Exploring the Scientific Method

Exploring the Scientific Method

CASES AND QUESTIONS

EDITED BY STEVEN GIMBEL

THE UNIVERSITY OF CHICAGO PRESS

Chicago & London

STEVEN GIMBEL is assistant professor of philosophy at Gettysburg College. He is the author of several journal articles and books, including *Defending Einstein: Hans Reichenbach's Writings on Space, Time, and Motion,* and *René Descartes: Profiles in Mathematics.*

The University of Chicago Press, Chicago 60637

The University of Chicago Press, Ltd., London

© 2011 by The University of Chicago

All rights reserved. Published 2011.

Printed in the United States of America

20 19 18 17 16 15 14 13 12 11   1 2 3 4 5

ISBN-13: 978-0-226-29481-0 (cloth)

ISBN-13: 978-0-226-29483-4 (paper)

ISBN-10: 0-226-29481-1 (cloth)

ISBN-10: 0-226-29483-8 (paper)

Library of Congress Cataloging-in-Publication Data

Exploring the scientific method : cases and questions / edited by Steven Gimbel.

   p. cm.

   Includes bibliographical references and index.

   ISBN-13: 978-0-226-29481-0 (cloth : alk. paper)

   ISBN-10: 0-226-29481-1 (cloth : alk. paper)

   ISBN-13: 978-0-226-29483-4 (pbk : alk. paper)

   ISBN-10: 0-226-29483-8 (pbk : alk. paper) 1. Science— Methodology. I. Gimbel, Steven, 1968–

   Q175.3.E97 2011

   507.2′1—dc22

2010039374

Contents

Introduction

What Is Science?

Many students in grade school science classes are made to memorize something the teacher calls "*the* scientific method." Usually a recipe including five
to seven very neatly laid-out steps, possibly in the form of a flowchart, the
so-called scientific method includes some sort of hypothesizing, an empirical testing regimen, and a vague notion of evidence. But if you were to walk
into the office of a theoretical physicist, the laboratory of biochemist, or find
an anthropologist in the field and ask one of these working scientists, "Which
step are you on?" you would likely receive a blank stare. Science just does
not work like that. But it does work. The question that this book is centered
around is "What actually is the scientific method?"

Philosophically minded scientists and scientifically minded philosophers
for centuries have tried to lay it out in clear terms, often relating it to their own
scientific endeavors or those that were most exalted at the time. We will examine several of these accounts with an eye toward answering three interrelated
questions about science:

(1) WHAT IS A SCIENTIFIC THEORY?

You can't talk meaningfully about a thing until you know what the thing is. In
order to see how science produces theories that we have good reason to believe, we first need to understand exactly what is meant by a theory. Confusion and misleading arguments result around issues like global warming and
the teaching of evolutionary theory in public schools when basic terms like
"scientific theory" are not well understood. Unfortunately, like much else in
philosophy, clearly working out the sense of even this basic notion can be
trickier than it appears.

(2) HOW DO SCIENTISTS COME UP WITH THEIR THEORIES?

Philosophers of science use the phrase "context of discovery" to talk about
the process by which scientists develop their theories. Is it the result of a strict
logic applied to undeniable first truths or to observations made in the lab or in
the world around us? Is it a matter of creativity and insight where any old idea
can be introduced as a scientific hypothesis no matter how outlandish and bizarre? What role do politics and the biases of the times play?

(3) WHAT MAKES SOME THEORIES BETTER THAN OTHERS?

Theories, in general, are easy to come by, but really good theories are a thing of beauty. Why are some theories worthy of belief and others not? What is the process by which we come to reject a theory we had held onto for centuries in favor of a new one? Is this a matter of observable evidence, the result of a logical inference, or do social factors play into it?

One way to evaluate proposed answers to these questions is to look at science itself, determining whether actual cases in the history of science mirror the methodological claims made by philosophers. Sadly, this is something that neither philosophy nor science students tend to do very often or very systematically. Since the middle of the twentieth century, the conversation in philosophy of science itself has become so rich that most courses in the philosophy of science can remain entirely within the philosophical literature and have sufficient issues and readings to take up several semesters' worth of work. On the other hand, science classes focus on training future scientists in the laboratory and in the mathematical techniques needed to do science and therefore tend to spend little, if any, time exploring the foundations and underlying methodological character of science or its history.

This is unfortunate for the effect it has on students. Those who go on to pursue scientific careers are given an impoverished picture of their science, devoid of the human element, the historical problems and connections with the times, and any chance to wrestle with the deeper questions about us and the world raised by the study. Where one cannot be a philosopher without having read Plato and Aristotle, you can be a working biologist without having read Darwin's own writings, or you can be a physicist who has never looked at Newton's *Principia* or Einstein's original papers. Chemists may recognize the name of Robert Boyle from the law named after him, but few learn anything about what he actually did to help shape chemistry as it exists today.

For the great majority of students who will not become scientists—indeed, who will very likely pursue scientific questions in the classroom no further than the minimum number of mandatory science courses—the standard approach to teaching both science and philosophy of science too often leaves them alienated from science. The many deep and interesting questions about the workings of the natural and human world they come in with, what draws them to science, will sadly all too often remain unaddressed.

This book is designed to fill that gap. In addition to readings that attempt to set out a version of the scientific method, there are nine parallel tracks that run through this book. These tracks correspond to fields of scientific inquiry: astronomy, physics, chemistry, geology, evolutionary biology, genetics, psychology, sociology, and economics. Students are asked to pick that science which most interests them. At six points in the text, students will (assisted with the resources in the book), develop a case study in the history of their science, that is, a story about an important episode in the development of that scientific field. The idea is to then see how the philosophers in that section would make sense of that episode and to evaluate whether they sufficiently describe the scientific method as it was used by real scientists in that real life situation. Maybe they did, maybe they didn't—that is for you to decide.

There may be no single correct answer about the degree of fit between the philosophical claims and the history of science, but there certainly are better and worse argued and informed positions. In explaining why the student decided what they did, they will need to acquire an understanding of the philosophical positions discussed in the readings and the major developments in the history of their science of interest. No previous background in any scientific field is required in order to work in these tracks, so if there is a science that seems interesting, but is one that the student has not studied, fear not. The resources will help them to understand each step. At the same time, if there is a science in which a student has done a lot of work, working through the history will deeply enrich the understanding of those studies.

This is a book intended for science and nonscience students alike. Anyone who is interested in the way science works will be able to think for him- or herself about what scientists do and how they ought to do it by looking at great thinkers' ideas about it and actual real-world examples. In a world of textbooks, we rarely work from scientists' own words. We lose their sense of style, their intellectual approach. This book is designed to allow students to connect with the history of science and engage it in a conversation about the foundational issues underlying scientific methodology, a conversation that is expected to include their voices.

How To Use This Book

For the Student

This book is unlike many philosophy textbooks in that it is not merely a collection of readings from which to receive the "wisdom of the elders"; rather, it is meant to provide a space in which the student actively engages with the history and philosophy of science, taking an interest in a particular science and critically examining the relative effectiveness of different views about the scientific method. Undergraduate philosophy students are often left with the nagging question, "How does this connect to the real world?" This book is intended to keep the conversation always focused on the real world and the ways in which important texts in philosophy do or do not account well for science as it is actually practiced.

Before beginning this book, please select one field from the following list:

- astronomy
- atomic physics
- chemistry
- genetics
- evolutionary biology
- geology
- psychology
- sociology
- economics

The science you pick will provide the world-pointing focus of your work, a set of historical episodes that you (with the help of resources in the text) will explain and use to evaluate the philosophical claims made about the working of science. The selection may be a subject in which you have undertaken significant coursework or simply a study that seems interesting to you even if you have no idea what it really is. Since the histories of the sciences are little covered in most traditionally taught science courses, don't worry if you want to select a subject about which you know nothing right now. The case-study exercises are constructed in a self-guiding fashion that presupposes no previous exposure. Pick something that you think will be interesting to learn about. The episodes are arranged chronologically so that by the end of the book, you will have a good sense of the stages of historical development that your science went through.

Just as the sciences evolved, so too did philosophical thinking about science. The book is divided into sections that pick out major trends and schools of thought about science and pair them with scientific advances. Your job will have four steps: (1) explain the scientific episode, (2) explain the philosophy of the section, (3) explain how someone with that philosophical inclination would make sense of the episode and how they would see the episode as supporting their picture of the scientific method, and (4) explain the ways they are successful and unsuccessful in fully capturing the type of reasoning that happens in real-life science. The key is that you have the last word. All of the thinkers included here are big names in the history of philosophy who said very smart things, but, of course, the conversation continues even after they spoke, so it is always interesting to see what they got right, what they got wrong, and why.

In addition to understanding some philosophy and coming to see the history of a given science, another thing this book is designed to do is help you develop the sort of skills used in philosophical research. For the first two episodes, the book includes the relevant selections from the scientists themselves. The third case-study exercise directs you (through a hint) to where in the scientist's writings the important work was done. Get the work and read at least the mentioned section. For the last three cases, you will have to find the resources. This incremental approach allows you to develop your own research skills in a step-wise fashion so that you will have experience with the sort of information you will be looking for.

One additional tip: There are many wonderful books written about the history of every science. It will be helpful to you, in addition to the resources given to you in the text, to go early in the term to your library and find a history of your science to use as a companion throughout these exercises. The cases selected for you to work on are all famous and important advances, and all will have many secondary conversations about them, that is, discussions by scholars where they explain and weigh in on those cases. It is always good to have someone else's take on the case while also looking at the documents produced by the people themselves. You may or may not agree with their assessment or understanding, but such differences are the stuff of interesting scholarly debate. A good essay will mix original sources and expert discussion while also advancing one's own view with textual support.

While it may not look like one, this is intended to be a workbook. This is not merely a text to be read and digested, but rather an occasion for you to actively engage with philosophical questions about the way scientists do and should think. This engagement will help you develop research skills and the ability to focus an argument supporting a point of view. Do not be worried if your point of view changes from paper to paper. Such development is natural in philosophy. The key is to understand the material well enough to form and express a cogent argument.

For the Instructor

The structure of this book is different from other texts in that it is designed to afford students a greater sense autonomy in selecting work to meet their interests and thereby create a deeper sense of ownership and investment in understanding the material and in the work produced. For this reason, care was taken to allow for a broad range of sciences to be represented. Should an instructor feel uncomfortable with such a broad range, the option is always available to circumscribe the number of tracks made available to students.

The selected case studies represent, of course, a very small slice of the entire history of these sciences and were selected both because they fit well into the conceptual scheme of the book and because they represent significant advances or turning points in the discipline. The effort was made to provide a full enough sketch of the episode in each paper prompt that would be sufficient to give an instructor (even one without a working background in the history of, say, economics or chemistry) a sufficient grasp of the relevant work discussed to be able to evaluate student work while still leaving room for the student to connect the dots and plug the gaps in their own words to fully explain the episode before analyzing it.

The papers not only advance historically, but with each assignment the student is made to take on a bit more of the tasks of research. For the first paper, an edited reading is provided; it requires only that students learn to quote and cite properly, to extract the argument contained in the excerpt, and to write a successful exposition. In the second paper, the selection is less selectively edited, forcing students to read more carefully. Starting with the third paper, the resources are no longer contained in the textbook; students are referred to the library. This allows them to become familiar with the stacks, how to check out books and locate articles on reserve, and perhaps even the use of interlibrary loan. The fourth and fifth papers also require library research, but with increasingly less guidance, allowing students to become more familiar with the text gathering process and the use of secondary sources.

As all of these case studies are significant episodes, there are easy to find and access web-based resources for each. It is up to the instructor to determine how much to expect from the students in terms of library and online sources. While all students today have an internalized habit to begin all projects with a Google search, educational contexts differ, and it is left up to the individual instructor to decide on the depth of detail desired in the sketch and the degree to which electronic sources are acceptable and desirable in the work of one's own students. Where Wikipedia entries may be appropriate in some contexts, limiting students to electronically posted journal articles may be appropriate for others. This book is designed to give the instructor the flexibility needed to conform expectations to the particular students and their skill sets.

Since the number of students will likely outnumber the number of tracks, one concern will be resource sharing. It would be helpful to the students if texts known to be needed, or at least the relevant excerpts, could be placed on reserve to protect against competition for needed works. The following are the books or articles that students will be utilizing (many of the articles are available online or reproduced in collections or source books):

- Galileo, A Dialogue on the Two Chief Systems of the Universe
- Ernest, Rutherford, "The Scattering of α and β Particles by Matter and the Structure of the Atom," *Philosophical Magazine*, May 1911 (series 6, vol. 21)
- Joseph Priestley, "Of Dephlogisticated Air, and of the Constitution of the Atmosphere," in *Experiments and Observations on Different Kinds of Air*
- Thomas Hunt Morgan, *The Theory of the Gene* and/or *Mendelian Heredity*
- Jean Baptiste Lamarck, introduction to part 2 of *Zoological Philosophy*
- Charles Lyell, "Causes which Have Retarded the Progress of Geology," in *Principles of Geology*, book 1, chapter 5
- Ivan Pavlov, "Technical Methods Employed in the Objective Investigation of the Functions of the Cerebral Hemispheres," in *Conditioned Reflexes*
- Max Weber, "Demonination and Social Stratification," in *The Protestant Ethic and the Spirit of Capitalism*
- Adam Smith, *The Wealth of Nations*, book 1, chapter 7

While this book has been designed to be teachable in a single semester, if an instructor wants to dwell longer on any given reading or set of readings, there is room to pick and choose essays without sacrificing coherence. Most sections contain several essays and instructors could have students work through all of them or choose a subset while maintaining the student's ability to complete the analysis of the section's case study. Again, flexibility to allow the instructor to mold this book to one's desired emphasis and educational context is something considered in the design of the book. The concept of the book is to allow for active engagement on the part of the student while also allowing maximum flexibility for the instructor.

SYNTACTIC VIEW OF THEORIES

DEDUCTIVISM

Science occupies a privileged place. The results of science are held up as the quintessential examples of reasonable belief. Why should this be true and how do scientists go about deriving and supporting their claims? The answers to these questions seem to lie in understanding the structure of scientific theories and the method used to generate and test them.

Major approaches to working out the details of scientific methodology can be divided into six categories, which make up the sections of this text. The first is called the "syntactic view of theories," and it holds that scientific theories are sets of sentences whose truth or falsity can each be tested by using some combination of logical reasoning and empirical observation. The key to scientific reasoning, syntactic advocates argue, is to find the logical mechanism by which each and every scientific claim is derived and individually evaluated.

Historically, the first attempt to work out this view is what we call "deductivism." In Greece, during the fourth century BC, when questions about the workings of the natural world were beginning to be answered in natural—instead of supernatural—terms, thinkers found an amazing model for acquiring knowledge in mathematics, specifically geometry. Starting with simple statements whose truth was so obvious as to be undeniable, they advanced step by indubitable step until they arrived with seemingly absolute certainty at geometric theorems that were so intricate and complex as to be anything but obvious.

This method of argument, then called "demonstration" and now called "deduction," is the logic that moves from general to specific propositions. Consider the following classic deductive argument:

All Greeks are mortal
Socrates is a Greek
Therefore, Socrates is mortal.

This is a deductive argument, or, a demonstration of the conclusion that Socrates is mortal because it is derived from the more general claim about all Greeks, a set to which Socrates belongs (or at least did before the hemlock).

Deductive logic has the great advantage that in moving from general to specific in a way that the information in the conclusion is contained within the supporting premises, if you know the premises to be absolutely true, then you can be completely assured that your new conclusion is as well. It derives certain truth from certain truth.

This property was so impressive that above the entryway to Plato's Academy was the inscription, "Let no one enter here who has not learned his geometry." In the *Republic,* Plato sets out the way philosophers (indeed, philosopher-kings) were to be trained; he argues that anyone who strives to be a serious thinker must learn mathematics and its deductive logic—not because all the world is mathematical, as the Pythagoreans just a few years earlier had held—but because the method used in geometry is the same that would be needed for the acquisition of knowledge of all other types.

But since this deductive logic takes us from general to specific, if we want general truths about our world, we must start with something even more general. Those truths that are bigger, broader, and more general than even the laws of physics are what we term "metaphysics." According to Plato, metaphysical reality consists of a set of nonmaterial "forms" that reside in their own eternal, unchanging realm, a separate more perfect reality. Material things are imperfect representations of the forms. With the eye of the mind, we could, if well trained, conceive of these forms and understand the more universal truths they could impart to us.

Plato's student Aristotle did not share his teacher's enthusiasm for worlds apart from our material surroundings, but he did agree that the deductive method is the best route to knowledge. In his book *Posterior Analytics,* Aristotle argues that scientific knowledge comes from deductive arguments called syllogisms. A syllogism is an argument with two premises. One, the minor premise, is a universal generalization, that is, a statement of the form "all As are Bs." For true scientific knowledge, "B" must be an essential property of anything that is "A," not a mere coincidence. For example, it is true that "all bachelors are unmarried," because being unmarried is an essential property of bachelorhood and we, according to Aristotle, can derive scientific knowledge from that fact. Whereas it may be true that "all bachelors are slobs," but because slovenliness is not a necessary component of being a bachelor, one cannot derive scientific understanding from such an accidental coordination.

But where are we to gain these general truths about the world? In *Physics,* Aristotle inquires into the nature of nature. Natural things, he argues, have inherent in them a nature and it is from this nature that we get our general truths about the world. But this nature has a four-fold structure, so scientific questions—questions about the causes of events—can be questions about any one of these four aspects of the nature of a thing. It may be about the thing's material cause, the stuff that makes it up. It may be about the thing's formal cause, its structure and means of internal organization. It may be about its efficient cause, in other words, that which brought it about. Or it may be a question about its final cause, that is, the ultimate goal that the thing is moving toward. If we consider an acorn, for example, the material cause is the matter making up the acorn, the formal cause is the internal structure of the acorn, the efficient cause is the parent oak tree that produced it, and the final cause is the mighty oak it is striving to become. When we understand these aspects of acorns, we may then use syllogisms to demonstrate scientific claims about oak trees.

A few generations after Aristotle, the deductive method found its greatest expression in the work of Euclid. His masterpiece, *Elements*, sets out as basic premises five simple axioms, statements like "given any two points, one can draw a line," and "given any point, we can draw a circle around it of any radius"; five simple common notions, such as "equals added to equals yield equals"; and twenty-three basic definitions of geometric terms. From these self-evidently true propositions derives what we know as plane geometry. Complex propositions, such as "if in a triangle the square on one of the sides equals the sum of the squares on the remaining two sides of the triangle, then the angle contained by the remaining two sides of the triangle is right," are shown to follow with absolute certainty from the few atomic claims that seemingly no one in his rational mind would dream of denying.

So impressive was his work that René Descartes thought it alone to be worthy as a template for the construction of all other knowledge, including scientific knowledge. As a young genius studying at the university in La Fleche, the finest school in all of Europe at the time, Descartes was disconcerted to find that in all fields other than geometry, there were no undisputed claims. No matter what a given smart person contended, there was an equally smart person who argued against it. But in geometry, Euclid's theorems were beyond challenge.

He sought to bring this same lack of doubt to all of science and knew that he needed the certainty that came with Euclid's deductive method. But he could not appeal to Aristotle's version, since Aristotle's arguments and the scientific worldview that followed from them (specifically his earth centered picture of the universe) were exactly the propositions being challenged at the time.

So Descartes sought his own Euclidean-type axioms, his own propositions that would be self-evident beyond any possible doubt. In order to deductive argue from general down to specific, where the specific in this case is the laws of physics, he must have absolute truth in metaphysics. The first undeniable truth Descartes found was his own existence: "I think, therefore I am." From this, he argued it must necessarily follow that an all-perfect God exists. From the existence of an all-perfect God, we get "clear and distinct" universal propositions from which to deductively derive the nature of the world around us.

The key to the scientific method, these theorists contend, is the use of deductive logic to give us beliefs about the world that are demonstrable and absolutely true.

Aristotle

Posterior Analytics

BOOK I

1.

All instruction given or received by way of argument proceeds from pre-existent knowledge. This becomes evident upon a survey of all the species of such instruction. The mathematical sciences and all other speculative disciplines are acquired in this way, and so are the two forms of dialectical reasoning, syllogistic and inductive; for each of these latter make use of old knowledge to impart new, the syllogism assuming an audience that accepts its premises, induction exhibiting the universal as implicit in the clearly known particular. Again, the persuasion exerted by rhetorical arguments is in principle the same, since they use either example, a kind of induction, or enthymeme, a form of syllogism.

The pre-existent knowledge required is of two kinds. In some cases admission of the fact must be assumed, in others comprehension of the meaning of the term used, and sometimes both assumptions are essential. Thus, we assume that every predicate can be either truly affirmed or truly denied of any subject, and that 'triangle' means so and so; as regards 'unit' we have to make the double assumption of the meaning of the word and the existence of the thing. The reason is that these several objects are not equally obvious to us. Recognition of a truth may in some cases contain as factors both previous knowledge and also knowledge acquired simultaneously with that recognition-knowledge, this latter, of the particulars actually falling under the universal and therein already virtually known. For example, the student knew beforehand that the angles of every triangle are equal to two right angles; but it was only at the actual moment at which he was being led on to recognize this as true in the instance before him that he came to know 'this figure inscribed in the semicircle' to be a triangle. For some things (viz. the singulars finally reached which are not predicable of anything else as subject) are only learnt in this way, i.e. there is here no recognition through a middle of a minor

From Aristotle, "Analytica Posteriora," in *Logic*, in *The Oxford Translation of Aristotle*, vol 1, ed. W. D. Ross, trans. G. R. G. Mure (Oxford: Oxford University Press, 1928), 70–72, 73–74. Reprinted by permission of Oxford University Press.

term as subject to a major. Before he was led on to recognition or before he actually drew a conclusion, we should perhaps say that in a manner he knew, in a manner not.

If he did not in an unqualified sense of the term know the existence of this triangle, how could he know without qualification that its angles were equal to two right angles? No: clearly he knows not without qualification but only in the sense that he knows universally. If this distinction is not drawn, we are faced with the dilemma in the Meno: either a man will learn nothing or what he already knows; for we cannot accept the solution which some people offer. A man is asked, 'Do you, or do you not, know that every pair is even?' He says he does know it. The questioner then produces a particular pair, of the existence, and so a fortiori of the evenness, of which he was unaware. The solution which some people offer is to assert that they do not know that every pair is even, but only that everything which they know to be a pair is even: yet what they know to be even is that of which they have demonstrated evenness, i.e. what they made the subject of their premiss, viz. not merely every triangle or number which they know to be such, but any and every number or triangle without reservation. For no premiss is ever couched in the form 'every number which you know to be such,' or 'every rectilinear figure which you know to be such': the predicate is always construed as applicable to any and every instance of the thing. On the other hand, I imagine there is nothing to prevent a man in one sense knowing what he is learning, in another not knowing it. The strange thing would be, not if in some sense he knew what he was learning, but if he were to know it in that precise sense and manner in which he was learning it.

2.

We suppose ourselves to possess unqualified scientific knowledge of a thing, as opposed to knowing it in the accidental way in which the sophist knows, when we think that we know the cause on which the fact depends, as the cause of that fact and of no other, and, further, that the fact could not be other than it is. Now that scientific knowing is something of this sort is evident—witness both those who falsely claim it and those who actually possess it, since the former merely imagine themselves to be, while the latter are also actually, in the condition described. Consequently the proper object of unqualified scientific knowledge is something which cannot be other than it is.

There may be another manner of knowing as well—that will be discussed later. What I now assert is that at all events we do know by demonstration. By demonstration I mean a syllogism productive of scientific knowledge, a syllogism, that is, the grasp of which is eo ipso such knowledge. Assuming then that my thesis as to the nature of scientific knowing is correct, the premises of demonstrated knowledge must be true, primary, immediate, better known than and prior to the conclusion, which is further related to them as effect to cause. Unless these conditions are satisfied, the basic truths will not be 'appropriate' to the conclusion. Syllogism there may indeed be without these conditions, but such syllogism, not being productive of scientific knowledge, will not be demonstration. The premises must be true: for that which is non-existent cannot be known—we cannot know, e.g. that the diagonal of a square is commensurate with its side. The premises must be primary and indemonstrable; otherwise they will require demonstration in order to be known, since to have knowledge, if it be not accidental knowledge, of things which are demonstrable, means precisely to have a demonstration of them. The premisses must be the causes of the conclusion, better known than it, and prior to it; its causes, since we possess scientific knowledge of a thing only when we know its cause; prior, in order to be causes; antecedently known, this antecedent knowledge being not our mere understanding of the meaning, but knowledge of the fact as well. Now 'prior' and 'better known' are ambiguous terms, for there is a difference between what is prior and better known in the order of being and what is prior and better known to man. I mean that objects nearer to sense are prior and better known to man; objects without qualification prior and better known are those further from sense. Now the most universal causes are furthest from sense and particular causes are nearest to sense, and they are thus exactly opposed to one another. In saying that the premisses of demonstrated knowledge must be primary, I mean that they must be the 'appropriate' basic truths, for I identify primary premiss and basic truth. A 'basic truth' in a demonstration is an immediate proposition. An immediate proposition is one which has no other proposition prior to it. A proposition is either part of an enunciation, i.e. it predicates a single attribute of a single subject. If a proposition is dialectical, it assumes either part indifferently; if it is demonstrative, it lays down one part to the definite exclusion of the other because that part is true. The term 'enunciation' denotes either part of a contradiction indifferently. A contradiction is an opposition which of its own nature excludes a middle. The part of a contradiction which conjoins a predicate with a subject is an affirmation; the part disjoining them is a negation. I call an immediate basic truth of syllogism a 'thesis' when, though it is not susceptible of proof by the teacher, yet ignorance of it does not constitute a total bar to progress on the part of the pupil: one which the pupil must know if he is to learn anything whatever is an axiom. I call it an axiom because there are such truths and we give them the name of axioms par excellence. If a thesis assumes one part or the other of an enunciation, i.e. asserts either the existence or the non-existence of a subject, it is a hypothesis; if it does not so assert, it is a definition. Definition is a 'thesis' or a 'laying something down,' since the arithmetician lays it down that to be a unit is to be quantitatively indivisible; but it is not a hypothesis, for to define what a unit is is not the same as to affirm its existence.

Now since the required ground of our knowledge—i.e. of our conviction—of a fact is the possession of such a syllogism as we call demonstration, and the ground of the syllogism is the facts constituting its premisses, we must not only know the primary premisses—some if not all of them—beforehand, but know them better than the conclusion: for the cause of an attribute's inherence in a subject always itself inheres in the subject more firmly than that attribute; e.g. the cause of our loving anything is dearer to us than the object of our love. So since the primary premisses are the cause of our knowledge—i.e. of our conviction—it follows that we know them better—that is, are more convinced of them—than their consequences, precisely because of our knowledge of the latter is the effect of our knowledge of the premisses. Now a man cannot believe in anything more than in the things he knows, unless he has either actual knowledge of it or something better than actual knowledge. But we are faced with this paradox if a student whose belief rests on demonstration has not prior knowledge; a man must believe in some, if not in all, of the basic truths more than in the conclusion. Moreover, if a man sets out to acquire the scientific knowledge that comes through demonstration, he must not only have a better knowledge of the basic truths and a firmer conviction of them than of the connexion which is being demonstrated: more than this, nothing must be more certain or better known to him than these basic truths in their character as contradicting the fundamental premisses which lead to the opposed and erroneous conclusion. For indeed the conviction of pure science must be unshakable.

4.

Since the object of pure scientific knowledge cannot be other than it is, the truth obtained by demonstrative knowledge will be necessary. And since demonstrative knowledge is only present when we have a demonstration, it follows that demonstration is an inference from necessary premisses. So we must consider what are the premisses of demonstration—i.e. what is their character: and as a preliminary, let us define what we mean by an attribute 'true in every instance of its subject,' an 'essential' attribute, and a 'commensurate and universal' attribute. I call 'true in every instance' what is truly predicable of all instances—not of one to the exclusion of others—and at all times, not at this or that time only; e.g. if animal is truly predicable of every instance of man, then if it be true to say 'this is a man,' 'this is an animal' is also true, and if the one be true now the other is true now. A corresponding account holds if point is in every instance predicable as contained in line. There is evidence for this in the fact that the objection we raise against a proposition put to us as true in every instance is either an instance in which, or an occasion on which, it is not true. Essential attributes are (1) such as belong to their subject as elements in its essential nature (e.g. line thus belongs to triangle, point to line; for the very being or 'substance' of triangle and line is composed of these elements, which are contained in the formulae defining triangle and line): (2) such that, while they belong to certain subjects, the subjects to which they belong are contained in the attribute's own defining formula. Thus straight and curved belong to line, odd and even, prime and compound, square and oblong, to number; and also the formula defining any one of these attributes contains its subject—e.g. line or number as the case may be.

Extending this classification to all other attributes, I distinguish those that answer the above description as belonging essentially to their respective subjects; whereas attributes related in neither of these two ways to their subjects I call accidents or 'coincidents'; e.g. musical or white is a 'coincident' of animal.

Further (a) that is essential which is not predicated of a subject other than itself: e.g. 'the walking [thing]' walks and is white in virtue of being something else besides; whereas substance, in the sense of whatever signifies a 'this somewhat,' is not what it is in virtue of being something else besides. Things, then, not predicated of a subject I call essential; things predicated of a subject I call accidental or 'coincidental.'

In another sense again (b) a thing consequentially connected with anything is essential; one not so connected is 'coincidental.' An example of the latter is 'While he was walking it lightened': the lightning was not due to his walking; it was, we should say, a coincidence. If, on the other hand, there is a consequential connexion, the predication is essential; e.g. if a beast dies when its throat is being cut, then its death is also essentially connected with the cutting, because the cutting was the cause of death, not death a 'coincident' of the cutting.

So far then as concerns the sphere of connexions scientifically known in the unqualified sense of that term, all attributes which (within that sphere) are essential either in the sense that their subjects are contained in them, or in the sense that they are contained in their subjects, are necessary as well as consequentially connected with their subjects. For it is impossible for them not to inhere in their subjects either simply or in the qualified sense that one or other of a pair of opposites must inhere in the subject; e.g. in line must be either straightness or curvature, in number either oddness or evenness. For within a single identical genus the contrary of a given attribute is either its privative or its contradictory; e.g. within number what is not odd is even, inasmuch as within this sphere even is a necessary consequent of not-odd. So, since any given predicate must be either affirmed or denied of any subject, essential attributes must inhere in their subjects of necessity.

Thus, then, we have established the distinction between the attribute which is 'true in every instance' and the 'essential' attribute.

I term 'commensurately universal' an attribute which belongs to every instance of its subject, and to every instance essentially and as such; from which it clearly follows that all commensurate universals inhere necessarily in their subjects. The essential attribute, and the attribute that belongs to its subject as such, are identical. E.g. point and straight belong to line essentially, for they belong to line as such; and triangle as such has two right angles, for it is essentially equal to two right angles.

An attribute belongs commensurately and universally to a subject when it can be shown to belong to any random instance of that subject and when the subject is the first thing to which it can be shown to belong. Thus, e.g. (1) the equality of its angles to two right angles is not a commensurately universal attribute of figure. For though it is possible to show that a figure has its angles equal to two right angles, this attribute cannot be demonstrated of any figure selected at haphazard, nor in demonstrating does one take a figure at random-a square is a figure but its angles are not equal to two right angles. On the other hand, any isosceles triangle has its angles equal to two right angles, yet isosceles triangle is not the primary subject of this attribute but triangle is prior. So whatever can be shown to have its angles equal to two right angles, or to possess any other attribute, in any random instance of itself and primarily-that is the first subject to which the predicate in question belongs commensurately and universally, and the demonstration, in the essential sense, of any predicate is the proof of it as belonging to this first subject commensurately and universally: while the proof of it as belonging to the other subjects to which it attaches is demonstration only in a secondary and unessential sense. Nor again (2) is equality to two right angles a commensurately universal attribute of isosceles; it is of wider application.

Physics

BOOK II

CHAPTER 1

Of things that exist, some exist by nature, some from other causes.

'By nature' the animals and their parts exist, and the plants and the simple bodies (earth, fire, air, water)—for we say that these and the like exist 'by nature.'

All the things mentioned present a feature in which they differ from things which are not constituted by nature. Each of them has within itself a principle of motion and of stationariness (in respect of place, or of growth and decrease, or by way of alteration). On the other hand, a bed and a coat and anything else of that sort, qua receiving these designations i.e. in so far as they are products of art—have no innate impulse to change. But in so far as they happen to be composed of stone or of earth or of a mixture of the two, they do have such an impulse, and just to that extent which seems to indicate that nature is a source or cause of being moved and of being at rest in that to which it belongs primarily, in virtue of itself and not in virtue of a concomitant attribute.

I say 'not in virtue of a concomitant attribute,' because (for instance) a man who is a doctor might cure himself. Nevertheless it is not in so far as he is a patient that he possesses the art of medicine: it merely has happened that the same man is doctor and patient—and that is why these attributes are not always found together. So it is with all other artificial products. None of them has in itself the source of its own production. But while in some cases (for instance houses and the other products of manual labour) that principle is in something else external to the thing, in others those which may cause a change in themselves in virtue of a concomitant attribute—it lies in the things themselves (but not in virtue of what they are).

'Nature' then is what has been stated. Things 'have a nature' which have a principle of this kind. Each of them is a substance; for it is a subject, and nature always implies a subject in which it inheres.

The term 'according to nature' is applied to all these things and also to

From Aristotle, "Physica," in *Philosophy of Nature*, in *The Oxford Translation of Aristotle*, vol. 2, ed. W. D. Ross, trans. R. P. Hardie and R. K. Gaye (Oxford: Oxford University Press, 1930), 192–95. Reprinted by permission of Oxford University Press.

attributes which belong to them in virtue of what they are, for instance property of fire to be carried upwards—which is not a 'nature' nor 'has a nature' but is 'by nature' or 'according to nature.'

What nature is, then, and the meaning of the terms 'by nature' and 'according to nature,' has been stated. *That* nature exists, it would be absurd to try to prove; for it is obvious that there are many things of this kind, and to prove what is obvious by what is not is the mark of a man who is unable to distinguish what is self-evident from what is not. (This state of mind is clearly possible. A man blind from birth might reason about colours. Presumably therefore such persons must be talking about words without any thought to correspond.)

Some identify the nature or substance of a natural object with that immediate constituent of it which taken by itself is without arrangement, e.g., the wood is the 'nature' of the bed, and the bronze the 'nature' of the statue.

As an indication of this Antiphon points out that if you planted a bed and the rotting wood acquired the power of sending up a shoot, it would not be a bed that would come up, but wood—which shows that the arrangement in accordance with the rules of the art is merely an incidental attribute, whereas the real nature is the other, which, further, persists continuously through the process of making.

But if the material of each of these objects has itself the same relation to something else, say bronze (or gold) to water, bones (or wood) to earth and so on, that (they say) would be their nature and essence. Consequently some assert earth, others fire or air or water or some or all of these, to be the nature of the things that are. For whatever any one of them supposed to have this character—whether one thing or more than one thing—this or these he declared to be the whole of substance, all else being its affections, states, or dispositions. Every such thing they held to be eternal (for it could not pass into anything else), but other things to come into being and cease to be times without number.

This then is one account of 'nature,' namely that it is the immediate material substratum of things which have in themselves a principle of motion or change.

Another account is that 'nature' is the shape or form which is specified in the definition of the thing.

For the word 'nature' is applied to what is according to nature and the natural in the same way as 'art' is applied to what is artistic or a work of art. We should not say in the latter case that there is anything artistic about a thing, if it is a bed only potentially, not yet having the form of a bed; nor should we call it a work of art. The same is true of natural compounds. What is potentially flesh or bone has not yet its own 'nature,' and does not exist until it receives the form specified in the definition, which we name in defining what flesh or bone is. Thus in the second sense of 'nature' it would be the shape or form (not separable except in statement) of things which have in themselves a source of motion. (The combination of the two, e.g. man, is not 'nature' but 'by nature' or 'natural.')

The form indeed is 'nature' rather than the matter; for a thing is more properly said to be what it is when it has attained to fulfillment than when it exists potentially. Again man is born from man, but not bed from bed. That is why people say that the figure is not the nature of a bed, but the wood is—if the bed sprouted not a bed but wood would come up. But even if the figure is art, then on the same principle the shape of man is his nature. For man is born from man.

We also speak of a thing's nature as being exhibited in the process of growth by which its nature is attained. The 'nature' in this sense is not like 'doctoring,' which leads not to the art of doctoring but to health. Doctoring must start from the art, not lead to it. But it is not in this way that nature (in the one sense) is related to nature (in the other). What grows qua growing grows from something into something. Into what then does it grow? Not into that from which it arose but into that to which it tends. The shape then is nature.

'Shape' and 'nature,' it should be added, are in two senses. For the privation too is in a way form. But whether in unqualified coming to be there is privation, i.e. a contrary to what comes to be, we must consider later.

CHAPTER 2

We have distinguished, then, the different ways in which the term 'nature' is used.

The next point to consider is how the mathematician differs from the physicist. Obviously physical bodies contain surfaces and volumes, lines and points, and these are the subject matter of mathematics.

Further, is astronomy different from physics or a department of it? It seems absurd that the physicist should be supposed to know the nature of sun or moon, but not to know any of their essential attributes, particularly as the writers on physics obviously do discuss their shape also and whether the earth and the world are spherical or not.

Now the mathematician, though he too treats of these things, nevertheless does not treat of them as the limits of a physical body; nor does he consider the attributes indicated as the attributes of such bodies. That is why he separates them; for in thought they are separable from motion, and it makes no difference, nor does any falsity result, if they are separated. The holders of the theory of Forms do the same, though they are not aware of it; for they separate the objects of physics, which are less separable than those of mathematics. This becomes plain if one tries to state in each of the two cases the definitions of the things and of their attributes. 'Odd' and 'even,' 'straight' and 'curved,' and likewise 'number,' 'line,' and 'figure,' do not involve motion; not so 'flesh' and 'bone' and 'man'—these are defined like 'snub nose,' not like 'curved.'

Similar evidence is supplied by the more physical of the branches of mathematics, such as optics, harmonics, and astronomy. These are in a way the converse of geometry. While geometry investigates physical lines but not qua physical, optics investigates mathematical lines, but qua physical, not qua mathematical.

Since 'nature' has two senses, the form and the matter, we must investigate its objects as we would the essence of snubness. That is, such things are neither independent of matter nor can be defined in terms of matter only. Here too indeed one might raise a difficulty.

Since there are two natures, with which is the physicist concerned? Or should he investigate the combination of the two? But if the combination of the two, then also each severally. Does it belong then to the same or to different sciences to know each severally?

If we look at the ancients, physics would to be concerned with the matter. (It was only very slightly that Empedocles and Democritus touched on the forms and the essence.)

But if on the other hand art imitates nature, and it is the part of the same discipline to know the form and the matter up to a point (e.g. the doctor has a knowledge of health and also of bile and phlegm, in which health is realized, and the builder both of the form of the house and of the matter, namely that it is bricks and beams, and so forth): if this is so, it would be the part of physics also to know nature in both its senses.

Again, 'that for the sake of which,' or the end, belongs to the same department of knowledge as the means. But the nature is the end or 'that for the sake of which.' For if a thing undergoes a continuous change and there is a stage which is last, this stage is the end or 'that for the sake of which.' (That is why the poet was carried away into making an absurd statement when he said 'he has the end for the sake of which he was born.' For not every stage that is last claims to be an end, but only that which is best.)

For the arts make their material (some simply 'make' it, others make it serviceable), and we use everything as if it was there for our sake. (We also are in a sense an end. 'That for the sake of which' has two senses: the distinction is made in our work On Philosophy.)

The arts, therefore, which govern the matter and have knowledge are two, namely the art which uses the product and the art which directs the production of it. That is why the using art also is in a sense directive; but it differs in that it knows the form, whereas the art which is directive as being concerned with production knows the matter. For the helmsman knows and prescribes what sort of form a helm should have, the other from what wood it should be made and by means of what operations. In the products of art, however, we make the material with a view to the function, whereas in the products of nature the matter is there all along.

Again, matter is a relative term: to each form there corresponds a special matter. How far then must the physicist know the form or essence? Up to a point, perhaps, as the doctor must know sinew or the smith bronze (i.e. until he understands the purpose of each): and the physicist is concerned only with things whose forms are separable indeed, but do not exist apart from matter. Man is begotten by man and by the sun as well. The mode of existence and essence of the separable it is the business of the primary type of philosophy to define.

CHAPTER 3

Now that we have established these distinctions, we must proceed to consider causes, their character and number. Knowledge is the object of our inquiry, and men do not think they know a thing till they have grasped the 'why' of (which is to grasp its primary cause).

So clearly we too must do this as regards both coming to be and passing away and every kind of physical change, in order that, knowing their principles, we may try to refer to these principles each of our problems.

In one sense, then, (1) that out of which a thing comes to be and which persists, is called 'cause,' e.g. the bronze of the statue, the silver of the bowl, and the genera of which the bronze and the silver are species.

In another sense (2) the form or the archetype, i.e. the statement of the essence, and its genera, are called 'causes' (e.g. of the octave the relation of 2:1, and generally number), and the parts in the definition.

Again (3) the primary source of the change or coming to rest; e.g. the man who gave advice is a cause, the father is cause of the child, and generally what makes of what is made and what causes change of what is changed.

Again (4) in the sense of end or 'that for the sake of which' a thing is done, e.g. health is the cause of walking about. ('Why is he walking about?' we say. 'To be healthy,' and, having said that, we think we have assigned the cause.) The same is true also of all the intermediate steps which are brought about through the action of something else as means towards the end, e.g. reduction of flesh, purging, drugs, or surgical instruments are means towards health. All these things are 'for the sake of' the end, though they differ from one another in that some are activities, others instruments.

This then perhaps exhausts the number of ways in which the term 'cause' is used. As the word has several senses, it follows that there are several causes of the same thing not merely in virtue of a concomitant attribute), e.g. both the art of the sculptor and the bronze are causes of the statue. These are causes of the statue qua statue, not in virtue of anything else that it may be—only not in the same way, the one being the material cause, the other the cause whence the motion comes. Some things cause each other reciprocally, e.g. hard work causes fitness and vice versa, but again not in the same way, but the one as end, the other as the origin of change. Further the same thing is the cause of contrary results. For that which by its presence brings about one result is sometimes blamed for bringing about the contrary by its absence. Thus we ascribe the wreck of a ship to the absence of the pilot whose presence was the cause of its safety.

All the causes now mentioned fall into four familiar divisions. The letters are the causes of syllables, the material of artificial products, fire, &c., of bodies, the parts of the whole, and the premises of the conclusion, in the sense of 'that from which.' Of these pairs the one set are causes in the sense of substratum, e.g. the parts, the other set in the sense of essence—the whole and the combination and the form. But the seed and the doctor and the adviser, and generally the maker, are all sources whence the change or stationariness originates, while the others are causes in the sense of the end or the good of the rest; for 'that for the sake of which' means what is best and the end of the things that lead up to it. (Whether we say the 'good itself or the 'apparent good' makes no difference.)

Such then is the number and nature of the kinds of cause.

René Descartes

. . . .

Discourse on Method

PART I

Good sense is, of all things among men, the most equally distributed; for
every one thinks himself so abundantly provided with it, that those even who
are the most difficult to satisfy in everything else, do not usually desire a larger
measure of this quality than they already possess. And in this it is not likely
that all are mistaken the conviction is rather to be held as testifying that the
power of judging aright and of distinguishing truth from error, which is properly what is called good sense or reason, is by nature equal in all men; and that
the diversity of our opinions, consequently, does not arise from some being
endowed with a larger share of reason than others, but solely from this, that
we conduct our thoughts along different ways, and do not fix our attention on
the same objects. For to be possessed of a vigorous mind is not enough; the
prime requisite is rightly to apply it. The greatest minds, as they are capable
of the highest excellences, are open likewise to the greatest aberrations; and
those who travel very slowly may yet make far greater progress, provided they
keep always to the straight road, than those who, while they run, forsake it.
For myself, I have never fancied my mind to be in any respect more perfect than those of the generality; on the contrary, I have often wished that I
were equal to some others in promptitude of thought, or in clearness and distinctness of imagination, or in fullness and readiness of memory. And besides
these, I know of no other qualities that contribute to the perfection of the
mind; for as to the reason or sense, inasmuch as it is that alone which constitutes us men, and distinguishes us from the brutes, I am disposed to believe
that it is to be found complete in each individual; and on this point to adopt
the common opinion of philosophers, who say that the difference of greater
and less holds only among the accidents, and not among the forms or natures
of individuals of the same species.

I will not hesitate, however, to avow my belief that it has been my singular good fortune to have very early in life fallen in with certain tracks which
have conducted me to considerations and maxims, of which I have formed

From René Descartes, *Discourse on Method*, trans. John Veitch (Chicago: Open Court, 1899),
1–4, 17–21, 34–69.

a method that gives me the means, as I think, of gradually augmenting my knowledge, and of raising it by little and little to the highest point which the mediocrity of my talents and the brief duration of my life will permit me to reach. For I have already reaped from it such fruits that, although I have been accustomed to think lowly enough of myself, and although when I look with the eye of a philosopher at the varied courses and pursuits of mankind at large, I find scarcely one which does not appear in vain and useless, I nevertheless derive the highest satisfaction from the progress I conceive myself to have already made in the search after truth, and cannot help entertaining such expectations of the future as to believe that if, among the occupations of men as men, there is any one really excellent and important, it is that which I have chosen.

After all, it is possible I may be mistaken; and it is but a little copper and glass, perhaps, that I take for gold and diamonds. I know how very liable we are to delusion in what relates to ourselves, and also how much the judgments of our friends are to be suspected when given in our favor. But I shall endeavor in this discourse to describe the paths I have followed, and to delineate my life as in a picture, in order that each one may also be able to judge of them for himself, and that in the general opinion entertained of them, as gathered from current report, I myself may have a new help towards instruction to be added to those I have been in the habit of employing.

My present design, then, is not to teach the method which each ought to follow for the right conduct of his reason, but solely to describe the way in which I have endeavored to conduct my own. They who set themselves to give precepts must of course regard themselves as possessed of greater skill than those to whom they prescribe; and if they err in the slightest particular, they subject themselves to censure. But as this tract is put forth merely as a history, or, if you will, as a tale, in which, amid some examples worthy of imitation, there will be found, perhaps, as many more which it were advisable not to follow, I hope it will prove useful to some without being hurtful to any, and that my openness will find some favor with all.

PART II

Among the branches of philosophy, I had, at an earlier period, given some attention to logic, and among those of the mathematics to geometrical analysis and algebra,—three arts or sciences which ought, as I conceived, to contribute something to my design. But, on examination, I found that, as for logic, its syllogisms and the majority of its other precepts are of avail—rather in the communication of what we already know, or even as the art of Lully, in speaking without judgment of things of which we are ignorant, than in the investigation of the unknown; and although this science contains indeed a number of correct and very excellent precepts, there are, nevertheless, so many others, and these either injurious or superfluous, mingled with the former, that it is almost quite as difficult to effect a severance of the true from the false as it is to extract a Diana or a Minerva from a rough block of marble. Then as to the analysis of the ancients and the algebra of the moderns, besides that they embrace only matters highly abstract, and, to appearance, of no use, the former is so exclusively restricted to the consideration of figures, that it can exercise the understanding only on condition of greatly fatiguing the imagination; and, in the latter, there is so complete a subjection to certain rules and formulas, that there results an art full of confusion and obscurity calculated to embarrass, instead of a science fitted to cultivate the mind. By these considerations I was induced to seek some other method which would comprise the advantages of the three and be exempt from their defects. And as a multitude of laws often only hampers justice, so that a state is best governed when, with few laws, these are rigidly administered; in like manner, instead of the great number of precepts of which logic is composed, I believed that the four following would prove perfectly sufficient for me, provided I took the firm and unwavering resolution never in a single instance to fail in observing them.

The *first* was never to accept anything for true which I did not clearly know to be such; that is to say, carefully to avoid precipitancy and prejudice, and to comprise nothing more in my judgment than what was presented to my mind so clearly and distinctly as to exclude all ground of doubt.

The *second*, to divide each of the difficulties under examination into as many parts as possible, and as might be necessary for its adequate solution.

The *third*, to conduct my thoughts in such order that, by commencing with objects the simplest and easiest to know, I might ascend by little and little, and, as it were, step by step, to the knowledge of the more complex; assigning in thought a certain order even to those objects which in their own nature do not stand in a relation of antecedence and sequence.

And the *last*, in every case to make enumerations so complete, and reviews so general, that I might be assured that nothing was omitted.

The long chains of simple and easy reasonings by means of which geometers are accustomed to reach the conclusions of their most difficult demonstrations, had led me to imagine that all things, to the knowledge of which man is competent, are mutually connected in the same way, and that there is nothing so far removed from us as to be beyond our reach, or so hidden that we cannot discover it, provided only we abstain from accepting the false for

the true, and always preserve in our thoughts the order necessary for the deduction of one truth from another. And I had little difficulty in determining
the objects with which it was necessary to commence, for I was already persuaded that it must be with the simplest and easiest to know, and, considering
that of all those who have hitherto sought truth in the sciences, the mathematicians alone have been able to find any demonstrations, that is, any certain
and evident reasons, I did not doubt but that such must have been the rule of
their investigations. I resolved to commence, therefore, with the examination
of the simplest objects, not anticipating, however, from this any other advantage than that to be found in accustoming my mind to the love and nourishment of truth, and to a distaste for all such reasonings as were unsound. But I
had no intention on that account of attempting to master all the particular sciences commonly denominated mathematics: but observing that, however different their objects, they all agree in considering only the various relations or
proportions subsisting among those objects, I thought it best for my purpose
to consider these proportions in the most general form possible, without referring them to any objects in particular, except such as would most facilitate
the knowledge of them, and without by any means restricting them to these,
that afterwards I might thus be the better able to apply them to every other
class of objects to which they are legitimately applicable. Perceiving further,
that in order to understand these relations I should sometimes have to consider them one by one and sometimes only to bear them in mind, or embrace
them in the aggregate, I thought that, in order the better to consider them individually, I should view them as subsisting between straight lines, than which
I could find no objects more simple, or capable of being more distinctly represented to my imagination and senses; and on the other hand, that in order to
retain them in the memory or embrace an aggregate of many, I should express
them by certain characters the briefest possible. In this way I believed that I
could borrow all that was best both in geometrical analysis and in algebra, and
correct all the defects of the one by help of the other.

PART IV

I am in doubt as to the propriety of making my first meditations in the place
above mentioned matter of discourse; for these are so metaphysical, and so
uncommon, as not, perhaps, to be acceptable to every one. And yet, that it
may be determined whether the foundations that I have laid are sufficiently
secure, I find myself in a measure constrained to advert to them. I had long
before remarked that, in relation to practice, it is sometimes necessary to
adopt, as if above doubt, opinions which we discern to be highly uncertain,

as has been already said; but as I then desired to give my attention solely to the search after truth, I thought that a procedure exactly the opposite was called for, and that I ought to reject as absolutely false all opinions in regard to which I could suppose the least ground for doubt, in order to ascertain whether after that there remained aught in my belief that was wholly indubitable. Accordingly, seeing that our senses sometimes deceive us, I was willing to suppose that there existed nothing really such as they presented to us; and because some men err in reasoning, and fall into paralogisms, even on the simplest matters of geometry, I, convinced that I was as open to error as any other, rejected as false all the reasonings I had hitherto taken for demonstrations; and finally, when I considered that the very same thoughts (presentations) which we experience when awake may also be experienced when we are asleep, while there is at that time not one of them true, I supposed that all the objects (presentations) that had ever entered into my mind when awake, had in them no more truth than the illusions of my dreams. But immediately upon this I observed that, whilst I thus wished to think that all was false, it was absolutely necessary that I, who thus thought, should be somewhat; and as I observed that this truth, I think, therefore I am, was so certain and of such evidence that no ground of doubt, however extravagant, could be alleged by the skeptics capable of shaking it, I concluded that I might, without scruple, accept it as the first principle of the philosophy of which I was in search.

In the next place, I attentively examined what I was and as I observed that I could suppose that I had no body, and that there was no world nor any place in which I might be; but that I could not therefore suppose that I was not; and that, on the contrary, from the very circumstance that I thought to doubt of the truth of other things, it most clearly and certainly followed that I was; while, on the other hand, if I had only ceased to think, although all the other objects which I had ever imagined had been in reality existent, I would have had no reason to believe that I existed; I thence concluded that I was a substance whose whole essence or nature consists only in thinking, and which, that it may exist, has need of no place, nor is dependent on any material thing; so that "I," that is to say, the mind by which I am what I am, is wholly distinct from the body, and is even more easily known than the latter, and is such, that although the latter were not, it would still continue to be all that it is.

After this I inquired in general into what is essential I to the truth and certainty of a proposition; for since I had discovered one which I knew to be true, I thought that I must likewise be able to discover the ground of this certitude. And as I observed that in the words I think, therefore I am, there is nothing at all which gives me assurance of their truth beyond this, that I see very clearly that in order to think it is necessary to exist, I concluded that I might take, as a general rule, the principle, that all the things which we very clearly and distinctly conceive are true, only observing, however, that there is some difficulty in rightly determining the objects which we distinctly conceive.

In the next place, from reflecting on the circumstance that I doubted, and that consequently my being was not wholly perfect (for I clearly saw that it was a greater perfection to know than to doubt), I was led to inquire whence I had learned to think of something more perfect than myself; and I clearly recognized that I must hold this notion from some nature which in reality was more perfect. As for the thoughts of many other objects external to me, as of the sky, the earth, light, heat, and a thousand more, I was less at a loss to know whence these came; for since I remarked in them nothing which seemed to render them superior to myself, I could believe that, if these were true, they were dependencies on my own nature, in so far as it possessed a certain perfection, and, if they were false, that I held them from nothing, that is to say, that they were in me because of a certain imperfection of my nature. But this could not be the case with—the idea of a nature more perfect than myself; for to receive it from nothing was a thing manifestly impossible; and, because it is not less repugnant that the more perfect should be an effect of, and dependence on the less perfect, than that something should proceed from nothing, it was equally impossible that I could hold it from myself: accordingly, it but remained that it had been placed in me by a nature which was in reality more perfect than mine, and which even possessed within itself all the perfections of which I could form any idea; that is to say, in a single word, which was God. And to this I added that, since I knew some perfections which I did not possess, I was not the only being in existence (I will here, with your permission, freely use the terms of the schools); but, on the contrary, that there was of necessity some other more perfect Being upon whom I was dependent, and from whom I had received all that I possessed; for if I had existed alone, and independently of every other being, so as to have had from myself all the perfection, however little, which I actually possessed, I should have been able, for the same reason, to have had from myself the whole remainder of perfection, of the want of which I was conscious, and thus could of myself have become infinite, eternal, immutable, omniscient, all-powerful, and, in fine, have possessed all the perfections which I could recognize in God. For in order to know the nature of God (whose existence has been established by the preceding reasonings), as far as my own nature permitted, I had only to consider in reference to all the properties of which I found in my mind some idea, whether their possession was a mark of perfection; and I was assured that no one which indicated any imperfection was in him, and that none of the rest was awanting. Thus I perceived that doubt, inconstancy, sadness, and such like, could not be found in God, since I myself would have been happy to be free from them. Besides, I had ideas of many sensible and corporeal things; for although I might suppose that I was dreaming, and that all which I saw or imagined was false, I could not, nevertheless, deny that the ideas were in reality in my thoughts. But, because I had already very clearly recognized in myself that the intelligent nature is distinct from the corporeal, and as I observed that all composition is an evidence of dependency, and that a state of dependency is manifestly a state of imperfection, I therefore determined that it could not be a perfection in God to be compounded of these two natures and that consequently he was not so compounded; but that if there were any bodies in the world, or even any intelligences, or other natures that were not wholly perfect, their existence depended on his power in such a way that they could not subsist without him for a single moment.

PART V

I would here willingly have proceeded to exhibit the whole chain of truths which I deduced from these primary but as with a view to this it would have been necessary now to treat of many questions in dispute among the earned, with whom I do not wish to be embroiled, I believe that it will be better for me to refrain from this exposition, and only mention in general what these truths are, that the more judicious may be able to determine whether a more special account of them would conduce to the public advantage. I have ever remained firm in my original resolution to suppose no other principle than that of which I have recently availed myself in demonstrating the existence of God and of the soul, and to accept as true nothing that did not appear to me more clear and certain than the demonstrations of the geometers had formerly appeared; and yet I venture to state that not only have I found means to satisfy myself in a short time on all the principal difficulties which are usually treated of in philosophy, but I have also observed certain laws established in nature by God in such a manner, and of which he has impressed on our minds such notions, that after we have reflected sufficiently upon these, we cannot doubt that they are accurately observed in all that exists or takes place in the world and farther, by considering the concatenation of these laws, it appears to me that I have discovered many truths more useful and more important than all I had before learned, or even had expected to learn.

But because I have essayed to expound the chief of these discoveries in a treatise which certain considerations prevent me from publishing, I cannot make the results known more conveniently than by here giving a summary of the contents of this treatise. It was my design to comprise in it all that, before I set myself to write it, I thought I knew of the nature of material objects. But like the painters who, finding themselves unable to represent equally well on a plain surface all the different faces of a solid body, select one of the chief, on which alone they make the light fall, and throwing the rest into the shade, allow them to appear only in so far as they can be seen while looking at the principal one; so, fearing lest I should not be able to compense in my discourse all that was in my mind, I resolved to expound singly, though at considerable length, my opinions regarding light; then to take the opportunity of adding something on the sun and the fixed stars, since light almost wholly proceeds from them; on the heavens since they transmit it; on the planets, comets, and earth, since they reflect it; and particularly on all the bodies that are upon the earth, since they are either colored, or transparent, or luminous; and finally on man, since he is the spectator of these objects. Further, to enable me to cast this variety of subjects somewhat into the shade, and to express my judgment regarding them with greater freedom, without being necessitated to adopt or refute the opinions of the learned, I resolved to leave all the people here to their disputes, and to speak only of what would happen in a new world, if God were now to create somewhere in the imaginary spaces matter sufficient to compose one, and were to agitate variously and confusedly the different parts of this matter, so that there resulted a chaos as disordered as the poets ever feigned, and after that did nothing more than lend his ordinary concurrence to nature, and allow her to act in accordance with the laws which he had established. On this supposition, I, in the first place, described this matter, and essayed to represent it in such a manner that to my mind there can be nothing clearer and more intelligible, except what has been recently said regarding God and the soul; for I even expressly supposed that it possessed none of those forms or qualities which are so debated in the schools, nor in general anything the knowledge of which is not so natural to our minds that no one can so much as imagine himself ignorant of it. Besides, I have pointed out what are the laws of nature; and, with no other principle upon which to found my reasonings except the infinite perfection of God, I endeavored to demonstrate all those about which there could be any room for doubt, and to prove that they are such, that even if God had created more worlds, there could have been none in which these laws were not observed. Thereafter, I showed how the greatest part of the matter of this chaos must, in accordance with these laws, dispose and arrange itself in such a way as to present the appearance of heavens; how in the meantime some of its parts must compose an earth and some planets and comets, and others a sun and fixed stars. And, making a digression at this stage on the subject of light, I expounded at considerable length what the nature of that light must be which is found in the sun and the stars, and how thence in an instant of time it traverses the immense spaces of the heavens, and how from the planets and comets it is reflected towards the earth. To this I likewise added much respecting the substance, the situation, the motions, and all the different qualities of these heavens and stars; so that I thought I had said enough respecting them to show that there is nothing observable in the heavens or stars of our system that must not, or at least may not appear precisely alike in those of the system which I described. I came next to speak of the earth in particular, and to show how, even though I had expressly supposed that God had given no weight to the matter of which it is composed, this should not prevent all its parts from tending exactly to its center; how with water and air on its surface, the disposition of the heavens and heavenly bodies, more especially of the moon, must cause a flow and ebb, like in all its circumstances to that observed in our seas, as also a certain current both of water and air from east to west, such as is likewise observed between the tropics; how the mountains, seas, fountains, and rivers might naturally be formed in it, and the metals produced in the mines, and the plants grow in the fields and in general, how all the bodies which are commonly denominated mixed or composite might be generated and, among other things in the discoveries alluded to inasmuch as besides the stars, I knew nothing except fire which produces light, I spared no pains to set forth all that pertains to its nature,—the manner of its production and support, and to explain how heat is sometimes found without light, and light without heat; to show how it can induce various colors upon different bodies and other diverse qualities; how it reduces some to a liquid state and hardens others; how it can consume almost all bodies, or convert them into ashes and smoke; and finally, how from these ashes, by the mere intensity of its action, it forms glass: for as this transmutation of ashes into glass appeared to me as wonderful as any other in nature, I took a special pleasure in describing it.

I was not, however, disposed, from these circumstances, to conclude that this world had been created in the manner I described; for it is much more likely that God made it at the first such as it was to be. But this is certain, and an opinion commonly received among theologians, that the action by which he now sustains it is the same with that by which he originally created it; so that even although he had from the beginning given it no other form than that of chaos, provided only he had established certain laws of nature, and had lent it his concurrence to enable it to act as it is wont to do, it may be believed, without discredit to the miracle of creation, that, in this way alone, things purely material might, in course of time, have become such as we observe them at present; and their nature is much more easily conceived when they are beheld coming in this manner gradually into existence, than when they are only considered as produced at once in a finished and perfect state.

From the description of inanimate bodies and plants, I passed to animals, and particularly to man. But since I had not as yet sufficient knowledge to enable me to treat of these in the same manner as of the rest, that is to say, by deducing effects from their causes, and by showing from what elements and in what manner nature must produce them, I remained satisfied with the supposition that God formed the body of man wholly like to one of ours, as well in the external shape of the members as in the internal conformation of the organs, of the same matter with that I had described, and at first placed in it no rational soul, nor any other principle, in room of the vegetative or sensitive soul, beyond kindling in the heart one of those fires without light, such as I had already described, and which I thought was not different from the heat in hay that has been heaped together before it is dry, or that which causes fermentation in new wines before they are run clear of the fruit. For, when I examined the kind of functions which might, as consequences of this supposition, exist in this body, I found precisely all those which may exist in us independently of all power of thinking, and consequently without being in any measure owing to the soul; in other words, to that part of us which is distinct from the body, and of which it has been said above that the nature distinctively consists in thinking, functions in which the animals void of reason may be said wholly to resemble us; but among which I could not discover any of those that, as dependent on thought alone, belong to us as men, while, on the other hand, I did afterwards discover these as soon as I supposed God to have created a rational soul, and to have annexed it to this body in a particular manner which I described.

PART VI

Three years have now elapsed since I finished the treatise containing all these matters; and I was beginning to revise it, with the view to put it into the hands of a printer, when I learned that persons to whom I greatly defer, and whose authority over my actions is hardly less influential than is my own reason over my thoughts, had condemned a certain doctrine in physics, published a short time previously by another individual to which I will not say that I adhered, but only that, previously to their censure I had observed in it nothing which I could imagine to be prejudicial either to religion or to the state, and nothing therefore which would have prevented me from giving expression to it in writing, if reason had persuaded me of its truth; and this led me to fear lest among my own doctrines likewise some one might be found in which I had departed from the truth, notwithstanding the great care I have always taken not to accord belief to new opinions of which I had not the most certain demonstrations, and not to give expression to aught that might tend to the hurt of any one. This has been sufficient to make me alter my purpose of publishing them; for although the reasons by which I had been induced to take this resolution were very strong, yet my inclination, which has always been hostile to writing books, enabled me immediately to discover other considerations sufficient to excuse me for not undertaking the task. And these reasons, on one side and the other, are such, that not only is it in some measure my interest here to state them, but that of the public, perhaps, to know them.

I have never made much account of what has proceeded from my own mind; and so long as I gathered no other advantage from the method I employ beyond satisfying myself on some difficulties belonging to the speculative sciences, or endeavoring to regulate my actions according to the principles it taught me, I never thought myself bound to publish anything respecting it. For in what regards manners, every one is so full of his own wisdom, that there might be found as many reformers as heads, if any were allowed to take upon themselves the task of mending them, except those whom God has constituted the supreme rulers of his people or to whom he has given sufficient grace and zeal to be prophets; and although my speculations greatly pleased myself, I believed that others had theirs, which perhaps pleased them still more. But as soon as I had acquired some general notions respecting physics, and beginning to make trial of them in various particular difficulties, had observed how far they can carry us, and how much they differ from the principles that have been employed up to the present time, I believed that I could not keep them concealed without sinning grievously against the law by which we are bound to promote, as far as in us lies, the general good of mankind. For by them I perceived it to be possible to arrive at knowledge highly useful in life; and in room of the speculative philosophy usually taught in the schools, to discover a practical, by means of which, knowing the force and action of fire, water, air the stars, the heavens, and all the other bodies that surround us, as distinctly as we know the various crafts of our artisans, we might also apply them in the same way to all the uses to which they are adapted, and thus render ourselves the lords and possessors of nature. And this is a result to be desired, not only in order to the invention of an infinity of arts, by which we might be enabled to enjoy without any trouble the fruits of the earth, and all its comforts, but also and especially for the preservation of health, which is without doubt, of all the blessings of this life, the first and fundamental one; for the mind is so intimately dependent upon the condition and relation of the organs of the body, that if any means can ever be found to render men wiser and more ingenious than hitherto, I believe that it is in medicine they must be sought for. It is true that the science of medicine, as it now exists, contains few things whose utility is very remarkable: but without any wish to depreciate it, I am confident that there is no one, even among those whose profession it is, who does not admit that all at present known in it is almost nothing in comparison of what remains to be discovered; and that we could free ourselves from an infinity of maladies of body as well as of mind, and perhaps also even from the debility of age, if we had sufficiently ample knowledge of their causes, and of all the remedies provided for us by nature. But since I designed to employ my whole life in the search after so necessary a science, and since I had fallen in with a path which seems to me such, that if any one follow it he must inevitably reach the end desired, unless he be hindered either by the shortness of life or the want of experiments, I judged that there could be no more effectual provision against these two impediments than if I were faithfully to communicate to the public all the little I might myself have found, and incite men of superior genius to strive to proceed farther, by contributing, each according to his inclination and ability, to the experiments which it would be necessary to make, and also by informing the public of all they might discover, so that, by the last beginning where those before them had left off, and thus connecting the lives and labours of many, we might collectively proceed much farther than each by himself could do.

I remarked, moreover, with respect to experiments, that they become always more necessary the more one is advanced in knowledge; for, at the commencement, it is better to make use only of what is spontaneously presented to our senses, and of which we cannot remain ignorant, provided we bestow on it any reflection, however slight, than to concern ourselves about more uncommon and recondite phenomena: the reason of which is, that the more uncommon often only mislead us so long as the causes of the more ordinary are still unknown; and the circumstances upon which they depend are almost always so special and minute as to be highly difficult to detect. But in this I have adopted the following order: first, I have essayed to find in general the principles, or first causes of all that is or can be in the world, without taking into consideration for this end anything but God himself who has created it, and without educing them from any other source than from certain germs of truths naturally existing in our minds In the second place, I examined what were the first and most ordinary effects that could be deduced from these causes; and it appears to me that, in this way, I have found heavens, stars, an earth, and even on the earth water, air, fire, minerals, and some other things of this kind, which of all others are the most common and simple, and hence the easiest to know. Afterwards when I wished to descend to the more particular, so many diverse objects presented themselves to me, that I believed it to be impossible for the human mind to distinguish the forms or species of bodies that are upon the earth, from an infinity of others which might have been, if it had pleased God to place them there, or consequently to apply them to our use, unless we rise to causes through their effects, and avail ourselves of many particular experiments. Thereupon, turning over in my mind I the objects that had ever been presented to my senses I freely venture to state that I have never observed any which I could not satisfactorily explain by the principles had discovered. But it is necessary also to confess that the power of nature is so ample and vast, and these principles so simple and general, that I have hardly observed a single particular effect which I cannot at once recognize as capable of being deduced in man different modes from the principles, and that my greatest difficulty usually is to discover in which of these modes the effect is dependent upon them; for out of this difficulty cannot otherwise extricate myself than by again seeking certain experiments, which may be such that their result is not the same, if it is in the one of these modes at we must explain it, as it would be if it were to be explained in the other. As to what remains, I am now in a position to discern, as I think, with sufficient clearness what course must be taken to make the majority those experiments which may conduce to this end: but I perceive likewise that they are such and so numerous, that neither my hands nor my income, though it were a thousand times larger than it is, would be sufficient for them all; so that according as henceforward I shall have the means of making more or fewer experiments, I shall in the same proportion make greater or less progress in the knowledge of nature. This was what I had hoped to make known by the treatise I had written, and so clearly to exhibit the advantage that would thence accrue to the public, as to induce all who have the common good of man at heart, that is, all who are virtuous in truth, and not merely in appearance, or according to opinion, as well to communicate to me the experiments they had already made, as to assist me in those that remain to be made.

Case Studies

. . . .

Astronomy Track

PAPER 1: ARISTOTLE ON THE MOTION OF THE HEAVENS

THE CASE

A central question in all of ancient thought was, "Why do things change?" It made sense why they would stay the same, but the first questions about the natural world had to do with things that changed. Among the changes that were most obvious to the ancients were the positions of objects in the sky. The sun, moon, and all the constellations of stars rise in the east and set in the west every day. While the exact point on the horizon where the sun rises, how long it is in the sky, and the path it takes are not the same day to day, they are periodic, repeating over and over again each year.

The fixed stars, those in the constellations, also rise and set, and are always in the same relative places. The constellations themselves remain stable, as do the distances in the sky between them. However, like the sun, their motion varies daily in a regular fashion. This motion, however, appears to be quicker than that of the sun so that the stars that appear on the horizon at midnight the night before will be a bit higher in the sky the next night, and higher the night after that, and so on, returning to the horizon at midnight 365 days later. Charted out, these stars rotate around the North Star.

These are well-established facts, and were for the ancients as well, but what accounts for the change and its periodic regularity?

THE SCIENTIST

Aristotle (384–322 BC) was the son of Nicomachus, the physician of King Amyntas of Macedonia. He studied with Plato at his school, the Academy, for twenty years, until Plato's death. He was invited to return to Macedonia by Philip in order to become the personal tutor of his son Alexander, who would later conquer much of the known world. Upon Philip's death, Aristotle returned to Athens and opened his own school, called the Lyceum. Upon the death of Alexander, Aristotle was chased from Athens to Chalcis, where he died soon after. His treatises on astronomy, physics, chemistry, biology, meteorology, and many other subjects are among the first systematic works on the natural world based on observation.

YOUR JOB

Ancient astronomers had to explain the periodicity of the motions of heavenly bodies. Your job is to read the excerpt of Aristotle's astronomical work (beginning on p. 327) and explain his answer. Explain the deductivist approach to science. How would a deductivist understand Aristotle's work setting out a picture of the heavens? Discuss in detail how Aristotle argues from metaphysical principles that the motion of the heavens must be eternal, circular, and regular. What are the metaphysical first principles he used and what are his deductive arguments from them to his resulting claims about heavenly motion? Does Aristotle's work remain truly within the deductivist method or is there another sort of reasoning also in play?

RESEARCH HINTS
- When reading the historical selection and the philosophical essays, underline passages that express the author's conclusion and supporting premises. Use these as direct quotations in your paper—type them out and then explain what the passage means and how it is used by the author in his argument.
- Structure is very important in a philosophical essay. Make clear what is being argued in each step of your essay by using strong thesis sentences and section headings.

Physics Track

PAPER 1: ANCIENT ATOMISM

THE CASE

Classical atomists occupied the middle ground between two other views. On the one hand, those following Parmenides argued that the entire universe was a single, indivisible, unchanging entity. There were no parts to reality to interact and all change, including motion, as Zeno famously argued, is illusory. On the other hand, those like Aristotle contended that matter is infinitely divisible. No matter how small the bit of matter, we could always cut it in half, and in half again, and in half again. . . . For Aristotle the ultimate components of matter were what we would now call states: air or gas, water or liquid, earth or solid, and fire or heat, which we now see as properties of the object and not separate states.

Atomists, differing from both, contend that all of reality is comprised of atoms, indivisible or uncuttable bits of matter bouncing around in the void, the pure emptiness of space. There is no such thing as a soul; rather, the universe is completely comprised of vacuous space and indivisible matter moving within it. These basic pieces interact with each according to well-defined, deterministic laws. The qualities of the individual atoms—particularly their shapes, their interactions, and contact between them as they move in the void—completely and uniquely determine the observable qualities of the resulting composite objects. Atoms of different shapes will fit together in different configurations and the falling and swerving of atoms moving in the void will bring them into contact. The world of the atomists was a tidy place.

THE SCIENTIST

Epicurus (341–270 BC) was a Greek thinker a generation after the death of Aristotle. Following the views of Democritus, the father of atomism, Epicurus argued that the world was comprised of atoms that interacted according to well-ordered laws. He believed that humans are capable of a well-lived life in such a completely material universe only if we free ourselves from anxiety and desire. For this reason, he contended that one should avoid politics, sex, religion, and marriage, focusing instead on friendship and questions we could answer.

YOUR JOB

Read Epicurus's letter to Herodotus (beginning on p. 330). What is Epicurus's argument for the finite divisibility of matter? Explain the deductivist approach to the scientific method in detail. On what metaphysical principles do Epicurus base his position? What aspects of Epicurus's work fit well into the deductivist scheme and what aspects deviate from it?

RESEARCH HINTS

- When reading the historical selection and the philosophical essays, underline passages that express the author's conclusion,and supporting premises. Use these as direct quotations in your paper—type them out and then explain what the passage means and how it is used by the author in his argument.
- Structure is very important in a philosophical essay. Make clear what is being argued in each step of your essay by using strong thesis sentences and section headings.

Chemistry Track

PAPER 1: HERMES TRISMEGISTUS, PARACELSUS, AND ALCHEMY

THE CASE

The first theories of matter come from the artisans who concerned themselves with metallurgy, dye making, beer brewing, and other practical uses of those substances found around them. They created lists of cookbook-type recipes telling how they took what they found in nature and used it to isolate other substances. Explanations for the successes of these methods led thinkers to believe that the substances we observe are in fact combinations of more basic substances—air, water, earth, and fire. Notice that these correspond to what we now call the states of matter. The ancients developed theories about the interconnection of these basic elements. We can see them organized, for example, in terms of their basic qualities of hot/cold and wet/dry. Fire is hot and dry while air is hot and wet. Earth is cold and dry while water is cold and wet. We can combine the elements and then operate upon them, heating or cooling, for example, in order to alter their properties and create new substances as a result. The system of such methods of affecting the essences of substances and thereby creating new ones, what was called "transmutation," was alchemy.

In the Renaissance, this system was revived. Out of fear that the knowledge could be misused by the greedy and in an attempt to make sure their knowledge was well-guarded, alchemists began to use a code in which to transmit their recipes. This code used mythological figures and could be interpreted as mystical stories as well as practical instructions. In this way, a mystique grew up around the practice. Adding to this reputation, the basic writings were claimed to have been rediscovered ancient Egyptian knowledge, that of Hermes Trismegistus. The Emerald Tablet was the touchstone of alchemy, providing the basic principles—the axioms of the system.

In addition to the transmutation of metals, medicine was also a significant interest of the alchemists. In this realm, the primary figure was Paracelsus, who may be considered the forefather of biochemistry. Following the traditional alchemical system, Paracelsus contended that one must understand the interactions of the basic substances—mercury, sulfur, and salt—in terms of the four elements. For the Christian Paracelsus, that the basic elements created by God would be three in number itself had mystical meaning, adding yet another level to the foundations of alchemy. Previous alchemists had held that everything was comprised of sulfur and mercury—the more sulfur, the more combustible the substance was, and relatively more mercury, the less so. Paracelsus added salt as the third element with the combustible sulfur representing the spirit, mercury the soul, and salt the body. Human beings as a combination of the three are a microcosm of the larger Creation and illnesses could be thought to be particular imbalances that could be cured by ingesting the correct restorative substance.

THE SCIENTIST

Theophrastus Bombastus von Hohenheim (1493–1541), better known as Paracelsus, was born in Zurich, Switzerland, the son of a doctor. As a boy, he worked in the mines and metallurgic shop of a noted alchemist and became well acquainted with the field before studying medicine. Upon becoming a professor of medicine, he became a contentious figure as the first university professor to lecture in German instead of Latin. He also began his courses by burning the books of Galen and Avicenna, the major authorities of medicine at the time, and inviting apothecaries and barber-surgeons—

common, nonacademic medical practitioners who were looked down upon by the professors—to give lectures at the university.

YOUR JOB
Read the excerpts of Paracelsus's *Hermetic and Alchemical Writings* (beginning on p. 332) and explain Paracelsus's doctrine of three ingredients and four elements. Discuss how this theory accounts for the claim by alchemists that all things can be transmuted into any other thing, if appropriately treated. Explain deductivism, making clear the role of metaphysical principles. How would a deductivist make sense of Paracelsus's alchemical approach? What are the metaphysical principles he makes use of? Why, for example, does he add a third basic ingredient? Is there any part of Paracelsus's work that does not fit well into the deductivists' picture of the scientific method?

RESEARCH HINTS

- When reading the historical selection and the philosophical essays, underline passages that express the author's conclusion and supporting premises. Use these as direct quotations in your paper—type them out and then explain what the passage means and how it is used by the author in his argument.
- Structure is very important in a philosophical essay. Make clear what is being argued in each step of your essay by using strong thesis sentences and section headings.

Genetics Track

PAPER 1: ARISTOTLE'S GERM THEORY

THE CASE

For the ancient Greeks, the nature of change was one of the central questions. We know why things stay the same, but why do they change? The most remarkable change is coming to be, the creation of a new thing. When an animal has an offspring, where did it come from and how does it know to develop in ways that make it the sort of thing that its parents are?

Aristotle's answer is his germ theory. Spending time on the island of Lesbos after leaving Athens following Plato's death, Aristotle conducted careful and detailed examinations of plants and animals. His meticulous documentation of the anatomical features of living things led him to a theory of animal reproduction. Males, according to this approach, provide sperm that carries the form of the offspring. It is the potential being that becomes actualized when it finds the matter in the female. The male provides the form, but the female provides the actual material that becomes the offspring. When form meets matter, a new entity is created.

The nature of the male is to be active and this is reflected in body and in the means of participation in the reproductive act. The nature of the female is passive; this, too, is reflected in the females' receptive role in the reproductive act. Since all things must have a form and matter, and the form is that which guides goal oriented change, it must be the active participant, or the male, that provides the form and it must be the passive participant that provides the matter. This, Aristotle contends, explains why it is women who give birth, but that birth is not possible without the semen of a male to originate the change from egg to embryo and to provide the internal instructions that guide the change and growth with the additional material from the mother's body.

THE SCIENTIST

Aristotle (384–322 BC) was the son of Nicomachus, the physician of King Amyntas of Macedonia. He studied with Plato at his school, the Academy, for twenty years, until Plato's death. He was invited to return to Macedonia by Philip in order to become the personal tutor of his son Alexander, who would later conquer much of the known world. Upon Philip's death, Aristotle returned to Athens and opened his own school, called the Lyceum. Upon the death of Alexander, Aristotle was chased from Athens to Chalcis, where he died soon after. His treatises on astronomy, physics, chemistry, biology, meteorology, and many other subjects, are among the first systematic works on the natural world based on observation.

YOUR JOB

Read the excerpt of Aristotle's *On the Generation of Animals* (beginning on p. 334) and explain his theory of reproduction. What parts of the offspring come from the mother? What parts come from the father? Why do the offspring and the father always have similar anatomical characteristics? Explain deductivism. How would a deductivist make sense of Aristotle's account of reproduction? What are the general principles he makes use of? Is there any part of this theory that does not fit in well with the deductivist approach?

RESEARCH HINTS

- When reading the historical selection and the philosophical essays, underline passages that express the author's conclusion and supporting premises. Use these as direct quotations in your paper—type them out and then explain what the passage means and how it is used by the author in his argument.
- Structure is very important in a philosophical essay. Make clear what is being argued in each step of your essay by using strong thesis sentences and section headings.

Evolutionary Biology Track

PAPER 1: ARISTOTLE'S SPECIES

THE CASE

For the ancients, the central question was the nature of change. It made sense why things stayed the same, but why and how they changed was something they deemed to be in need of explanation. Of all the changes they could observe, little was as mysterious or as prevalent as birth. Living things, things that had not previously existed, would suddenly come into being. It was truly a marvel and something that had to be accounted for.

For Aristotle, the cause of offspring could be explained by answering four questions, each corresponding to one of his four notions of cause. Spending time on the island of Lesbos, Aristotle conducted careful and detailed examinations of plants and animals. His meticulous documentation of the anatomical features of living things led him to recognize all sorts of regularities and allowed him to group them according to these characteristics. Aristotle was the first naturalist to devise a complex and extensive catalog of species.

While he thought that some reproduce themselves, some arise from other sorts of animals (for example, sterile hybrids like the mule), and others arise spontaneously (for example, maggots in rotting flesh), he believed all living things must possess a soul that serves to guide its growth. The soul contains the potentiality that is the final cause of the growth. An acorn, for example, is a potential oak tree and the soul of the oak tree that the acorn is to become resides from the start within the seed. The soul is not material and cannot be made from nothing, and so would have to preexist and reside within the seed.

In this way, reproductive beings give rise to their own sort. The result is that while individuals are corruptible and doomed to perish, the species may live on. Species are then categorized and capable of being ranked in a ladder of nature from the inanimate elements, up through plants, to animals, that are capable of motion and perception, and up to humans, who have reason.

THE SCIENTIST

Aristotle (384–322 BC) was the son of Nicomachus, the physician of King Amyntas of Macedonia. He studied with Plato at his school, the Academy, for twenty years, until Plato's death. He was invited to return to Macedonia by Philip in order to become the personal tutor of his son Alexander, who would later conquer much of the known world. Upon Philip's death, Aristotle returned to Athens and opened his own school, called the Lyceum. Upon the death of Alexander, Aristotle was chased from Athens to Chalcis, where he died soon after. His treatises on astronomy, physics, chemistry, biology, meteorology, and many other subjects, are among the first systematic works on the natural world based on observation.

YOUR JOB

In *On the Generation of Animals,* Aristotle argues that individuals are members of species that existed before and will continue on after them. Each species is distinguished by a type of soul. Read the excerpt (beginning on p. 336) and explain the role of the soul in reproduction according to Aristotle. How does he rank the species? Explain the deductivist approach to science. How would an advocate of the deductivist system make sense of Aristotle's biology? What are the general principles he makes use of? Are there any parts of this work that seem not to fit in well with the deductivist approach? If so, what are they and why don't they fit?

RESEARCH HINTS

- When reading the historical selection and the philosophical essays, underline passages that express the author's conclusion and supporting premises. Use these as direct quotations in your paper—type them out and then explain what the passage means and how it is used by the author in his argument.
- Structure is very important in a philosophical essay. Make clear what is being argued in each step of your essay by using strong thesis sentences and section headings.

Geology Track

PAPER 1: JOHN WOODWARD, CATASTROPHISM, AND THE GREAT FLOOD

Low Soil
bis changed

THE CASE

When geological theory first began to devise accounts of the earth's formation, several prominent phenomena cried out for explanation. Miners tunneling into the earth noticed that there were fairly well-defined strata of rock, layers where one type of rock would sit above another type of rock with a sharp demarcation between them. Further, there was, for the most part, a regular order in which these layers would appear.

A second noticeable phenomenon came to light as more and more scientists explored the tops of mountains and examined the rock strata at the top. It baffled many to find the fossilized shells of animals that were clearly sea dwellers. How had the remains of aquatic beings ended up in a place so incredibly far removed from any available body of water?

John Woodward was part of a movement termed "catastrophism," in which geological features were held to be the result of major earth-changing events. These events were not regularly recurring, but occasional, such as major floods or earthquakes. In Woodward's case, it was water that was held to be operative force in creating the observed structure of the world.

Strata result from settling during and following periods of flooding, and the unexpected placement of sea fossils follow from the fact that the water level had once been higher so that some part of the mountains were indeed below sea level, only to have the waters later recede. This view fit in nicely with the biblical story of the Great Flood, allowing contemporary religious and scientific views to dovetail. If the features of the face of the earth could be explained by water, then there would be scientific reason to suppose that theological accounts were true.

THE SCIENTIST

John Woodward (1665–1722) trained as a biologist but became fascinated by fossils. This triggered an interest in geology more generally and he was among the first to undertake it as a serious study. He amassed one of the greatest collections of fossils among early scientists and it remains to this day at Cambridge University.

YOUR JOB

Read the excerpt of *An Essay towards a Natural History of the Earth* (beginning on p. 339) and explain Woodward's catastrophism. How did this view explain the observed geological structure of the world? Explain deductivism. How would a deductivist account for Woodward's work? What metaphysical postulates does Woodward make use of? Are there aspects to Woodward's work that the deductivist cannot account for?

RESEARCH HINTS

- When reading the historical selection and the philosophical essays, underline passages that express the author's conclusion and supporting premises. Use these as direct quotations in your paper—type them out and then explain what the passage means and how it is used by the author in his argument.
- Structure is very important in a philosophical essay. Make clear what is being argued in each step of your essay by using strong thesis sentences and section headings.

Psychology Track

PAPER 1: HIPPOCRATES, HUMOURS, AND MENTAL ILLNESS

THE CASE

The greatest of the ancient Greek physicians was Hippocrates. Modeling the understanding of the human body on the ancient understanding of the universe, which is made up of four elements—earth, air, fire, and water—Hippocrates argued that the body has four humours, or vital fluids—blood, phlegm, yellow bile, and black bile. In line with the Greek love of symmetry, good health was seen as the result of a proper balance of these humours and illness a result of imbalance.

Hippocrates was the first to realize that problems like epilepsy and depression (which he termed melancholia) were diseases of the brain. Just as the elements were warm and cold and moist and dry (fire = dry and warm; earth = dry and cold; air = moist and warm; water = moist and cold), so too were the humours (yellow bile = dry and warm; black bile = dry and cold; blood = moist and warm; phlegm = moist and cold). The heat of the elements, Hippocrates argued, had an effect on the working of the brain and therefore the mind. An excess of heat and the result is fear—think of the effects of being scared: sweating, red face, and feeling flushed. Indeed, this accounted for seasonal changes in mood. Why are we sadder in winter and happier with the arrival of spring? Providing the first ever account of seasonal affective disorder, Hippocrates argued that it was because of the change from heat to cold.

But excess heat could be the result of internal imbalances as well. An excess of black bile, for example, would cause the brain to heat up beyond its natural state, resulting in mental illness, specifically melancholia. Depression, according to Hippocrates, is a disease caused by the imbalance of natural substances, a disturbance in the natural harmony of the healthy body.

THE SCIENTIST

Hippocrates (460–377 BC) was born in Cos, which was home of a sanctuary for the guild of Asclepius, the Greek god of medicine. While Asclepius is mentioned along with Apollo in the famous oath that bears Hippocrates' name, his approach to medicine was systematic rather than mystical. He devoted significant effort to understanding the causes of epilepsy and pneumonia. Hippocrates was a traveling teacher and healer; his writings are likely the work of students in the Hippocratic School rather than that of Hippocrates himself, although the views expressed and aphorisms contained are considered to be reflective of his views.

YOUR JOB

Read the excerpts of *The Nature of Man* and *The Sacred Disease* (beginning on pp. 342 and 343, respectively) and explain Hippocrates' explanation of mental illness. How do mental states relate to physical states? Explain deductivism. How would a deductivist make sense of Hippocrates' work? What metaphysical principles are employed? Are there aspects of Hippocrates' account of mental illness that do not fit in well with the deductivists' approach?

- When reading the historical selection and the philosophical essays, underline passages that express the author's conclusion and supporting premises. Use these as direct quotations in your paper—type them out and then explain what the passage means and how it is used by the author in his argument.
- Structure is very important in a philosophical essay. Make clear what is being argued in each step of your essay by using strong thesis sentences and section headings.

Sociology Track

PAPER 1: HOBBES, COMMONWEALTH, AND LAWS OF NATURE

THE CASE

Thomas Hobbes was born during some of the most tumultuous fighting of the British Civil War. The experiences of seeing social order collapse and reassert itself left a mark on him, and he sought to understand the structure of society, a structure that he thought must follow from reason and the circumstances in which humans find themselves.

If humans are now in society, there must have been an earlier time when it first formed, and, indeed, a state before that. It is this development of civilization that separated humans from the rest of the animal kingdom, so without civil society humans would be in a state of nature like everything else. Given that one always has the innate imperative to do whatever one believes to be in the best interest of personal survival, and given that others may pose a threat to one's survival, the state of nature necessarily is a state of war.

But the state of war is itself a threat to our survival, and so it is a natural law that we must desire peace. Peace requires that we cease being threats to one another, but if someone were to be the only one to let down his or her guard, then peace would not result, only unfortunate bodily circumstances for that one individual. For peace to be possible, we must all renounce our rights to everything that can help us survive (including and specifically our right to kill those who could harm us) at the same time. Further, there must be something that makes sure we abide by this promise to one another.

It is in this way that society forms necessarily and as a result of natural law. The commonwealth, Hobbes argues, is a result of our dictated desire to seek and maintain peace for our own self-interest, something we must follow.

Thomas Hobbes (1588–1679) was a child of conflict who was born as the Spanish Armada approached England. His father was a strong-headed vicar who lost his job fighting with another vicar from an adjacent congregation. Hobbes's dark sense of the world pervaded his worldview and he was a strong and vocal opponent of democracy, a view that led to his exile during the English Civil War. This exile, however, put him in touch with some of the leading intellectuals in Europe, allowing him to meet Galileo and to be one of the few to read and comment upon a copy of Descartes' *Meditations* before publication.

YOUR JOB

Read the excerpt from *Leviathan* (beginning on p. 344) and explain in detail how society forms according to Hobbes. How do the laws of nature guide human beings, as rational animals, to have to form civilized society in certain necessary ways? Explain deductivism. How would a deductivist make sense of Hobbes's work? What metaphysical principles is he making use of? Are there any aspects to Hobbes's system that do not fit in well with this view?

RESEARCH HINTS

- When reading the historical selection and the philosophical essays, underline passages that express the author's conclusion and supporting premises. Use these as direct quotations in your paper—type them out and then explain what the passage means and how it is used by the author in his argument.
- Structure is very important in a philosophical essay. Make clear what is being argued in each step of your essay by using strong thesis sentences and section headings.

Economics Track

PAPER 1: ARISTOTLE AND CURRENCY

THE CASE

The ancient Greeks were fascinated by change. The point of science was to be able to understand the notion of change. This could be change in the heavens, in the growth of a plant or animal, or within a state or household. In this way, to understand economics, one must understand the final cause of exchange, that for the sake of which it occurs. Household management is the art of providing an environment for human flourishing.

Since one will not naturally have all one needs in one's household, some sense of exchange and wealth acquisition must be a natural part of life and therefore liable to scientific understanding. But Aristotle saw two arts of wealth-getting: one natural and therefore a science, but the second, which is subtly different, unnatural and not a science.

While acquisition of wealth is part of nature, coin, that is, money, is an artificial human invention, and the art of maximizing one's monetary wealth, that is, retail trade, fails to focus on the conditions that would allow for human flourishing. Indeed, to be concerned with amassing wealth without limit is to turn away from those things that would create a life well lived.

If we truly want to understand the way exchange works, we must understand the role and limits of property in human well-being. It is too easy to think of wealth-getting as a good in itself, instead of money as a good for something else and this takes us away from science and nature.

THE SCIENTIST

Aristotle (384–322 BC) was the son of Nicomachus, the physician of King Amyntas of Macedonia. He studied with Plato at his school, the Academy, for twenty years, until Plato's death. He was invited to return to Macedonia by Philip in order to become the personal tutor of his son Alexander, who would later conquer much of the known world. Upon Philip's death, Aristotle returned to Athens and opened his own school, called the Lyceum. Upon the death of Alexander, Aristotle was chased from Athens to Chalcis, where he died soon after. His treatises on astronomy, physics, chemistry, biology, meteorology, and many other subjects, are among the first systematic works on the natural world based on observation.

YOUR JOB

Read the excerpt from the *Politics* (beginning on p. 347) and explain Aristotle's view of exchange. Explain the deductivist approach to science. How would a deductivist makes sense of Aristotle's approach to economics? What metaphysical principles is he making use of? Are there aspects of Aristotle's approach that do not fit in well with this view?

RESEARCH HINTS

- When reading the historical selection and the philosophical essays, underline passages that express the author's conclusion and supporting premises. Use these as direct quotations in your paper—type them out and then explain what the passage means and how it is used by the author in his argument.
- Structure is very important in a philosophical essay. Make clear what is being argued in each step of your essay by using strong thesis sentences and section headings.

INDUCTIVISM

While the reliance on demonstration gave the deductivists the advantage of a scientific method that gave results with guaranteed certain truth, deductivism struck many as deeply problematic. By setting up deduction as the central form of inference employed in scientific reasoning, it forced all scientific laws to be derived down from higher-level truths: metaphysical claims whose meaning and verification required discussing topics far beyond the world we see, feel, hear, taste, and smell. Indeed, observation has little role in science at all on this view. Maybe as a mere check to make sure your metaphysical deductions are correct, but no formative role. _observn not formal role_

But surely observable evidence, the set of statements about the way we see the world behaving, is more central to the way science is actually done. So believed a number of thinkers who sought to replace the top-down inference of deductivism with a different logical mechanism, one that included experiment and experience in a more pivotal role. This new means of reasoning is what we call "induction." _Inductn -new way observ math_

Where deductive inferences reason from the general to the specific, inductive inferences are what logicians call "ampliative," that is, their conclusions go beyond the scope of the supporting premises. Consider the following argument:

> The first swan we saw was white.
> The second swan we saw was white.
> The third swan we saw was white.
>
> . . .
>
> The 10,473,411th swan we saw was white.
> The list contains all the swans we've seen.
> Therefore, all swans are white.

The conclusion is a general proposition about all swans, that is, every single swan everywhere, at any time. The list of observed instances, while extremely large, does not include statements about visual inspections of the color of all possible swans. The inference, therefore, is one in which the conclusion is broader than the premises, it is inductive. Unlike the broad to narrow, top-down sort of reasoning in deduction, this proceeds from narrow to broad, bottom-up.

The power of inductive inferences is that from a finite set of observations, one can infer something about a potentially infinite set. Inductive inferences enhance knowledge; they take one beyond what is already known to bigger, more general truths. But, as with the rest of life, in logic there is no free lunch. This ampliative nature comes with a price. Inductive inferences, even seemingly very good ones like that above, come with some risk that the conclusion may be wrong. Indeed, the argument above was considered cogent by many until Captain Cook went to Australia and discovered black swans. Where deductive arguments, because they have their conclusions contained within their premises, give certainty, inductive arguments, because their conclusions move beyond their premises, give a high probability of truth.

The view that inductive inferences are the central logical step in the scientific method is called "inductivism." Inductivists contend that the deductivist approach that starts from high-minded, out-of-this world claims and then reasons down to the laws of nature has it completely wrong. To describe how the world behaves and the details of how its systems work, one needs to start with the world. Step one in science, according to the inductivists, is to look at the world.

The first major advocate of this view was Francis Bacon. In his work *Novum Organum,* Bacon sets out to keep the philosophers—the metaphysicians—from undermining experimental science. While not a scientist himself, Bacon first laid out in detail the approach to science that begins not with ideas, but with sense perception, with observation. This method, he readily admits, is not infallible. It will lead, especially in its earliest stages, to errors. But over time, he contends, it progressively gives results that are more and more certain. With more and more observation, our inductive inferences get stronger and stronger and our results more and more likely to be true.

While Bacon is the forefather of inductivism, it is Isaac Newton who is its great champion. In his masterwork, *The Mathematical Principles of Natural Philosophy,* the book in which Newton lays out his three laws of motion and his law of universal gravitation, he also provides us with an insight to how he came up with them. In the very middle of the book that gives us the greatest theory in the history of science, Newton also gives us a short lesson on scientific methodology by setting out his four "Rules of Reasoning." The first two provide us with versions of Occam's razor, which implores us to commit to the existence of as few entities as possible in explaining the world. The second two, however, are explicitly inductive in nature. Indeed, Newton makes perfectly clear that "we are to look upon propositions inferred by general induction from phenomena as accurately or very nearly true."

John Stuart Mill agrees with Newton that inductive inference is the logical heart of the scientific method but contends that the notion of induction must be expanded. Newton's form of inductive argument is what we call universal generalization where we move from a complete list of observed instances, "A_1 is B," "A_2 is B," "A_3 is B," ... "A_n is B," to the general conclusion "all As are B." Mill argues that working scientists need other more intricate inductive tools in their toolboxes to conduct the complex work of real science.

To that end, he proposes his four "methods of experimental inquiry," each being a new form of inductive inference. The method of agreement allows a scientist who, when setting up causes A, B, and C, observes effects a, b, and c, and when setting up causes A, D, and E, observes effects a, d, and e, to infer that A causes a. The fact that the effect a resulted each time and the only thing in common was the causal factor A means that the agreement gives us good reason to think that it was A that brought about a in both cases.

The method of difference is the inference in which a scientist who sets up causes A, B, and C and observes effects a, b, and c and then sets up only A and B and fails to observe c can infer that C is a cause of c.

The method of concomitant variations takes a situation in which causes A, B, and C bring about effects a, b, and c. If we increase the intensity of cause C, leaving the other two constant, and we see proportionately greater amounts of effect c, then we can conclude that C is a cause of c.

The method of residues lets us argue inductively by elimination. If causes A, B, and C bring about effects a, b, and c, and we know that A causes a but not c, and B causes b but not c, then we can be confident in our belief that C causes c because it is what is left over.

With these methods, Mill has sketched out a more sophisticated view of inductive inference to provide a fuller picture of how real science is done.

Francis Bacon

Novum Organum

PREFACE

Those who have taken upon them to lay down the law of nature as a thing already searched out and understood, whether they have spoken in simple assurance or professional affectation, have therein done philosophy and the sciences great injury. For as they have been successful in inducing belief, so they have been effective in quenching and stopping inquiry; and have done more harm by spoiling and putting an end to other men's efforts than good by their own. Those on the other hand who have taken a contrary course, and asserted that absolutely nothing can be known—whether it were from hatred of the ancient sophists, or from uncertainty and fluctuation of mind, or even from a kind of fullness of learning, that they fell upon this opinion—have certainly advanced reasons for it that are not to be despised; but yet they have neither started from true principles nor rested in the just conclusion, zeal and affectation having carried them much too far. The more ancient of the Greeks (whose writings are lost) took up with better judgment a position between these two extremes—between the presumption of pronouncing on everything, and the despair of comprehending anything; and though frequently and bitterly complaining of the difficulty of inquiry and the obscurity of things, and like impatient horses champing at the bit, they did not the less follow up their object and engage with nature, thinking (it seems) that this very question—viz., whether or not anything can be known—was to be settled not by arguing, but by trying. And yet they too, trusting entirely to the force of their understanding, applied no rule, but made everything turn upon hard thinking and perpetual working and exercise of the mind.

Now my method, though hard to practice, is easy to explain; and it is this. I propose to establish progressive stages of certainty. The evidence of the sense, helped and guarded by a certain process of correction, I retain. But the mental operation which follows the act of sense I for the most part reject; and instead of it I open and lay out a new and certain path for the mind

From Francis Bacon, *The Works of Francis Bacon,* ed. James Spedding, Robert Leslie Ellis, and Douglas Denon Heath, trans. James Spedding (New York: Hurd and Houghton, 1877), 59–64, 372–74.

to proceed in, starting directly from the simple sensuous perception. The necessity of this was felt, no doubt, by those who attributed so much importance to logic, showing thereby that they were in search of helps for the understanding, and had no confidence in the native and spontaneous process of the mind. But this remedy comes too late to do any good, when the mind is already, through the daily intercourse and conversation of life, occupied with unsound doctrines and beset on all sides by vain imaginations. And therefore that art of logic, coming (as I said) too late to the rescue, and no way able to set matters right again, has had the effect of fixing errors rather than disclosing truth. There remains but one course for the recovery of a sound and healthy condition—namely, that the entire work of the understanding be commenced afresh, and the mind itself be from the very outset not left to take its own course, but guided at every step; and the business be done as if by machinery. Certainly if in things mechanical men had set to work with their naked hands, without help or force of instruments, just as in things intellectual they have set to work with little else than the naked forces of the understanding, very small would the matters have been which, even with their best efforts applied in conjunction, they could have attempted or accomplished. Now (to pause a while upon this example and look in it as in a glass) let us suppose that some vast obelisk were (for the decoration of a triumph or some such magnificence) to be removed from its place, and that men should set to work upon it with their naked hands, would not any sober spectator think them mad? And if they should then send for more people, thinking that in that way they might manage it, would he not think them all the madder? And if they then proceeded to make a selection, putting away the weaker hands, and using only the strong and vigorous, would he not think them madder than ever? And if lastly, not content with this, they resolved to call in aid the art of athletics, and required all their men to come with hands, arms, and sinews well anointed and medicated according to the rules of the art, would he not cry out that they were only taking pains to show a kind of method and discretion in their madness? Yet just so it is that men proceed in matters intellectual—with just the same kind of mad effort and useless combination of forces—when they hope great things either from the number and cooperation or from the excellency and acuteness of individual wits; yea, and when they endeavor by logic (which may be considered as a kind of athletic art) to strengthen the sinews of the understanding, and yet with all this study and endeavor it is apparent to any true judgment that they are but applying the naked intellect all the time; whereas in every great work to be done by the hand of man it is manifestly impossible, without instruments and machinery, either for the strength of each to be exerted or the strength of all to be united.

Upon these premises two things occur to me of which, that they may not be overlooked, I would have men reminded. First, it falls out fortunately as I think for the allaying of contradictions and heartburnings, that the honor and reverence due to the ancients remains untouched and undiminished, while I may carry out my designs and at the same time reap the fruit of my modesty. For if I should profess that I, going the same road as the ancients, have something better to produce, there must needs have been some comparison or rivalry between us (not to be avoided by any art of words) in respect of excellency or ability of wit; and though in this there would be nothing unlawful or new (for if there be anything misapprehended by them, or falsely laid down, why may not I, using a liberty common to all, take exception to it?) yet the contest, however just and allowable, would have been an unequal one perhaps, in respect of the measure of my own powers. As it is, however (my object being to open a new way for the understanding, a way by them untried and unknown), the case is altered: party zeal and emulation are at an end, and I appear merely as a guide to point out the road—an office of small authority, and depending more upon a kind of luck than upon any ability or excellency. And thus much relates to the persons only. The other point of which I would have men reminded relates to the matter itself.

Be it remembered then that I am far from wishing to interfere with the philosophy which now flourishes, or with any other philosophy more correct and complete than this which has been or may hereafter be propounded. For I do not object to the use of this received philosophy, or others like it, for supplying matter for disputations or ornaments for discourse—for the professor's lecture and for the business of life. Nay, more, I declare openly that for these uses the philosophy which I bring forward will not be much available. It does not lie in the way. It cannot be caught up in passage. It does not flatter the understanding by conformity with preconceived notions. Nor will it come down to the apprehension of the vulgar except by its utility and effects.

Let there be therefore (and may it be for the benefit of both) two streams and two dispensations of knowledge, and in like manner two tribes or kindreds of students in philosophy—tribes not hostile or alien to each other, but bound together by mutual services; let there in short be one method for the cultivation, another for the invention, of knowledge.

And for those who prefer the former, either from hurry or from considerations of business or for want of mental power to take in and embrace the other (which must needs be most men's case), I wish that they may succeed to their desire in what they are about, and obtain what they are pursuing. But if there be any man who, not content to rest in and use the knowledge which has already been discovered, aspires to penetrate further; to overcome, not an adversary in argument, but nature in action; to seek, not pretty and probable conjectures, but certain and demonstrable knowledge—I invite all such to join themselves, as true sons of knowledge, with me, that passing by the outer courts of nature, which numbers have trodden, we may find a way at length into her inner chambers. And to make my meaning clearer and to familiarize the thing by giving it a name, I have chosen to call one of these methods or ways *Anticipation of the Mind,* the other *Interpretation* of Nature.

Moreover, I have one request to make. I have on my own part made it my care and study that the things which I shall propound should not only be true, but should also be presented to men's minds, how strangely soever preoccupied and obstructed, in a manner not harsh or unpleasant. It is but reasonable, however (especially in so great a restoration of learning and knowledge), that I should claim of men one favor in return, which is this: if anyone would form an opinion or judgment either out of his own observation, or out of the crowd of authorities, or out of the forms of demonstration (which have now acquired a sanction like that of judicial laws), concerning these speculations of mine, let him not hope that he can do it in passage or by the by; but let him examine the thing thoroughly; let him make some little trial for himself of the way which I describe and lay out; let him familiarize his thoughts with that subtlety of nature to which experience bears witness; let him correct by seasonable patience and due delay the depraved and deep-rooted habits of his mind; and when all this is done and he has begun to be his own master, let him (if he will) use his own judgment.

ON THE INTERPRETATION OF NATURE, OR THE REIGN OF MAN

1.

To generate and superinduce a new nature or new natures, upon a given body, is the labor and aim of human power: whilst to discover the form or true difference of a given nature, or the nature to which such nature is owing, or source from which it emanates (for these terms approach nearest to an explanation of our meaning), is the labor and discovery of human knowledge . . .

2.

The unhappy state of man's actual knowledge is manifested even by the common assertions of the vulgar. It is rightly laid down that true knowledge is that which is deduced from causes. The division of four causes also is not amiss: matter, form, the efficient, and end or final cause. Of these, however, the latter is so far from being beneficial, that it even corrupts the sciences, except in the intercourse of a man with a man. The discovery of form is considered desperate. As for the efficient cause and matter (according to the present system of inquiry and the received opinions concerning them, by which they are placed remote from, and without any latent process towards form), they are but desultory and superficial, and of scarcely any avail to real and active knowledge. Nor are we unmindful of our having pointed out and corrected above the error of the human mind, in assigning the first qualities of essence to forms. For although nothing exists in nature except individual bodies, exhibiting clear individual effects according to particular laws, yet in each branch of learning, that very law, its investigation, discovery, and development, are the foundation both of theory and practice. This law, therefore, and its parallel in each science, is what we understand by the term form, adopting that word because it has grown into common use, and is of familiar occurrence.

3.

He who has learnt the cause of a particular nature (such as whiteness or heat), in particular subjects only, among those which are susceptible of it, has acquired but an imperfect power. But he who has learnt the efficient and material cause (which causes are variable and mere vehicles conveying form to particular substances) may perhaps arrive at some new discoveries in matters of a similar nature, and prepared for the purpose, but does not stir the limits of things which are much more deeply rooted: whilst he who acquainted with forms, comprehends the unity of substances apparently most distinct from each other. He can disclose and bring forward, therefore, (though it has never yet been done) things which neither the vicissitudes of nature, nor the industry of experiment, nor chance itself, would have ever brought about, and which would forever have escaped man's thoughts; from the discovery of forms, therefore, results genuine theory and free practice.

4.

Although there is a most intimate connection, and almost an identity between the ways of human power and human knowledge, yet, on account of the pernicious and inveterate habit of dwelling upon abstractions, it is by far the safest method to commence and build up the sciences from those foundations which bear a relation to the practical division, and to let them mark out and limit the theoretical. We must consider, therefore, what precepts, or what direction or guide, a person would most desire, in order to generate and su-perinduce any nature upon a given body: and this not in abstruse, but in the plainest language . . .

5.
But the rule or axiom for the transformation of bodies is of two kinds. The first regards the body as an aggregate or combination of the simple natures. Thus, in gold are united the following circumstances: it is yellow, heavy, of a certain weight, malleable and ductile to a certain extent; it is not volatile, loses part of its substance by fire, melts in a particular manner, is separated and dis-solved by particular methods, and so of the other natures observable in gold. An axiom, therefore, of this kind deduces the subject from the forms of simple natures . . .

The second kind of axiom (which depends on the discovery of the latent process) does not proceed by simple natures, but by concrete bodies, as they are found in nature and in its usual course. For instance, suppose the inquiry to be, from what beginnings, in what manner, and by what process gold or any metal or stone is generated from the original menstruum, or its elements, up to the perfect mineral: or, in like manner, by what process plants are gen-erated, from the first concretion of juices in the earth, or from seeds, up to the perfect plant, with the whole successive motion, and varied and uninter-rupted efforts of nature; and the same inquiry be made as to a regularly de-duced system of generation of animals from coition to birth, and so on of other bodies . . .

9.
From the two kinds of axioms above specified, arise the two divisions of phi-losophy and the sciences, and we will use the commonly adopted terms which approach the nearest to our meaning, in our own sense. Let the investigation of forms, which (in reasoning at least, and after their own laws), are eternal and immutable, constitute metaphysics, and let the investigation of the effi-cient cause of matter, latent process, and latent conformation (which all relate merely to the ordinary course of nature, and not her fundamental and eternal laws), constitute physics . . .

10.
The object of our philosophy being thus laid down, we proceed to precepts, in the most clear and regular order. The signs for the interpretation of nature comprehend two divisions, the first regards the eliciting or creating of axioms

from experiment, the second the deducing or deriving of new experiments from axioms. The first admits of three subdivisions: 1. To the senses. 2. To the memory. 3. To the mind or reason.

For we must first prepare as a foundation for the whole, a complete and accurate natural and experimental history. We must not imagine or invent, but discover the acts and properties of nature.

But natural and experimental history is so varied and diffuse that it confounds and distracts the understanding unless it be fixed and exhibited in due order. We must, therefore, form tables and co-ordinations of instances, upon such a plan, and in such order, that the understanding may be enabled to act upon them.

Even when this is done, the understanding, left to itself and to its own operation, is incompetent and unfit to construct its axioms without direction and support. Our third ministration, therefore, must be true and legitimate induction, the very key of interpretation.

Isaac Newton

. . . .

Principia
Rules of Reasoning in Philosophy

RULE I

We are to admit no more causes of natural things than such as are both true and sufficient to explain their appearances.

To this purpose the philosophers say that Nature does nothing in vain, and more is in vain when less will serve; for Nature is pleased with simplicity, and affects not the pomp of superfluous causes.

RULE II

Therefore to the same natural effects we must, as far as possible, assign the same causes.

As to respiration in a man and a beast; the descent of stones in *Europe* and in *America;* the light of our culinary fire and of the sun; the reflection of light in the earth, and in the planets.

RULE III

The qualities of bodies, which admit neither intensification nor remission of degrees, and which are found to belong to all bodies within the reach of our experiments, are to be esteemed the universal qualities of all bodies whatsoever.

For since the qualities of bodies are only known to us by experiments, we are to hold for universal all such as universally agree with experiments; and such as are not liable to diminution can never be quite taken away. We are certainly not to relinquish the evidence of experiments for the sake of dreams and vain fictions of our own devising; nor are we to recede from the analogy of Nature, which is wont to be simple, and always consonant to itself. We no other way know the extension of bodies than by our senses, nor do these reach it in all bodies; but because we perceive extension in all that

From Isaac Newton, *Sir Isaac Newton's Mathematical Principles of Natural Philosophy and His System of the World,* trans. Andrew Motte and Florian Cajori (Berkeley: University of California Press, 1934), 398–400. Reprinted with the permission of the University of California Press.

are sensible, therefore we ascribe it universally to all others also. That abundance of bodies are hard, we learn by experience; and because the hardness of the whole arises from the hardness of the parts, we therefore justly infer the hardness of the undivided particles not only of the bodies we fell but of all others. That all bodies are impenetrable, we gather not from reason, but from sensation. The bodies which we handle we find impenetrable, and thence conclude impenetrability to be an universal property of all bodies whatsoever. That all bodies are movable, and endowed with certain powers (which we call the inertia) of preserving in their motion, or in their rest, we only infer from the extension, hardness, impenetrability, and inertia of the whole, result from the extension, hardness, impenetrability, and inertia of the parts; and hence we conclude the least particles of all bodies to be also all extended, and hard, and impenetrable, and movable, and endowed with their proper inertia. And this is the foundation of all philosophy. Moreover, that the divided but contiguous particles of bodies may be separated from one another, is a matter of observation; and, in the particles that remain undivided, our minds are able to distinguish yet lesser parts, as is mathematically demonstrated. But whether the parts so distinguished, and not yet divided, may, by the powers of Nature, be actually divided and separated from one another, we cannot certainly determine. Yet, had we the proof of but one experiment that any undivided particle, in breaking a hard and solid body, suffered a division, we might by virtue of this rule conclude that the undivided as well as the divided particles may be divided and actually separated to infinity.

Lastly, if it universally appears, by experiments and astronomical observations, that all bodies about the earth gravitate towards the earth, and that in proportion to the quantity of matter which they severally contain; that the moon likewise, according to the quantity of its matter, gravitates towards the earth; that, on the other hand, our sea gravitates towards the moon; and all the planets towards one another; and the comets in like manner towards the sun; we must, in consequence of this rule, universally allow that all bodies whatsoever are endowed with a principle of mutual gravitation. For the argument from the appearances concludes with more force for the universal gravitation of all bodies than for their impenetrability; of which, among those in the celestial regions, we have no experiments, nor any manner of observation. Not that I affirm gravity to be essential to bodies: by their *vis insita* I mean nothing but their inertia. This is immutable. Their gravity is diminished as they recede from the earth.

RULE IV

In experimental philosophy we are to look upon propositions inferred by general induction from phenomena as accurately or very nearly true, notwithstanding any contrary hypotheses that may be imagined, till such time as other phenomena occur, by which they may either be made more accurate, or liable to exceptions.

This rule we must follow, that the argument of induction may not be evaded by hypotheses.

John Stuart Mill

· ·

A System of Logic
Of Induction

PRELIMINARY OBSERVATIONS ON INDUCTION IN GENERAL

§1.

The portion of the present inquiry upon which we are now about to enter may be considered as the principle, both from its surpassing in intricacy all the other branches, and because it relates to a process which has been shown in the preceding Book to be that in which the investigation of nature essentially consists. We have found that all Inference, consequently all Proof, and all discovery of truths not self-evident, consists of inductions, and the interpretations of inductions; that all our knowledge, not intuitive, comes to us exclusively from that source. What Induction is, therefore, and what conditions render it legitimate, cannot but be deemed the main question of the science of logic—the question which includes all others. It is, however, one which professed writers on logic have almost entirely passed over. The generalities of the subject have not been altogether neglected by metaphysicians; but, for want of sufficient acquaintance with the processes by which science has actually succeeded in establishing general truths, their analyses of inductive operation, even when unexceptional as to correctness, has not been specific enough to be made the foundation of practical rules, which might be for induction itself what the rules of the syllogism are for the interpretation of induction; while those by whom physical science has been carried to its present state of improvement—and who, to arrive at a complete theory of the process, needed only to generalize, and adapt to all varieties of problems, the methods which they themselves employed in their habitual pursuits—never until very lately made any serious attempt to philosophise on the subject, nor regarded the mode in which they arrived at their conclusions as deserving study, independently of the conclusion themselves.

From John Stuart Mill, *A System of Logic* (London: Longman, 1906), 185–86, 253–56, 258–62, 263–64, 299–304.

§2.

For the purposes of the present inquiry, Induction may be defined, the operation of discovering and proving general propositions. It is true that (as already
shown) the process of indirectly ascertaining individual facts is as truly inductive as that by which we establish general truths. But it is not a different kind
of induction; it is a form of the very same process: since, on the one hand,
generals are but collections of particulars, definite in kind but indefinite in
number; and on the other hand, whenever the evidence which we derive from
observation of known cases justifies us in drawing a similar inference with respect to a whole class of cases. The inference either does not hold at all, or it
holds in all cases of a certain description; in all cases which, in certain defin-
able respects, resemble those we have observed. truth

OF THE FOUR METHODS OF EXPERIMENTAL INQUIRY

§1.

The simplest and most obvious modes of singling out from among the circumstances which precede or follow a phenomenon those with which it is really
connected by an invariable law are two in number. One is, by comparing together different instances in which the phenomenon occurs. The other is, by
comparing instances in which the phenomenon does occur, with instances in
other respects similar in which it does not. These two methods may be respectively denominated the Method of Agreement and the Method of Difference.

In illustrating these methods, it will be necessary to bear in mind the two-
fold character of inquiries into the laws of phenomena, which may be either
inquiries into the cause of a given effect, or into the effects or properties of a
given cause. We shall consider the methods in their application to either order
of investigation, and shall draw our examples equally from both.

We shall denote antecedents by the large letters of the alphabet, and the
consequents corresponding to them by the small. Let A, then, be an agent
or cause, and let the object of our inquiry be to ascertain what are the effects
of this cause. If we can either find or produce the agent A in such varieties of
circumstances that the different cases have no circumstance in common except A, then whatever effect we find to be produced in all our trials is indicated as the effect of A. Suppose, for example, that A is tried along with B and
C, and that the effect is *a b c*; and suppose that A is next tried with D and E,
but without B and C, and that the effect is *a d e*. Then we may reason thus: *b*
and *c* are not effects of A, for they were not produced by it in the second ex-

periment; nor are *d* and *e*, for they were not produced in the first. Whatever is really the effect of A must have been produced in both instances; now this condition is fulfilled by no circumstance except *a*. The phenomenon *a* could not have been the effect of B or C, since it was not produced where they were not. Therefore it is the effect of A.

For example, let the antecedent A be the contact of an alkaline substance and an oil. This combination being tried under several varieties of circumstances, resembling each other in nothing else, the results agree in the production of a greasy and detersive or saponaceous substance: it is therefore concluded that the combination of an oil and an alkali causes the production of a soap. It is thus we inquire, by the Method of Agreement, into the effect of a given cause.

In a similar manner we may inquire into the cause of a similar effect. Let *a* be the effect. Here, as shown in the last chapter, we have only the resource of observation without experiment: we cannot take a phenomenon of which we know not the origin, and try to find its mode of production by producing it: if we succeeded in such a random trial it could only be by accident. But if we can observe *a* in two different combinations, *a b c* and *a d e*; and if we know, or can discover, that the antecedent circumstances in these cases respectively were A B C and A D E, we may conclude by a reasoning similar to that in the preceding example, that A is the antecedent connected with the consequent *a* by a law of causation. B and C, we may say, cannot be causes of *a*, since on its second occurrence they were not present; nor are D and E, for they were not present on its first occurrence. A, alone of the five circumstances, was found among the antecedents of *a* in both instances.

For example, let the effect *a* be crystallisation. We compare instances in which bodies are known to assume crystalline structure, but which have no other point of agreement; and we find them to have one, and, as far as we know only one, antecedent in common: the deposition of a solid matter from a liquid state, either a state of fusion or solution. We conclude, therefore, that the solidification of a substance from a liquid state is an invariable antecedent of its crystallisation.

In this example we may go farther, and say, it is not only the invariable antecedent, but the cause, or at least the proximate event which completes the cause. For in this case we are able, after detecting the antecedent A, to produce it artificially, and by finding that *a* follows it, verify the result of our induction. The importance of thus reversing the proof was strikingly manifested when by keeping a phial of water charged with siliceous particles undisturbed for years, a chemist (I believe Dr. Wollaston) succeeded in obtaining crystals of quartz;

and in the equally interesting experiment in which Sir James Hall produced artificial marble by the cooling of its materials from fusion under immense pressure; two admirable examples of the light which may be thrown upon the most secret processes of Nature by well-contrived interrogation of her.

But if we cannot artificially produce the phenomenon A, the conclusion that it is the cause of *a* remains subject to very considerable doubt. Though an invariable, it may not be the unconditional antecedent of *a*, but may precede it as day precedes night and night day. This uncertainty arises from the impossibility of assuring ourselves that A is the *only* immediate antecedent common to both the instances. If we could be certain of having ascertained all the invariable antecedents, we might be sure that the unconditional invariable antecedent or cause must be found somewhere among them. Unfortunately it is hardly ever possible to ascertain all of the antecedents, unless the phenomenon is one which we can produce artificially. Even then, the difficulty is merely lightened, not removed: men knew how to raise water in pumps long before they adverted to what was really the operating circumstance in the means they employed, namely, the pressure of the atmosphere, on the open surface of the water. It is, however, much easier to analyse a set of arrangements made by ourselves, than the whole complex mass of the agencies which nature happens to be exerting at the moment of production of a given phenomenon. We may overlook some of the material circumstances in an experiment with an electrical machine; but we shall, at the worst, be better acquainted with them than with those of a thunderstorm.

The mode of discovering and proving laws of nature, which we have now examined, proceed on the following axiom. Whatever circumstances can be excluded, without prejudice to the phenomenon, or can be absent notwithstanding its presence, is not connected with it in the way of causation. The causal circumstances being thus eliminated, if only one remains, that one is the cause which we are in search of: if more than one, they either are, or contain among them, the cause; and so *mutadis mutandis,* of the effect. As this method proceeds by comparing different instances to ascertain in what they agree, I have termed it the Method of Agreement; and we may adopt as its regulating principle the following canon:—

First Canon.

If two or more instances of the phenomenon under investigation have only one circumstance in common, the circumstance in which alone all the instances agree is the cause (or effect) of the given phenomenon.

Quitting for the present the Method of Agreement, to which we shall almost immediately return, we proceed to a still more potent instrument of the investigation of nature, the Method of Disagreement.

§2.

In the Method of Agreement, we endeavored to obtain instances which agreed in the given circumstance but differed in every other: in the present method we require, on the contrary, two instances resembling one another in every possible respect, but differing in the presence or absence of the phenomenon we wish to study. If our object be to discover the effects of an agent A, we must procure A in some set of ascertained circumstances, as A B C, and having noted the effects produced, compare them with the effects of the remaining circumstances B C, when A is absent. If the effect of A B C is *a b c*, and the effect of B C, *b c*, it is evident that the effect of A is *a*. So again, if we begin at the other end, and desire to investigate the cause of an effect *a*, we must select an instance, as *a b c*, in which the effect occurs, and in which the antecedents were A B C, and we must look out for another instance in which the remaining circumstances, *b c*, occur without *a*. If the antecedents, in that instance, are B C, we know that the cause of *a* must be A: either A alone, or A in conjunction with some of the other circumstances present.

It is scarcely necessary to give examples of a logical process to which we owe almost all the inductive conclusions we draw in early life. When a man is shot through the heart, it is by this method we know that it was the gunshot which killed him: for he was in the fullness of life immediately before, all the circumstances being the same except the wound.

The axioms implied in this method are evidently the following. Whatever antecedent cannot be excluded without preventing the phenomenon, is the cause, or a condition of that phenomenon: Whatever consequent can be excluded, with no other difference in the antecedents than the absence of a particular one, is the effect of that one. Instead of comparing different instances of a phenomenon, to discover in what they agree, this method compares an instance of its occurrence with an instance of its non-occurrence, to discover in what they differ. The canon which is the regulating principle of the Method of Difference may be expressed as follows:—

Second Canon.

If an instance in which the phenomenon under investigation occurs, and an instance in which it does not occur, have every circumstance in common save one, that one occurring only in the former; the circumstance in which alone the two instances differ is the effect, or the cause, or an indispensable part of the cause, of the phenomenon.

§3.

The two methods which we have now stated have many features of resemblance, but there are also many distinctions between them. Both are methods of *elimination*. This term (employed in the theory of equations to denote the process by which one after another of the elements of a question is excluded, and the solution made to depend on the relation between the remaining elements only) is well suited to express the operation, analogous to this, which has been understood since the time of Bacon to be the foundation of experimental inquiry, namely, the successive exclusion of the various circumstances which are found to accompany a phenomenon in a given instance, in order to ascertain what are those among them which can be absent consistently with the existence of the phenomenon. The Method of Agreement stands on the ground that whatever can be eliminated is not connected with the phenomenon by any law. The Method of Difference has for its foundation, that whatever cannot be eliminated is connected with the phenomenon by a law.

Of these methods, that of Difference is more particularly a method of artificial experiment; while that of Agreement is more especially the resource employed where experimentation is impossible . . .

§4.

There are, however, many cases in which, though our power of producing the phenomenon is complete, the Method of Difference either cannot be made available at all, or not without a previous employment of the Method of Agreement. This occurs when the agency by which we can produce the phenomenon is not that of a single antecedent, but a combination of antecedents which we have no power of separating from each other and exhibiting apart. For instance, suppose the subject of inquiry to be the cause of the double refraction of light. We can produce this phenomenon at pleasure by employing any one of the substances which are known to refract light in that particular manner. But if, taking one of those substances, as Iceland spar, for example, we wish to determine on which of the properties of Iceland spar this remarkable property depends, we can make no use for that purpose of the Method of Difference; for we cannot find another substance precisely resembling Iceland spar except in some one property. The only mode, therefore, of prosecuting this inquiry is that afforded by the Method of Agreement; by which, in fact, through a comparison of all known substances which have the property of doubly refracting light, it was ascertained that they agree in the circumstances of being crystalline substances; and though the converse does not hold, though all crystalline substances have not the property of double refraction, it was concluded with reason, that there is a real connection between these two properties; that either crystalline structure, or the cause which gives rise to that structure, is one of the conditions of double refraction.

Out of this employment of the Method of Agreement arises a peculiar modification of that method, which is sometimes of great avail in the investigation of nature. In cases similar to the above, in which it is not possible to obtain the precise pair of instances which our second canon requires—instances agreeing in every antecedent except A, or in every consequent except *a*—we may yet be able, by a double employment of the Method of Agreement, to discover in what the instances which contain A or *a* differ from those which do not.

If we compare various instances in which *a* occurs, and find that they all have in common the circumstance A, and (as far as can be observed) no other circumstance, the Method of Agreement, so far, bears testimony to a connection between A and *a*. In order to convert this evidence of connection into proof of causation by the direct Method of Difference, we ought to be able, in some of these instances, as, for example, A B C, to leave out A, and observe whether by doing so, *a* is prevented. Now supposing (what is often the case) that we are not able to try this decisive experiment, yet, provided we can by any means discover what would be its result if we could try it, the advantage will be the same. Suppose, then, as we previously examined, a variety of instances in which *a* occurred, and found them to agree in containing A, so we now observe a variety of instances in which *a* does not occur, and find them agree in not containing A; which establishes, by the Method of Agreement, the same connection between the absence of A and the absence of *a*, which was before established between their presence. As, then, it had been shown that whenever A is present *a* is present, so it being now shown that when A is taken away *a* is removed along with it, we have by the one proposition A B C, *a b c,* by the other, B C, *b c,* the positive and negative instances which the Method of Difference requires.

This method may be called the Indirect Method of Difference, or the Joint Method of Agreement and Difference, and consists n a double employment of the Method of Agreement, each proof being independent of the other, and corroborating it. But it is not equivalent to a proof by the direct Method of Difference. For the requisitions of the Method of Difference are not satisfied unless we can be quite sure either that the instances affirmative of *a* agree in no antecedent whatever but A, or that the instances negative of *a* agree in nothing But the negation of A. Now if it were possible, which it never is, to have this assurance, we should not need the joint method; for either of the two sets of instances separately would then be sufficient to prove causation.

This indirect method, therefore, can only be regarded as a great extension and improvement of the Method of Agreement, but not as participating in the more cogent nature of the Method of Difference. The following may be stated as its canon:—

Third Canon.

If two or more instances in which the phenomenon occurs have only one circumstance in common, while two or more instances in which it does not occur have nothing in common save the absence of that circumstance, the circumstance in which alone the two sets of instances differ is the effect, or the cause, or an indispensable part of the cause, of the phenomenon.

We shall presently see that the Joint Method of Agreement and Difference constitutes, in another respect not yet adverted to, an improvement upon the common Method of Agreement, namely, in being unaffected by a characteristic imperfection of that method, the nature of which still remains to be pointed out. But as we cannot enter into this exposition without introducing a new element of complexity into this long and intricate discussion, I shall postpone it to a subsequent chapter, and shall at once proceed to a statement of two other methods, which will complete the enumeration of the means which mankind possesses for exploring the laws of nature by specific observation and experience.

§5.

The first of these has been aptly denominated the Method of Residues. Its principle is very simple. Subducting from any given phenomenon all the portions which, by virtue of preceding inductions, can be assigned to known causes, the remainder will be the effect of the antecedents which had been overlooked, or of which the effect was as yet an unknown quantity.

Suppose, as before, that we have the antecedents, A B C, followed by the consequents *a b c,* and that by previous inductions (founded, we will suppose on the Method of Difference) we have ascertained the causes of some of these effects, or the effects of some of these causes; and are thence apprised that the effect of A is *a,* and that the effect of B is *b.* Subtracting the sum of these effects from the total phenomenon, there remains *c,* which now without any fresh experiments, we may know to be the effect of C. This Method of Residues is in truth a peculiar modification of the Method of Difference. If the instance A B C, *a b c,* could have been compared with a single instance A B, *a b,* we should have proved C to be the cause of *c,* by the common process of the Method of Difference. In the present case, however, instead of a single instance A B, we have has to study separately the causes A and B, and to infer from the effects which they produce separately what effect they must produce in the case A B C where they act together. Of the two instances, therefore, which the Method of Difference requires,—the one positive, the other negative,—the negative one, or that in which the given phenomena is absent, is not the result of observation and experiment, but has been arrived at by deduction. As one of the forms of the Method of Difference, the Method of Residues partakes of its rigourous certainty, provided the previous inductions, those which gave the effects of A and B, were obtained by the same infallible method, and provided we are certain that C is the *only* antecedent to which the residual phenomenon *c* can be referred; the only agent of which we had not already calculated and subducted the effect. But as we can never be quite certain of this, the evidence derived from the Method of Residues is not complete unless we can obtain C artificially and try it separately, or unless its agency, when once suggested, can be accounted for, and proved deductively, from known laws.

Even with these reservations, the Method of Residues is one of the most important among our instruments of discovery. Of all the methods of investigating laws of nature, this is the most fertile in unexpected results: often informing us of sequences in which neither the cause nor the effect were sufficiently conspicuous to attract of themselves the attention of observers. The agent C may be an obscure circumstance, not likely to have been perceived unless sought for, nor likely to have been sought for until attention had been awakened by the insufficiency of the obvious causes to account for the whole of the effect. And *c* may be so disguised by its intermixture with *a* and *b*, that it would scarcely have presented itself spontaneously as a subject of separate study. Of these uses of the method we shall presently cite some remarkable examples. The canon of the Method of Residues is as follows:—

Fourth Canon.

Subduct from any phenomenon such part as is known by previous inductions to be the effect of certain antecedents, and the residue of the phenomenon is the effect of the remaining antecedents.

§6.

There remains a class of laws which it is impracticable to ascertain by any of the three methods which I have attempted to characterize, namely, the laws of those Permanent Causes, or indestructible natural agents, which it is impossible either to exclude or to isolate; which we can neither hinder from being present, nor contrive that they shall be present alone. It would appear at first sight that we could by no means separate the effects of these agents from the effects of those other phenomena with which they cannot be prevented from co-existing. In respect, indeed, to most of the permanent causes, no such difficulty exists; since, though we cannot eliminate them as co-existing facts, we can eliminate them as influencing agents, by simply trying our experiment in a local situation beyond the limits of their influence. The pendulum, for example, has its oscillations disturbed by the vicinity of a mountain: we remove the pendulum to a sufficient distance from the mountain, and the disturbance ceases: from these data we can determine by the Method of Difference the amount of effect due to the mountain; and beyond a certain distance everything goes on precisely as it would do if the mountain exercised no influence whatever, which, accordingly, we, with sufficient reason, conclude to be the fact.

The difficulty, therefore, in applying the methods already treated of to determine the effects of Permanent Causes, is confined to the cases in which it is impossible for us to get out of the local limits of their influence. The pendulum can be removed from the influence of the mountain, but it cannot be removed from the influence of the earth: we cannot take away the earth from the pendulum, nor the pendulum from the earth, to ascertain whether it would continue to vibrate if the action which the earth exerts upon it were withdrawn. On what evidence, then, do we ascribe its vibration to the earth's influence? Not on any sanctioned by the Method of Difference; for one of the two instances, the negative instance, is wanting. Nor by the Method of Agreement; for though all pendulums agree in this, that during their oscillations the earth is always present, why may we not as well ascribe the phenomena to the sun, which is equally a co-existent fact in all the experiments? It is evident that to establish even so simple a fact of causation as this, there was required some method over and above those which we have yet examined.

As another example, let us take the phenomenon Heat. Independently of all hypothesis as to the real nature of the agency so called, this fact is certain, that we are unable to exhaust any body of the whole of its heat. It is equally certain that no one ever perceived heat not emanating from a body. Being unable, then, to separate Body and Heat, we cannot effect such a variation of circumstances as the foregoing three methods require; we cannot ascertain, by those methods, what portion of the phenomena exhibited by any body is due to the heat contained in it. If we could observe a body with its heat, and the same body entirely divested of heat, the Method of Difference would show the effect due to the heat, apart from that due to the body. If we could observe heat under circumstances agreeing in nothing but heat, and therefore not characterised also by the presence of a body, we could ascertain the effects of heat, from an instance of heat without a body and an instance of heat with a body, by the Method of Agreement; or we could determine by the Method of Difference what effect was due to the body, when the remainder which was due to the heat would be given by the Method of Residues. But we can do none of these things; and without them the application of any of the three methods to the solution of this problem would be illusory. It would be idle, for instance, to attempt to ascertain the effect of heat by subtracting from the phenomena exhibited by a body all that is due to its other properties; for as we have never been able to observe any bodies without a portion of heat in them, effects due to that heat might form a part of the very results which we were affecting to subtract in order that the effect of heat might be shown by the residue.

If, therefore, there were no other methods of experimental investigation than these three, we should be unable to determine the effects due to heat as a cause. But we have still a resource. Though we cannot exclude an antecedent altogether, we may be able to produce, or nature may be able to produce for us, some modification in it. By a modification is here meant a change in it, not amounting to its total removal. If some modification in the antecedent A is always followed by a change in the consequent *a,* the other consequents *b* and *c* remaining the same; or *vice versa,* if every change in *a* is found to have been preceded by some modification in A, none being observable in any of the other antecedents; we may safely conclude that *a* is wholly or in part, an effect traceable to A, or at least in some way connected with it through causation. For example, in the case of heat, though we cannot expel it altogether from any body, we can modify it in quantity, we can increase or diminish it; and doing so, we find by the various methods of experimentation or observation already treated of, that such increase or diminution of heat is followed by expansion or contraction of the body. In this manner we arrive at the conclusion, otherwise unattainable by us, that one of the effects of heat is to enlarge the dimensions of bodies; or what is the same thing in other words, to widen the distances between the particles.—

A change in a thing, not amounting to its total removal, that is, a change which leaves it still the same thing it was, must be a change either in its quantity, or in some of its variable relations to other things, of which variable relations the principle is its position in space. In the previous example, the modification which was produced in the antecedent was an alteration in its quantity. Let us now suppose the question to be, what influence the moon exerts on the surface of the earth. We cannot try an experiment in the absence of the moon, so as to observe what terrestrial phenomena her annihilation would put an end to; but when we find that all the variations in the *position* of the moon are followed by corresponding variations in the time and place of high water, the place being always either the part of the earth which is nearest to, or that which is the most remote from, the moon, we have ample evidence that the moon is wholly, or partially, the cause which determines the tides. It very commonly happens, as it does in this instance, that the variations of an effect are correspondent, or analogous, to those of its cause; as the moon moves farther towards the east, the high water point does the same: but this is not an indispensable condition, as may be seen in the same example; for along with that high-water point there is at the same instant another high-water point diametrically opposed to it, and which, therefore, of necessity, moves towards the west, as the moon, followed by the nearer of the tide-waves, advances towards the east: and yet both of these motions are equally effects of the moon's motion . . .

The method by which these results were obtained may be termed the Method of Concomitant Variations: it is regulated by the following canon:—

Fifth Canon.

Whatever phenomenon varies in any manner whenever another phenomenon varies in some particular manner, is either a cause or an effect of that phenomenon, or is connected with it through some fact of causation.

This last clause is subjoined because it by no means follows, when two phenomena accompany each other in their variations, that the one is cause and the other effect. The same thing may, and indeed must happen, supposing them to be two different effects of a common cause: and by this method alone it would never be possible to ascertain which of the suppositions is the true one. The only way to solve the doubt would be that which we have so often adverted to, viz. by endeavoring to ascertain whether we can produce the one set of variations by means of the other. In the case of heat, for example, by increasing the temperature of a body we increase its bulk, but by increasing its bulk we do not increase its temperature; on the contrary, (as in the rarefaction of air under the receiver of an air pump, we generally diminish it: therefore heat is not an effect, but a cause, of increase of bulk. If we cannot ourselves produce the variations, we must endeavor, though it is an attempt which is seldom successful, to find them produced by nature in some case in which the pre-existing circumstances are perfectly known to us.

It is scarcely necessary to say, that in order to ascertain the uniform concomitants of variations in the effect with variations in the cause, the same precautions must be used as in any other case of the determination of an invariable sequence. We must endeavour to retain all the other antecedents unchanged, while that particular one is subjected to the requisite series of variations; or, in other words, that we may be warranted in inferring causation from concomitance of variations, the concomitance itself must be proved by the Method of Difference.

It might first appear that the Method of Concomitant Variations assumes a new axiom, or law of causation in general, namely, that every modification of the cause is followed by a change in the effect. And it does usually happen that when a phenomenon A causes a phenomenon *a*, any variation in the quantity or in the various relations of A is uniformly followed by a variation in the quantity or relations of *a*. To take a familiar instance, that of gravitation. The sun causes a certain tendency to motion in the earth; here we have cause and effect; but that tendency is *towards* the sun, and therefore varies in direction as the sun varies in the relation of position; and moreover the tendency varies in intensity, in a certain numerical correspondence to the sun's distance from the earth, that is, according to another relation of the sun. Thus we see that there is not only an invariable connection between the sun and the earth's gravitation, but that two of the relations of the sun, its position with respect to the earth and its distance from the earth, are invariably connected as antecedents with the quantity and direction of the earth's gravitation. The cause of the earth's gravitating at all is simply the sun; but the cause of its gravitating with a given intensity and in a given direction is the existence of the sun in a given direction and at a given distance. It is not strange that a modified cause, which is in truth a different cause, should produce a different effect.

Although it is for the most part true that a modification of the cause is followed by a modification of the effect, the Method of Concomitant Variations, does not, however, presuppose an axiom. It only requires the converse proposition, that anything on whose modifications, modifications of an effect are invariably consequent, must be the cause (or connected with the cause) of that effect; a proposition, the truth of which is evident; for if the thing itself had no influence on the effect, neither could the modifications of the thing have any influence. If the stars have no power over the fortunes of mankind, it is implied in the very terms that the conjunctions or oppositions of different stars can have no such power.

Although the most striking applications of the Method of Concomitant Variations take place in the cases in which the Method of Difference, strictly so called, is impossible, its use is not confined to those cases; it may often usefully follow after the Method of Difference, to give additional precision to a solution which that has found. When by the method of difference it has first been ascertained that a certain object produces a certain effect, the Method of Concomitant Variations may be usefully called in to determine according to what law the quantity or the different relations of the effect follow those of the cause.

§ 7.

The case in which this method admits of the most extensive employment is that in which the variations of the cause are variations of quantity. Of such variations we may in general affirm with safety that they will be attended not only with variations, but similar variations of the effect: the proposition, that more of the cause is followed by more of the effect, being a corollary from the principle of the Composition of Causes, which, as we have seen, is the general rule of causation; cases of the opposite description, in which causes change their properties on being conjoined with one another, being, on the contrary, special and exceptional. Suppose, then, that when A changes in quantity, and in such a manner that we can trace the numerical relation which the changes of the one bear to such changes of the other as take place within our limits of observation. We may then, with certain precautions, safely conclude that the same numerical relation will hold beyond those limits. If, for instance, we find that when A is double, a is double; that when A is treble or quadruple, a is treble or quadruple; we may conclude that if A were a half or a third, a would be a half or a third; and finally, that if A were annihilated, a would be annihilated; and that a is wholly the effect of A, or wholly the effect of the same cause with A. And so with any other numerical relation according to which A and a would vanish simultaneously; as, for instance, if a were proportional to the square of A. If, on the other hand, a is not wholly the effect of A, but yet varies when A varies, it is probably a mathematical function not of A alone, but of A and something else; its changes, for example, may be such as would occur if part of it remained constant, or varied on some other principle, and the remainder varied in some numerical relation to the variations of A. In that case, when A diminishes, a will be seen to approach not towards zero, but towards some other limit; and when the series of variations is such as to indicate what the limit is, if constant, or the law of its variation if variable, the limit will exactly measure how much of a is the effect of some other and independent cause, and the remainder will be the effect of A (or the cause of A) . . .

OF THE DEDUCTIVE METHOD

§ 1.

The mode of investigation which, from the proved inapplicability of direct methods of observation and experiment, remains to us as the main source of the knowledge we possess or can acquire respecting the conditions and laws of recurrence of the more complex phenomena, is called, in its most general expression, the Deductive Method, and consists of three operations—the first, one of direct induction; the second, of ratiocination; the third, of verification.

I call the first step in the process an inductive operation, because there must be a direct induction as the basis of the whole, though in many particular investigations the place of the induction may be supplied by a prior deduction; but the premises of the prior deduction must have been derived from induction.

The problem of the Deductive Method is to find the law of an effect from the laws of the different tendencies of which it is the joint result. The first requisite, therefore, is to know the laws of those tendencies—the law of each of the concurrent causes; and this supposes previous process of observation or experiment upon which each cause separately, or else a previous deduction, which also must depend for its ultimate premises on observation or experiment. Thus, if the subject be social or historical phenomena, the premises of the Deductive Method must be the laws of the causes which determine that class of phenomena; and those causes are human actions, together with the general outward circumstances under the influence of which mankind are placed, and which constitute man's position on the earth. The Deductive Method applied to social phenomena must begin, therefore, by investigating, or must suppose to have been already investigated, the laws of human action, and those properties of outward things by which the actions of human beings in society are determined. Some of these general truths will naturally be obtained by observation and experiment, others by deduction; the more complex laws of human action, for example, may be deduced from the simpler ones, but the simple or elementary laws will always and necessarily have been obtained by a directly inductive process.

To ascertain, then, the laws of each separate cause which takes a share in producing the effect is the first desideratum of the Deductive Method. To know what the causes are which must be subjected to this process of study may or may not be difficult. In the case last mentioned, this first condition is of easy fulfillment. That social phenomena depend on the acts and mental impressions of human beings never could have been a matter of any doubt, however imperfectly it may have been known either by what laws those impressions and actions are governed, or to what social consequences their laws naturally lead. Neither, again, after physical science had attained a certain development, could there be any real doubt where to look for the laws on which the phenomena of life depend, since they must be the mechanical and chemical laws of the solid and fluid substances composing the organized body and the medium in which it subsists, together with the particular vital laws of the different tissues constituting the organic structure. In other cases really far more simple than these, it was much less obvious in what quarter the causes were to be looked for, as in the case of the celestial phenomena. Until by combining the laws of certain causes, it was found that those laws explained all the facts which experience had proved concerning the heavenly motions, and led to predictions which it always verified, mankind never knew that those *were* the causes. But whether we are able to put the question before or not until after we have become capable of answering it, in either case it must be answered; the laws of the different causes must be ascertained before we can proceed to deduce from them the conditions of the effect . . .

In general, the laws of the causes on which the effect depends may be obtained by an induction from comparatively simple instances, or, at the worst, by deduction from the laws of simpler causes, so obtained. By simple instances are meant, of course, those in which the actions of each cause was not intermixed or interfered with, or not to any great extent, by other causes whose laws were unknown; and only when the induction which furnished the premises to the Deductive Method rested on such instances has the application of such a method to the ascertainment of the laws of a complex effect been attended with brilliant results.

§2.

When the laws of the causes have been ascertained, and the first stage of the great logical operation now under discussion satisfactorily accomplished, the second part follows; that of determining from the laws of the causes what effect any given combination of those causes will produce. This is a process of calculation, in the wider sense of the term, and very often involves processes of calculation in the narrowest sense.

It is a ratiocination; and when our knowledge of the causes is so perfect as to extend to the exact numerical laws which they observe in producing their effects, the ratiocination may reckon among its premises the theorems of the science of number, in the whole immense extent of that science. Not only are the most advanced truths of mathematics often required to enable us to compute an effect the numerical law of which we already know, but, even by the aid of those most advanced truths, we can go but a little way. In so simple a case as the common problem of three bodies gravitating towards one another, with a force directly as their mass and inversely as the square of their distance, all the resources of the calculus have not hitherto sufficed to obtain any general solution but an approximate one. In a case a little more complex, but still one of the simplest which arise in practice, that of the motion of a projectile, the causes which affect the velocity and range (for example) of a cannon-ball may be all known and estimated; the force of the gunpowder, the angle of elevation, the density of the air, the strength and direction of the wind; but it is one of the most difficult of mathematical problems to combine all these, so as to determine the effect resulting from their collective action.

Besides the theorems of number, those of geometry also come in as premises, where the effects take place in space, and involve motion and extension, as in mechanics, optics, acoustics, astronomy. But when the complications increase, and the effects are under the influences of so many and such shifting causes as to give no room either for fixed numbers or for straight lines and regular curves, (as in the case of physiological, to say nothing of mental and social phenomena,) the laws of number and extension are applicable, if at all, only on that large scale on which the precision of details becomes unimportant. Although these laws play a conspicuous part in the most striking examples of the investigation of nature by the Deductive Method, as, for example, in the Newtonian theory of the celestial motions, they are by no means an indispensable part of every such process. All that is essential in it is reasoning from a general law to a particular case, that is, determining by means of the particular circumstances of that case what result is required in that instance to fulfill the law. Thus in the Torricellian experiment, if the fact that air has weight had been previously known, it would have been easy, without any numerical data, to deduce from the general law of equilibrium that the mercury would stand in the tube at such a height that the column of mercury would exact balance a column of the atmosphere of equal diameter; because, otherwise, equilibrium would not exist.

By such ratiocinations from the separate laws of the causes we may, to a certain extent, succeed in answering either of the following questions: Given a certain combination of causes, what effect will follow? and, What combination of causes, if it existed, would produce a given effect? In the one case, we determine the effect to be expected in any complex circumstances of which the different elements are known: in the other case we learn, according to what law—under what antecedent conditions—a given complex will occur.

§3.
But (it may be asked) are not the same arguments by which the methods of direct observation and experiment were set aside as illusory when applied

to the laws of complex phenomena, applicable with equal force against the Method of Deduction? When in every single instance a multitude, often an unknown multitude, of agencies are clashing and combining, what security have we that in our computation *a priori* we have taken all these into our reckoning? How many must we not generally be ignorant of? Among those which we know, how probable that some have been overlooked; and, even were all included, how vain the pretense of summing up the effects of many causes unless we know accurately the numerical law of each,—a condition in most cases not to be fulfilled; and even when it is fulfilled, to make the calculation transcends, in any but very simple cases, the utmost power of mathematical science with all its most modern improvements.

These objections have real weight, and would altogether be unanswerable, if there were no test by which, when we employ the Deductive Method, we might judge whether an error of any of the above descriptions had been committed or not. Such a test, however, there is; and its application forms, under the name of Verification, the third essential component part of the Deductive Method, without which all the results it can give have little other value than that of conjecture. To warrant reliance on the general conclusions arrived at by deduction, these conclusions must be found, on careful comparison, to accord with the results of direct observation wherever it can be had. If, when we have experience to compare with them, this experience confirms them, we may safely trust to them in other case of which our specific experience is yet to come. But if our deductions have led to the conclusion that from a particular combination of causes a given effect would result, then in all known cases where that combination can be shown to have existed, and where the effect has not followed, we must be able to show (or at least to make a probable surmise) what frustrated it: if we cannot, the theory is imperfect, and not yet to be relied upon. Nor is the verification complete, unless some of the cases in which the theory is borne out by the observed result, are of at least equal complexity with any other cases in which its application could be called for.

If direct observation and collation of instance have furnished us with any empirical laws of the effect, (whether true in all observable cases, or only true for the most part,) the most effectual verification of which the theory could be susceptible would be, that it led deductively to those empirical laws; that the uniformities, whether complete or incomplete, which were observed to exist among the phenomena were accounted for by the laws of the causes—were such as could not but exist if those be really the causes by which the phenomena are produced. Thus it was very reasonably deemed an essential requisite of any true theory of the causes of the celestial motions, that it should lead by deduction to Kepler's laws; which accordingly the Newtonian theory did.

To the Deductive Method, thus characterized in its three constituent parts, Induction, Ratiocination, and Verification, the human mind is indebted for its most conspicuous triumphs in its investigation of nature. To it we owe all the theories by which vast and complicated phenomena are embraced under a few simple laws, which, considered as the laws of those great phenomena, could never have been detected by their direct study.

Case Studies

. . . .

Astronomy Track

PAPER 2: PTOLEMY ON THE MOTION OF THE PLANETS

THE CASE

Aristotle's account of the heavens could explain the regularity of observable phenomena, but was not entirely able to give a complete accounting for everything that could be seen in the night sky. Other scientists worked hard to do so, including Aristotle's contemporary Eudoxus and astronomers who followed like Callippus, Apollonius, and Hipparchus. These later thinkers suggested alterations of the Aristotelian system designed to account for some of the observed deviations.

The ancients knew that the motions of astronomical bodies deviated from regularity. The moon, for example, rises and sets at slightly different points on the horizon each night. Eudoxus accounted for this complex motion by contending that the sphere attached to the moon was itself attached at the poles to another sphere slightly smaller that fit inside of it like Russian nesting dolls. This inner sphere rotated about its axis while the outer sphere rotated along its axis. The combination of these rotations would produce an orbit around the Earth that itself moved back and forth on the horizon. Eudoxus repeated this process so that the moon used four nested spheres in order to account for the appearance of the moon's actual orbit.

But the moon was the easy case. The tougher one was that of the so-called wandering stars, or planets as we now call them, which followed even stranger paths. While nightly the planets usually moved across the heavens in the same direction as everything else, when one found a certain planet in the night sky, one night it would be to the east of a given constellation, then a period of time later it would be in the constellation, finally appearing to the west of the constellation. Nothing bizarre there, they just moved a little faster around the Earth. But occasionally, they would exhibit what is called retrograde motion, appearing one night farther east, or behind, the position they had occupied in the sky the night before. It was as if the planet was moving backwards. But a short while later, the planet would once again move in its "normal" direction.

Apollonius, about a century and a half after Eudoxus, introduced two additional pieces of machinery to the Aristotelian system to account for these astronomical observations. The first included a new type of sphere, one that was not nested like Eudoxus's, but rather, one that rolled itself along the larger moving sphere. The motion of such a sphere rolling on top of a sphere was called an epicycle. Such motions were needed to explain the retrograde motions of planets. The second adjustment was to make the point of rotation a point other than the center of the sphere. Eccentric spheres rotated with a wobble like a flat tire and that wobble allowed for astronomers to account for some deviations from perfectly uniform motion.

Hipparchus was perhaps the greatest of the ancient astronomers, producing both a great catalog of astronomical observations and a complex system of eccentric spheres and epicycles to account for his data. Unfortunately, much of Hipparchus's writings have been lost. Fortunately, they were quoted widely by an astronomer who followed a couple hundred years later. Claudius Ptolemy, in his masterwork the *Almagest*, expands on Hipparchus's work and provides us with a nearly perfectly predictive system of eccentric spheres and epicycles to account for all the motions that could be seen in the heavens. He does so using eccentric orbits, epicycles, and one more bit of machinery, the equant, a point within the sphere that is different from the center such that from the point of view of someone sitting at the equant, the planet rolling around on the surface of the sphere would be seen to always be moving at the same speed. From the actual center, however, the planet would seem to speed up and slow down. Using these three methods to augment the circular orbits, Ptolemy was able to create a system of the heavens that was remarkably accurate for nearly every body in the night sky.

THE SCIENTIST

Claudius Ptolemy (85–165 AD) was an astronomer and mathematician at the museum at Alexandria in Egypt, the center of learning established in honor of Alexander. Conducting his astronomical observation close by at Canopus, Ptolemy acquired a great amount of data concerning the locations of heavenly bodies. Although these data were slightly less accurate than those of his predecessor Hipparchus, Ptolemy was able to utilize them. He employed the notions of ecliptics (circles at angles to each other) and epicycles (circles on circles) used by Hipparchus and Apollonius before him to construct a system that accounted for virtually all astronomical observations that can be made without the use of a telescope. This system is worked out in great detail in his book the *Almagest*, a title derived from the Arabic for "the greatest one," which is the earliest complete work on astronomy from the ancient world.

YOUR JOB

Ptolemy came out of the Aristotelian tradition, but knew that Aristotle's system was inadequate. Read the excerpt from the *Almagest* (beginning on p. 350) and explain why Aristotle could not have been exactly right and why his view would need to be adjusted. Show how Ptolemy's account of planetary motion works and why it does things that Aristotle's does not. The *Almagest* is one of the great scientific advances of the ancient world. Does Ptolemy's departure from the Aristotelian system signal a difference in scientific methodology? Explain how inductivism differs from deductivism, and point out the elements in Ptolemy's writings that would lead one to believe whether he was more of an inductivist or a deductivist. Are there any places where Mill's methods seem to be employed? Which ones? Does his work more clearly follow one or the other methodology? Is it a combination of the two? Support your claim with references to the *Almagest*.

RESEARCH HINT

- The scientific episode is a famous case that many writers have discussed in great detail. For the section of your paper in which you explain the case, find a secondary source (that is, a history of your science, a textbook, or a biography of the scientist) that explains it. Read that before working through the primary source (the scientist's own words in the selection in the back of this book). This will help you figure out the function of each part of the selection in the author's larger argument. Quote both the primary and the secondary sources (with proper citations, of course!) in the expository section of your paper.

Physics Track

PAPER 2: MAXWELL'S KINETIC THEORY OF HEAT

THE CASE

The move from the classical to the modern theory of the atom in physics came by way of thermodynamics. Since the time of Aristotle, heat was thought to be a substance unto itself. The heat substance flows, according to this view, and seeks equilibrium just as a liquid or gas does. When chemistry began to show some problems with this view, physicists took note and a new picture of heat that was more sympathetic to the Newtonian mechanical picture of the world began to be developed.

On this view, matter was made up of basic bits called "atoms" by some and "molecules" by others. The modern sense of molecules being made up of atoms was slow in coming, but up to Maxwell's time, the term "atom" carried with it the Greek sense of being indivisible, whereas a molecule was simply the smallest amount of a substance that had the same properties as a larger sample.

A number of physicists explored the possibility of accounting for thermodynamic properties of a substance in terms of the motion of constitutive molecules of that substance. They understood well the notions of speed, energy, and elastic collisions from Newton, and the hope was that they could now explain temperature, pressure, and volume in the same terms. Relations discovered by chemists like Robert Boyle, Joseph Louis Gay-Lussac, and Jacques Charles provided a sense that this was the proper line of reasoning, but physicists were loathe to accept the reality of something like atoms they could not observe, especially if the evidence came from *chemists*.

To give good reason to believe that atoms were real, the measurable properties of the atoms would have to be determined through calculations from physical theory. Such work was undertaken by some very important names in the history of physics: Daniel Bernoulli, Rudolf Clausius, and Ludwig Boltzmann are but three. But the thinker who pulled it all together into a single coherent kinetic, or motion-based, theory of heat was James Clerk Maxwell.

In a series of papers, Maxwell developed an increasingly complex theory of molecular motion that was able to account for observations in both physics and chemistry. From this work, Maxwell was able both to explain observations and derive facts about the nature of these molecules, for example, their size, their speed, and the length they travel before likely bumping into another molecule. Since these measures are quantified properties of atoms, there must be atoms to have the properties. In this way, Maxwell argued that the directly unobservable atoms we first read about in Democritus and Epicurus are in fact real.

THE SCIENTIST

James Clerk Maxwell (1831–1879) was born in Scotland and educated at Cambridge. Picking up on the work of Michael Faraday, a great physicist who worked in a pictorial/analogical fashion, he provided a complete, simple, and elegant mathematical structure unifying electricity and magnetism. His strength was applying mathematics to complex physical problems. After taking a position in Scotland to be close to his ailing father, Maxwell turned his attention to the rings of Saturn—the problem set out for the Adams Prize, a contest at Cambridge. Maxwell showed that to be stable, Saturn's ring was not a single solid feature but rather a collection of very small individual particles, a result that was confirmed 122 years later by the spacecraft *Voyager*. This work naturally led to questions about the interactions of small particles and that in turn helped give rise to Maxwell's kinetic theory of gasses which reduces much of thermodynamics to statistical statements.

YOUR JOB

Maxwell argued for the reality of molecules based not on metaphysical reasoning but upon scientific results. Many scientists at the time thought that belief in atoms was unscientific because one could never observe them. Read the excerpt from Maxwell's paper "Molecules" (beginning on p. 357) and explain his argument for the reality of atoms, citing a couple of his pieces of experimental evidence. Explain the inductivists' approach to the scientific method. Were the opponents of early-modern atomists being inductivists? How is Maxwell's argument inductivist? Is it completely inductivist or are there aspects that an advocate of inductivism would have a hard time accounting for?

- The scientific episode is a famous case that many writers have discussed in great detail. For the section of your paper in which you explain the case, find a secondary source (that is, a history of your science, a textbook, or a biography of the scientist) that explains it. Read that before working through the primary source (the scientist's own words in the selection in the back of this book). This will help you figure out the function of each part of the selection in the author's larger argument. Quote both the primary and the secondary sources (with proper citations, of course!) in the expository section of your paper.

Chemistry Track

PAPER 2: BOYLE'S LAW AND OBSERVATION

THE CASE

Objecting to the Aristotelian view in which all was composed of earth, air, fire, and water, and to Paracelsus's view that everything is comprised of mercury, sulfur, and salt, Robert Boyle posited that the different observed chemical properties came from the combination of a number of different ultimate constituents. As twenty-six letters can be used to create a range of words, so a collection of a multiplicity of atoms arranged in different ways can give rise to different substances with different observable qualities that interact in a number of different ways. While not surrendering the old view altogether—earth, air, fire, and water were still considered elemental substances in their own right—he augmented the world with additional atoms and therefore is considered the father of the modern notion of "element."

Boyle rejected the older, more philosophical approach to chemistry, stressing experimentation and observation, especially when the initial conditions and the result could be quantified. To observe numerical regularities in measurable properties was seen to be a mark of truth and rigor in a field that was largely qualitative before him.

Boyle contended that unlike the alchemists who guard their methods, chemists need to make their methods and results public. He was the first to stress the reproducibility of scientific results. This especially included their failures, as one can learn as much from the inability to do something as from succeeding in showing something new.

In this way, Boyle argued, the relation between fire and air could be truly investigated. By testing flammable substances in a number of ways under a number of different circumstances, reasonable beliefs could be formed about how air and fire were related and this would allow a deeper understanding of the constituent parts of other substances when altered by burning.

Robert Boyle (1627–1691) is best known for Boyle's law, which states that the pressure of a gas is inversely proportional to its volume. This resulted from his adherence to Rene Descartes' mechanical philosophy in which all of nature is to be explained through mechanical interactions. As such, Boyle considered gasses to be collections of particles that interacted with each other and their container. Descartes himself, however, denied the possibility of such an explanation, because an atomistic point of view required there to be material atoms in a vacuous void and he argued that a vacuum was a logical impossibility. To save his Cartesian view from Descartes' own objections, Boyle became interested in vacuum pumps, and this gave rise not only to his statement of the relation between thermodynamic properties but also to his objections to Aristotelian and alchemical explanations for chemical phenomena. His most famous work, *The Skeptical Chymist,* argues why Aristotle's picture of four elements and Paracelsus's picture of three ingredients was incapable of accounting for observed phenomena.

YOUR JOB

Read the excerpt from *The Skeptical Chymist* (beginning on p. 362) and explain Boyle's experimental approach to finding the relation between fire and air. How did he create experiments and by what means did he draw conclusions from them? Explain inductivism, taking care to clearly discuss Mill's methods. How would Boyle's work with fire and air fit into Mill's approach to the scientific method? Is there any aspect of Boyle's work that deviates from the inductivist scheme?

RESEARCH HINT

- The scientific episode is a famous case that many writers have discussed in great detail. For the section of your paper in which you explain the case, find a secondary source (that is, a history of your science, a textbook, or a biography of the scientist) that explains it. Read that before working through the primary source (the scientist's own words in the selection in the back of this book). This will help you figure out the function of each part of the selection in the author's larger argument. Quote both the primary and the secondary sources (with proper citations, of course!) in the expository section of your paper.

Genetics Track

THE CASE

Farmers and botanists had been artificially fertilizing plants for quite a while in order to create new varieties with desirable properties. From this clue, Gregor Mendel set out to find regularities in the way hybrid plants expressed the characteristics of their parents.

Being trained in physics, Mendel hoped to find easily expressed numerical relationships attached to properties as their relative frequency could be observed generation after generation. The trouble was that each plant had many properties and not all of them were quantifiable, but rather matters of degree, of more or less, that could not really be put into the sort of clean relation he hoped for.

He was looking for a plant with three specific properties: observable characteristics that could be differentiated, hybrids that were easily guarded against foreign pollen, and hybrids that suffered no significant disturbance in fertility in subsequent generations. Carefully searching, Mendel settled on the sweet pea (*Pisum sativum*) and considered seven properties: the form of the ripe seeds, the form of the pods, the color of the ripe seeds, the color of the seed-coat, the color of the unripe pods, the position of the flowers, and the length of the stem.

Breeding plants over eight years, observing all of these properties in thousands of plants and tens of thousands of seeds, Mendel arrived at the result that all of the properties could be divided into those that were dominant and those that were recessive. In all cases, the ratio of dominant to recessive properties observed came out almost exactly to 3:1.

THE SCIENTIST

Gregor Mendel (1822–1884) was raised on a farm and studied physics and mathematics at the University of Vienna with the great physicist Christian Doppler before becoming a monk. The monastery's garden allowed him to pursue his interests in botany, particularly with respect to heredity. Cleverly using the monastery's greenhouse, he was able both to control for unwanted cross-pollinations and to allow the plants to grow several seasons for each year, speeding up the data collection immensely.

YOUR JOB

Read the excerpt from *Experiments in Plant Hybridization* (beginning on p. 365) and explain Mendel's experiments with pea plants. What did he show by demonstrating that in all seven cases of the observable characteristics of the plants he grew and made into hybrids, there was a constant ratio of dominant to recessive traits? Explain inductivism, taking care to clearly describe Mill's methods. How might an inductivist explain Mendel's approach? Which of Mill's methods would he or she see as employed by Mendel? Is there any part of Mendel's work that deviates from the inductivist approach to science?

RESEARCH HINT

- The scientific episode is a famous case that many writers have discussed in great detail. For the section of your paper in which you explain the case, find a secondary source (that is, a history of your science, a textbook, or a biography of the scientist) that explains it. Read that before working through the primary source (the scientist's own words in the selection in the back of this book). This will help you figure out the function of each part of the selection in the author's larger argument. Quote both the primary and the secondary sources (with proper citations, of course!) in the expository section of your paper.

Evolutionary Biology Track

PAPER 2: LINNAEUS AND THE TAXONOMY OF LIFE

THE CASE

Aristotle had followed his teacher Plato in arranging all things, living and nonliving, into a linear hierarchy, a ladder of nature. This was adopted by later Christian theologians who transformed it into the Great Chain of Being, which ordered living things according to their metaphysical perfection, humans almost at the top, just below divinities like angels and God Himself.

This linear structure was deeply embedded in the scientific approach to nature until Carolus Linnaeus, a Swedish naturalist, proposed a new, intricate, and more complex structure by which to categorize living things. Linnaeus did not object to the theological basis of the linear Great Chain model, he contended that the divine plan was to be seen in the intricacies of nature, in the relationships and anatomical similarities that can be found in different species. For this reason, he constructed a new way of classifying the different forms of life, a system of classification, he contended, that made God's plan clearer to eye and mind of a rational observer.

A "balance of nature" was essential to the stability of Creation. Species had to be understood as pieces in a bigger puzzle, their perfection understood in terms of their roles in a harmonious working system. Every species fit neatly into its niche, maintaining the health of the whole. As such, the pieces could not be understood apart from the whole, and also the whole could not be fully appreciated independent of the pieces. In this way, Linnaeus's taxonomy was intended to provide a map, a key to Creation.

A generation before Linnaeus, botanists Nehemiah Grew and Rudolf Jacob Camerarius had shown that plants reproduced sexually. Linnaeus was therefore able to follow Aristotle in arguing that reproduction is the central task of life, and therefore, the natural way to begin to organize the groups of living things was by the type of reproductive organs. From there, other observed anatomical regularities became the means by which groups and subgroups were determined.

While modern biology made some changes, the current means of classification derives from the work of Linnaeus and from it concepts like ecological balance would find their way into evolutionary theory generations later.

THE SCIENTIST

Carolus Linnaeus (1707–1778) was the son of a Lutheran pastor who had a penchant for gardening. This early exposure left Linnaeus with a love of plants. It led him to study medicine in which the medicinal use of plants was a significant field of inquiry. He left Sweden for Holland and traveled widely, observing and collecting life forms wherever possible. His masterwork, *Systema Naturae*, began as a small pamphlet, but over Linnaeus's lifetime he continued to edit and expand it until, at his death, it was a multivolume encyclopedic work.

YOUR JOB

Read the excerpt from *Systema Naturae* (beginning on p. 371) and explain Linnaeus's taxonomy and the foundational beliefs behind it. Explain the inductivist approach to science. How would an inductivist make sense of Linnaeus's work? Which aspects fit well into the inductivists' picture of scientific methodology and which aspects do not?

RESEARCH HINT

- The scientific episode is a famous case that many writers have discussed in great detail. For the section of your paper in which you explain the case, find a secondary source (that is, a history of your science, a textbook, or a biography of the scientist) that explains it. Read that before working through the primary source (the scientist's own words in the selection in the back of this book). This will help you figure out the function of each part of the selection in the author's larger argument. Quote both the primary and the secondary sources (with proper citations, of course!) in the expository section of your paper.

Geology Track

THE CASE

Woodward's theologically charged approach eventually lost its religious edge while scientists still considered water to be the operative cause in giving rise to geological features. This view, termed "Neptunism," which continued to argue that great oceans had covered mountains at one time only to later recede, was successfully advanced by many thinkers, most notably, Abraham Werner, because it accounted well for sedimentary rock strata and the existence of marine fossils in areas that were far removed from water.

But not everyone bought Werner's Neptunist explanation. A competing approach looked at the fossils of aquatic organisms in the cliffs far away from the sea and posited the opposite approach. Instead of the water receding from the mountains, it was argued that the mountains rose up out of the water. But a mechanism was needed to fuel the process.

It was well known that granite lay at the heart of many mountains and an alternate theory contended that instead of settling out as sediment, as Werner would have it, the granite had come from beneath the surface of the earth as molten rock, cooled, and solidified.

This view that a subterranean fire was driving the uplift of mountains, termed "Plutonism," was famously championed by James Hutton. He argued that in great fits, the fire would create simultaneous and catastrophic upheavals around the planet, raising mountains. Over time, the weather would erode the mountains, creating soil fit for life. But this soil, too, would erode, getting washed into tributaries and ultimately out to sea, where it would settle as lower layers would again be melted in the core, collecting and awaiting the next great upheaval. In this way, mountains grew and eroded in a great repeating cycle.

Hutton had held and developed this view for many years but lacked the empirical evidence to support it. If it was true, he contended, then there ought to be empirical observations that support the view. Experiments are virtually impossible in geology, but new data can still be collected through observation.

Such evidence presented itself, and in his *System of the Earth,* Hutton argued that Werner's Neptunism must be wrong and his view must be correct, as only heat can explain several observed phenomena. First, the formulation of strata requires heat, he contends, as water may be able to dissolve minerals but lacks the ability to reconsolidate them. Further, the existence of crystals and veins, intrusions of minerals into strata to which they do not belong, shows evidence of formation by heat. He also cited as evidence what we now call angular unconformities, that is, places where strata of rock fold over onto themselves or rise up at strange angles. Such slanted strata could only be the result, he argued, of land lifting up in places causing parts to shift in awkward ways. Finally, if there was a subterranean fire responsible for geological formation, we would expect to find evidence of it and we do, he contends, in the form of volcanoes. These observations are to be expected on his view, but not so with Werner's competing alternative.

THE SCIENTIST

James Hutton (1726–1797) was educated as a doctor before becoming a farmer. With his training in chemistry, he opened a small business manufacturing agricultural chemicals, and this provided him with sufficient funds to give up farming, move to Edinburgh, and pursue his interests in science full-time. Presenting his findings as one of the initial papers before the newly formed Royal Society, Hutton became the leading geological thinker of his day in Britain.

YOUR JOB

Read the excerpt from *System of the Earth* (beginning on p. 375) and explain Hutton's Plutonism. How does this lead to the prediction of intrusions of granite or angular uncomformities? Explain the inductivist approach to science, especially Mill's methods. How would an inductivist make sense of Hutton's work? Which of Mill's methods did Hutton seem to be employing? Are there any aspects that do not fit in well with inductivism?

RESEARCH HINT

- The scientific episode is a famous case that many writers have discussed in great detail. For the section of your paper in which you explain the case, find a secondary source (that is, a history of your science, a textbook, or a biography of the scientist) that explains it. Read that before working through the primary source (the scientist's own words in the selection in the back of this book). This will help you figure out the function of each part of the selection in the author's larger argument. Quote both the primary and the secondary sources (with proper citations, of course!) in the expository section of your paper.

Psychology Track

THE CASE

Hippocrates and Galen after him located the effects of the mind in the brain, making psychology a branch of medicine. But the brain was all but impossible to study at the time. The human mind remained clouded in mystery; it was open to contemplative philosophy, but there was little in terms of scientific study.

Psychology made a significant turn away from the speculative and toward the scientific in the nineteenth century when attention was paid to measuring perception. Physiologists had become skilled at devising experiments and opportunities to observe the body at work and, beginning with Ernst Weber, these methods were used to test the ability and limits of perception. Weber set up a series of experiments wherein subjects were asked to differentiate between the strength of stimuli in order to determine the sensitivity of the mind. For example, a person would be asked which of two seemingly identical weights was heavier. The muscles would respond, but the question was at what point the person would acknowledge sensing a difference in what the muscles were experiencing. How small of a difference does there have to be between two stimuli before the brain is unable to resolve the difference?

Weber concluded from his research that it was a constant ratio and not the absolute strength of the stimulus that was important. The mind could sense difference up to one part in forty when sensing weight. Whether it was conducted with one ounce or half-ounce weights, the sensitivity of the mind could only resolve differences greater than 1/40th of the original stimulus intensity. This was true not only for holding the weights but also when the weights were left to rest on the skin. He compared it also in the cases of visually determining whether one line was longer and whether one could hear the difference in pitch between two tones.

Weber's experiments and the results he reported changed the direction of psychology. Gustav Fechner was so impressed with Weber's work that he extended it and tried to derive a mathematical description relating the physical and the mental. Wilhelm Wundt picked up on Fechner's work and created the first psychological laboratory. From the observations of Weber, experimental psychology was born.

THE SCIENTIST

Ernst Heinrich Weber (1795–1878) was a physiologist at the University of Leipzig. Best known for his work on just noticeable differences, he not only tested all the senses but was interested in how small of a separation there could be between two objects such that the mind can still determine that it was being simultaneously touched by two very close objects. He tested this in various places on the human body, including the tongue, the check, and the back, discovering a difference in sensitivity.

YOUR JOB
Read the excerpt from "The Sense of Touch and the Common Feeling" (beginning on p. 378) and explain Weber's various experiments and the results he achieved. Explain inductivism, taking care to outline Mill's methods. How would an inductivist make sense of Weber's work? In particular, which of Mill's methods, if any, best describe Weber's approach? Are there aspects of Weber's work that do not fit in well with the inductivist approach?

RESEARCH HINT
- The scientific episode is a famous case that many writers have discussed in great detail. For the section of your paper in which you explain the case, find a secondary source (that is, a history of your science, a textbook, or a biography of the scientist) that explains it. Read that before working through the primary source (the scientist's own words in the selection in the back of this book). This will help you figure out the function of each part of the selection in the author's larger argument. Quote both the primary and the secondary sources (with proper citations, of course!) in the expository section of your paper.

Sociology Track

PAPER 2: DURKHEIM, SOCIAL FACTS, AND SUICIDE

THE CASE
The question that interested Hobbes—how society is formed and the large-scale rules by which it develops—was picked up in the centuries that followed. Positivists like Auguste Comte and the Young Hegelian Karl Marx sketched out broad-stroke, step-wise developmental structures that society as a whole was meant to follow. Herbert Spencer argued that social institutions evolved just as species did under selection pressures akin to Darwin's natural selection.

Sociology came into its own with the work of Émile Durkheim when the field was taken beyond the generalities of its predecessors and began to concern itself with social facts. Social facts, Durkheim argued, were not psychological facts about the mindset of an individual and were not biological facts about the structure of humans; rather, they were socially imposed and enforced structures that direct human actions. We internalize these rules and they then guide us invisibly, but they may be studied by social scientists as facts by observing their effects upon behavioral data and trends.

One of Durkheim's most famous studies focuses on suicide, an epidemic of which was sweeping Europe at the time. While the decision to kill oneself is surely connected to the details of personal circumstances in each case, by finding a relatively constant rate of suicides per capita among "principal European countries," including a general decrease in rates among almost all such countries in 1848, Durkheim argued that the phenomenon of suicide may be studied on the sociological level, looking at those social facts that make the option more or less available to members of society. By distinguishing different types of suicide, the social factors that lead to increases and decreases in the rates of these acts may be determined. Age, sex, and time of year of the suicide all turn out to relate to structural aspects that can be teased out.

THE SCIENTIST

Émile Durkheim (1858–1917) was born to a French Jewish family. His father, grandfather, and great-grandfather were all rabbis. Sent to rabbinic school himself, he opted for an entirely secular life. Teaching at the secondary level, he came to occupy a governmental post that allowed him to introduce sociology into the French academic curriculum. He became the first French professor of sociology at the Sorbonne. He is widely considered to be the father of sociology because he produced not only notable sociological research but also deep and careful discussions of sociological methodology.

YOUR JOB

Read the excerpt from *Suicide* (beginning on p. 380) and explain why Durkheim sees suicide as a subject for sociology and not just psychology. Explain the inductivist approach to science, focusing especially on Mill's methods. How would an inductivist make sense of Durkheim's work? How does Durkheim argue that that the reason for suicide is not a sociologically relevant factor, but that religion is? Which of Mill's methods might he have been using? Are there any aspects of Durkheim's work that do not fit well into this account?

RESEARCH HINT

- The scientific episode is a famous case that many writers have discussed in great detail. For the section of your paper in which you explain the case, find a secondary source (that is, a history of your science, a textbook, or a biography of the scientist) that explains it. Read that before working through the primary source (the scientist's own words in the selection in the back of this book). This will help you figure out the function of each part of the selection in the author's larger argument. Quote both the primary and the secondary sources (with proper citations, of course!) in the expository section of your paper.

Economics Track

THE CASE

In France, a group of thinkers called the physiocrats were the first to put forward detailed discussions about the ways in which fiscal policies of the government affected the larger economy of the nation. Seeing that Great Britain was far more prosperous than their own country, these French thinkers sought to account for the differences between the two economies. They found them in the tax structures of the nations, where the British had a much more hands-off, or *laissez-faire*, approach, and this was correlated with overall wealth.

François Quesnay, the most famous of the physiocrats, argued that the view Aristotle had rejected, considering wealth to be a quantity of money, was indeed wrong. Wealth was not found in the hording of silver and gold coinage, but rather, he argued, in production of goods and the resulting trade that connected the people of the community. Farmers, he argued, provided the original stuff that allows trade to be possible, and therefore, all taxes levied on trade ended up being indirect taxes on agriculture that hampers the initial creation of wealth for the nation.

Agriculture creates value and surplus, whereas manufacturing and commerce create false wealth that is detrimental to the health of the nation. Yet, the tax structure encourages trade and discourages farming, causing peasants to leave the countryside for the factories of the city. This causes there to be less food to feed the same number of people, makes it more expensive, and drains the nation of its wealth instead of creating more valuable land and less expensive necessities, something that would enrich and strengthen the country.

THE SCIENTIST

François Quesnay (1694–1774) was born to a working-class family and did not learn to read until age eleven, when he became obsessed with books. He was a surgeon, a professor of medicine, and a member of the court of Louis XV (as the personal physician to the King's mistress). At that point, he decided to undertake a study of the economy along with other notable thinkers who made up the group that would come to be called the physiocrats. The group opposed the wealthy merchants who surrounded the king, and the physiocrats' doctrine explicitly aimed at undermining their mercantilist views. Quesnay's most famous contribution is his *Tableau Économique*, a visual representation of his view of the structure of the economy.

Read Quesnay's essay "Farmers" (beginning on p. 383) and explain why Quesnay contends that agriculture and not trade is the basis of the economy. How do the relative prosperity of France and England support his views on tax policy? Explain inductivism, taking care to sketch out Mill's methods. How would an inductivist make sense of Quesnay's work? Which of Mill's methods does he employ in comparing the economic success of France and England? Are there aspects of the work that do not fit in well with inductivism?

RESEARCH HINT

- The scientific episode is a famous case that many writers have discussed in great detail. For the section of your paper in which you explain the case, find a secondary source (that is, a history of your science, a textbook, or a biography of the scientist) that explains it. Read that before working through the primary source (the scientist's own words in the selection in the back of this book). This will help you figure out the function of each part of the selection in the author's larger argument. Quote both the primary and the secondary sources (with proper citations, of course!) in the expository section of your paper.

HYPOTHETICO-DEDUCTIVISM

While deductivism and inductivism disagree upon the central logical mechanism in the scientific method, they do agree on a foundational element in their construction of a model of the scientific method. Both contend that the purported laws of nature given to us in our best scientific theories are the result of logical inferences. They may disagree on the type of logical inference, but both agree that there is what philosophers of science call a "logic of discovery." In other words, the way we discover new scientific principles is the result of a logical procedure. Sch cosy in loose

This stands in stark contrast to what was called the "method of hypothesis," where logic need not play any role whatsoever in the initial phase of the scientific process; rather, there is room left for the scientist to guess, to be creative, to think outside of the box. William Whewell, a scientist, historian of science, and philosopher of science, argued that this creative space was not only a real feature of science as it is actually done but imperative for progress in science. Where opponents of the method of hypothesis, like John Stuart Mill, argued that allowing blind guesses into the scientific game would introduce unfathomable degrees of error, error that could be avoided if only we stuck fast to our logical machinery, Whewell retorted that we learn valuable lessons from our mistakes, that without the mistakes, great advances might not be possible.

Scientists (and Whewell was the first person in history to use that word) do not sit down with observations of the nice A, B, C and *a, b, d* form that Mill envisioned. There are a potentially infinite number of factors that could be looked at in any number of ways. The scientist, in an act of creativity, selects a set of facts about the world and decides to try to correlate them—"colligation" in Whewell's terminology—with another set of facts about the world. Now maybe those facts are related in that way, maybe they are not. But even if they are not correlated in exactly the way the scientist hypothesized, he may still be right that those facts are in some way actually related to one another. Some of the greatest scientific discoveries were the result of guesses that missed the mark but shot at the right target.

The scientific method, Whewell argued, is as much art as it is science. There must be room for the creativity of the scientist. What made the great geniuses so magnificent was not that they followed a turn-the-crank, plug-and-chug, logical methodology—anyone could do that. No, they had creative insights into the workings of the world. The scientist's mind is an active component in scientific discovery. By substituting a free context of discovery for the deductivists' and inductivists' logic of discovery, Whewell contended that he is showing us why the great figures of science are, in fact, great.

But science is not a free-for-all. Surely, there is rigor there somewhere. But if not in the context of discovery, the means by which we first come up with hypotheses that may or may not be true, then where? It is in the context of justification, that is, in the means of testing those hypotheses. Scientists are free to use whatever mechanism they choose in order to derive hypotheses to be tested. They could make educated guesses, but they could also use astrological star charts or a Ouija board, they could pull letters from a Scrabble bag and arrange them randomly into equations, or they could follow the chemist Kekulé, who discovered the geometric molecular structure of benzene in a dream. But framing a hypothesis does not mean there is any reason to believe that hypothesis. For that, the hypothesis must go through rigorous testing.

It is here that we reintroduce logic into the scientific method. As Rudolf Carnap and R. B. Braithwaite argued, the hypotheses proposed as laws of nature will be general and abstract, but from them we may deductively derive testable consequences. That is, from the highest level hypotheses, we may make specific predictions about what will happen if a specific situation were to be arranged.

Next, we arrange for such a situation to actually be brought about and see whether our predicted result is observed. If it is not, Braithwaite contended, we know that our hypotheses was false by employing a deductive inference known as *modus tollens*.

Modus tollens is the argument form:

If H is true, then O is true
O is false
Therefore, H is false.

We know that if our hypothesis H were true, then O, our observable consequence, would also be true. But since we did not observe O, that is, we know O to be false, it must necessarily follow then that H is false as well.

On the other hand, if we do observe O, this does not mean that H must be the case. Something other than H may be the reason for O. But the fact that O was predicted H does seem to convey some reason to believe that H may be true. H is not proven by O, but it is, in Braithwaite's terminology, established by it. In other words, we argue inductively to the likely truth of H from the observation of the predicted result H. This picture of the scientific method, what is called the hypothetico-deductive method can be sketched out in this way:

1. Find a natural phenomenon to be accounted for.
2. Frame a hypothesis that if true would account for the phenomenon in question.
3. Derive a directly observable result from the hypothesis, given a set of initial conditions.
4. Set up an experiment that brings about those initial conditions and check to see if the observable result occurs.
5. If the observable result does not occur, reject the hypothesis deductively by *modus tollens;* go back to step 2 and frame a new hypothesis.
6. If the observable result does occur, the hypothesis receives inductive support; go back to step 3 and test again. — how to do this method

Step 2 is the creativity that Whewell demanded we account for. Step 4 is the connection with observation we intuitively demand be a part of science. Steps 3 and 5 are deductive and clearly justified. Step 6 is an inductive inference that seems to be Bacon's universal generalization from instances.

Hypothetico-deductivism marries the insights of deductivism and inductivism, using the strength of each to seemingly overcome the weaknesses of the other. Science, they argue, is a combination of deductive derivation, focused observation, and inductive extrapolation.

William Whewell

Novum Organum Renovatum
Of Certain Characteristics of Scientific Induction

APHORISM X

The process of scientific discovery is cautious and rigorous, not by abstaining from hypothesis, but by rigorously comparing hypotheses with facts, and by resolutely rejecting all which the comparison does not confirm.

APHORISM XI

Hypotheses may be useful, though involving much that is superfluous, and even erroneous: for they may supply the true bond of connexion of the facts; and the superfluity and errour may afterwards be pared away.

APHORISM XII

It is a test of true theories not only to account for, but to predict phenomena.

APHORISM XIII

Induction *is a term applied to describe the* process *of a true Colligation of Facts by means of an exact and appropriate Conception. An* Induction *is also employed to denote the* proposition *which results from this process.*

APHORISM XIV

The Consilience of Inductions takes place when an Induction, obtained from one class of facts, coincides with an Induction, obtained from another different class. This Consilience is a test of the truth of the Theory in which it occurs.

APHORISM XV

An induction is not the mere sum of the Facts which are colligated. The Facts are not only brought together, but seen in a new point of view. A new mental Element is superinduced; *and a peculiar constitution and discipline of mind are requisite in order to make this Induction.*

From William Whewell, *Novum Organum Renovatum* (London: J. W. Parker & Son, 1858), 70–87.

APHORISM XVI

Although in Every Induction a new conception is superinduced upon the Facts; yet this once effectually done, the novelty of the conception is overlooked, and the conception is considered as a part of the fact.

SECT. I INVENTION A PART OF INDUCTION

1.

The two operations spoken of in the preceding chapters,—the Explication of the Conceptions of our own minds, and the Colligation of observed Facts by the aid of such Conceptions,—are, as we have just said, inseparably connected with each other. When united, and employed in collecting knowledge from the phenomena which the world presents to us, they constitute the mental process of *Induction;* which is usually and justly spoken of as the genuine source of all our *real general knowledge* respecting the external world. And we see, from the preceding analysis of this process into its two constituents, from what origin it derives each of its characters. It is *real,* because it arises from the combination of real facts, but it is *general,* because it implies the possession of General Ideas. Without the former, it would not be knowledge of the External World, without the latter, it would not be Knowledge at all. When Ideas and Facts are separated from each other, the neglect of Facts gives rise to empty speculations, idle subtleties, visionary inventions, false opinions concerning the laws of phenomena, disregard of the true aspect of nature: while the want of Ideas leaves the mind overwhelmed, bewildered, and stupefied by particular sensations, with no means of connecting the past with the future, the absent with the present, the example with the rule; open to the impression of all appearances, but capable of appropriating none. Ideas are the *Form,* facts are the *Material* of our structure. Knowledge does not consist in the empty mould, or in the brute mass of matter, but in the rightly-moulded substance. Induction gathers general truths from particular facts;—and in her harvest, the corn and the reaper, the solid ears and the binding band, are alike requisite. All our knowledge of nature is obtained by Induction; the term being understood according to the explanation we have now given. And our knowledge is then most complete, then most truly deserves the name of Science, when both its elements are perfect;—when the Ideas which have been concerned in its formation have, at every step, been clear and consistent; and when they have, at every step also, been employed in binding together real and certain Facts.

2.

Induction is familiarly spoken of as a process by which we collect a *General Proposition* from a number of *Particular Cases:* and it appears to be frequently imagined that the general proposition results from a mere juxta-position of the cases, or at most, from merely conjoining them and extending them. But if we consider the process more closely, as exhibited in the cases lately spoken of, we shall perceive that this is an inadequate account of the matter. The particular facts are not merely brought together, but there is a New Element added to the combination by the very act of thought by which they are combined. There is a Conception of the mind introduced in the general proposition, which did not exist in any of the observed facts. When the Greeks, after long observing the motions of the planets, saw that these motions might be rightly considered as produced by the motion of one wheel revolving in the inside of another wheel, these Wheels were Creations of their Minds, added to the facts which they perceived by sense. And even if the wheels were no longer to be material, but were reduced to mere geometrical spheres or circles, they were not the lesser products of the mind alone,—something additional to the facts observed. The same is the case in all other discoveries. The facts are known, but they are insulated and unconnected, till the discoverer supplies from his own stores a Principle of Connexion. The pearls are there, but they will not hang together till someone provides the String. The distances and periods of the planets were all so many separate facts; by Kepler's Third Law they are connected into a single truth: but the Conceptions which this law involves were supplied by Kepler's mind, and without these, the facts were of no avail. The planets described ellipses round the sun, in the contemplation of others as well as of Newton; but Newton conceived the deflection from the tangent in these elliptical motions in a new light,—as the effect of a Central Force following a certain law; and then it was, that such a force was discovered truly to exist.

Thus in each inference made by Induction, there is introduced some General Conception, which is given, not by the phenomenon, but by the mind. The conclusion is not contained in the premises, but includes them by the introduction of a New Generality. In order to obtain our inferences, we travel beyond the cases which we have before us; we consider them as mere exemplifications of some Ideal Case in which the relations are complete and intelligible. We take a Standard, and measure the facts by it; and this Standard is constructed by us, not offered by Nature. We assert, for example, that a body left to itself will move on with unaltered velocity; not because our senses ever disclosed to us a body doing this, but because (taking this as our Ideal

Case) we find that all actual cases are intelligible and explicable by means of the Conception of *Forces,* causing change and motion, and exerted by surrounding bodies. In like manner, we see bodies striking each other, and thus moving and stopping, accelerating and retarding each other: but in all this, we do not perceive by our senses that abstract quantity, *Momentum,* which is always lost by one body as it is gained by another. This Momentum is a creation of the mind, brought in among the facts, in order to convert their apparent confusion into order, their seeming chance into certainty, their perplexing variety into simplicity. This the *Conception of Momentum gained and lost* does: and in like manner, in any other case in which a truth is established by Induction, some Conception is introduced, some Idea is applied, as the means of binding together the facts, and thus producing the truth.

3.

Hence in every inference by Induction, there is some Conception *superinduced* upon the Facts: and we may henceforth conceive this to be the peculiar import of the term *Induction.* I am not to be understood as asserting that the term was originally or anciently employed with this notion of its meaning; for the peculiar feature just pointed out in Induction has generally been overlooked. This appears by the accounts generally given of Induction. 'Induction' says Aristotle,[1] 'is when by means of one extreme term[2] we infer the other extreme term to be true of the middle term.' Thus, (to take such exemplifications as belong to our subject,) from knowing that Mercury, Venus, Mars, describe ellipses about the Sun, we infer that all Planets describe ellipses about the Sun. In making this inference syllogistically, we assume that the evident proposition, 'Mercury, Venus, Mars, do what all plants do,' may be taken *conversely,* 'All Planets do what Mercury, Venus, Mars, do.' But we may remark that, in this passage, Aristotle (as was his natural line of discussion) turns his attention entirely to the *evidence* of the inference; and overlooks a step which is of far more importance to our knowledge, namely the *invention* of the second extreme term. In the above instance, the particular luminaries, Mercury, Venus, Mars, are one logical *Extreme;* the general designation Planets is the *Middle Term;* but having these before us, how do we come to think of the *description of ellipses,* which is the other Extreme of the syllogism? When we have once invented this 'Second Extreme Term,' we may, or may not, be satisfied with the evidence of the syllogism; we may, or may not, be convinced that so far as this property goes, the extremes are co-extensive with the middle term; but the *statement* of the syllogism is the important step in science. We know how long Kepler laboured, how hard he

fought, how many devices he tried, before he hit upon this *Term,* the Elliptical Motion. He rejected, as we know, many other 'second extreme Terms,'
for example, various combinations of epicyclical constructions, because they
did not represent with sufficient accuracy the special facts of observation.
When he had established his premises, that 'Mars does describe an ellipse
about the Sun,' he does not hesitate to *guess* at least that, in this respect, he
might *convert* the other premises, and assert that 'All the Planets do what
Mars does.' But the main business was, the inventing and verifying the proposition respecting the Ellipse. The Invention of the Conception was the great
step in the *discovery;* the Verification of the Proposition was the great step in
the *proof* of the discovery. If Logic consists in pointing out the conditions of
proof, the Logic of Induction must consist in showing what are the conditions of proof, in such inferences as this: but this subject must be pursued in
the next chapter; I now speak of the principally of the act of *Invention,* which
is requisite in every inductive inference.

4.

Although in every inductive inference, an act of invention is requisite, the act
soon slips out of notice. Although we bind together facts by superinducing
upon them a new Conception, this Conception, once introduced and applied,
is looked upon as inseparably connected with the facts, and necessarily implied in them. Having once had the phenomena bound together in their minds
in virtue of the Conception, men can no longer easily restore them back to the
detached and incoherent condition in which they were before they were thus
combined. The pearls once strung, they seem to form a chain by their nature.
Induction has given them a unity which it is so far from costing us an effort to
imagine it dissolved. For instance, we usually represent to ourselves the Earth
as *round,* the Earth and the Planets as *revolving* about the Sun, and as *drawn* to
the Sun by a Central Force; we can hardly understand how it could cost the
Greeks, and Copernicus, and Newton, so much pains and trouble to arrive at
a view which to us is so familiar. These are no longer to us Conceptions caught
hold of and kept hold of by a severe struggle; they are the simplest modes of
conceiving the facts: they are really Facts. We are willing to *own* our obligation
to those discoverers, but we hardly *feel* it: for in what other manner (we ask in
our thoughts) could we represent the facts to ourselves?

Thus we see why it is that this step of which we now speak, the Invention
of a new Conception in every inductive inference, is so generally overlooked
that it has hardly been noticed by preceding philosophers. When once per-

formed by the discoverer, it takes a fixed and permanent place in the understanding of every one. It is a thought which, once breathed forth, permeates all men's minds. All fancy they nearly or quite knew it before. It oft was thought, or almost thought, though never till now expressed. Men accept it and retain it, and know it cannot be taken from them, and look upon it as their own. They will not and cannot part with it, even though they may deem it trivial and obvious. It is a secret, which once uttered, cannot be recalled, even though it be despised by those to whom it is imparted. As soon as the leading term of a new theory has been pronounced and understood, all the phenomena change their aspect. There is a standard to which we cannot help referring them. We cannot fall back into the helpless and bewildered state in which we gazed at them when we possessed no principle which gave them unity. Eclipses arrive in mysterious confusion: the notion of a *Cycle* dispels the mystery. The Planets perform a tangled and mazy dance; but *Epicycles* reduce the maze to order. The Epicycles themselves run into confusion; the Conception of an *Ellipse* makes all clear and simple. And thus, from stage to stage, new elements of intelligible order are introduced. But this intelligible order is so completely adopted by the human understanding, as to seem part of its texture. Men ask Whether Eclipses follow a Cycle; Whether the Planets describe Ellipses; and they imagine that so long as they do not *answer* such questions rashly, they take nothing for granted. They do not recollect how much they assume in *asking* the question:—how far the conceptions of Cycle and Ellipses are beyond the surface of the celestial phenomena:— how many ages elapsed, how much thought, how much observation, were needed, before men's thoughts were fashioned into the words which they now so familiarly use. And thus they treat the subject, as we have seen Aristotle treating it; as if it were a question, not of invention, but of proof; not of substance, but of form: as if the main thing were not *what* we assert, but *how* we assert it. But for our purpose, it is requisite to bear in mind the feature we have thus attempted to mark; and to recollect that, in every inference by induction, there is a Conception supplied by the mind and superinduced upon the Facts.

5.

In collecting scientific truths by Induction, we often find (as has already been observed) a Definition and a Proposition established at the same time,—introduced together, and mutually dependent on each other. The combination of the two constitutes the Inductive act; and we may consider the Definition as representing the superinduced Conception, and the Proposition as exhibiting the Colligation of Facts.

SECT. II. USE OF HYPOTHESES

6.

To discover a Conception of the mind which will justly represent a train of observed facts is, in some measure, a process of conjecture, as I have stated already; and as I then observed, the business of conjecture is commonly conducted by called up before our minds several suppositions, and selecting that one which most agrees with what we know of observed facts. Hence he who has to discover the laws of nature may have to invent many suppositions before he hits upon the right one; and amongst the endowments which lead to his success, we must reckon that fertility of invention which ministers to him such imaginary schemes, till at last he finds the one which conforms to the true order of nature. A facility in devising hypotheses, therefore, is so far from being a fault in the intellectual character of a discoverer, that it is, in truth, a faculty indispensable to his task. It is, for his purpose, much better that he should be too ready in contriving, too eager in pursuing systems which promise to introduce law and order among a mass of unarranged facts, than that he should be barren of such inventions and hopeless of such success. Accordingly, as we have already noticed, great discoverers have often invented hypotheses which would not answer to all the facts, as well as those which would; and have fancied themselves to have discovered laws, which a more careful examination of the facts overturned.

The tendencies of our speculative nature, carrying us onwards in pursuit of symmetry and rule, and thus producing all true theories, perpetually show their vigour by overshooting the mark. They obtain something, by aiming at much more. They detect the order and connexion which exist, by conceiving imaginary relations of order and connexion which have no existence. Real discoveries are thus mixed with baseless assumptions; profound sagacity is combined with fanciful conjecture; not rarely, or in peculiar instances, but commonly, and in most cases; probably in all, if we could read the thoughts of discoverers as we read the books of Kepler. To try wrong guesses is, with most persons, the only way to hit upon right ones. The character of the true philosopher is, not that he never conjectures hazardously, but that his conjectures are clearly conceived, and brought into rigid contact with the facts. He sees and compares distinctly the Ideas and the Things;—the relations of his notions to each other and to phenomena. Under these conditions, it is not only excusable, but necessary for him, to snatch at every semblance of general rule,—to try all promising forms of simplicity and symmetry.

Hence advances in knowledge are not commonly made without the previous exercise of some boldness in license in guessing. The discovery of new truths requires, undoubtedly, minds careful and scrupulous in examining what is suggested; but it requires no less, such as are quick and fertile in suggesting. What is Invention, except the talent of rapidly calling before us the many possibilities, and selecting the appropriate one? It is true, that when we have rejected all the inadmissible suppositions, they are often quickly forgotten; and few think it necessary to dwell on these discarded hypotheses, and on the process by which they were condemned. But all who discover truths, must have reasoned upon many errours to obtain each truth; every accepted doctrine must have been one chosen out of many candidates. If many of the guesses of philosophers of bygone times now appear fanciful and absurd, because time and observation have refuted them, others, which were at the time equally gratuitous, have been confirmed in a manner which makes them appear marvelously sagacious. To form hypotheses, and then to employ much labour and skill in refuting them, if they do not succeed in establishing them, is a part of the usual process of inventive minds. Such a proceeding belongs to the *rule* of the genius of discovery, rather than (as have often been thought in modern times) to the *exception*.

7.

But if it be an advantage for the discoverer of truth that he be ingenious and fertile in inventing hypotheses which may connect the phenomena of nature, it is indispensably requisite that he be diligent and careful in comparing his hypotheses with the facts, and ready to abandon his invention as soon as it appears that it does not agree with the course of actual occurrences. This constant comparison of his own conceptions and supposition with observed facts under all aspects, forms the leading employment of the discoverer: this candid and simple love of truth, which makes him willing to suppress the most favourite production of his own ingenuity as soon as it appears to be at variance with realities, constitutes the first characteristic of his temper. He must have neither the blindness which cannot, nor the obstinacy which will not, perceive the discrepancy of his fancies and his facts. He must allow no indolence, or partial views, or self-complacency, or delight in seeming demonstration, to make him tenacious of the schemes which he devises, any further than they are confirmed by their accordance with nature. The framing of hypotheses is, for the inquirer after truth, not the end, but the beginning of his work. Each of his systems is invented, not that he may admire it and follow it into all its consistent consequences, but that he may make it the occasion of a course of active experiment and observation. And if the results of this process contradict his fundamental assumptions, however ingenious, however symmetrical, however elegant his system may be, he rejects it without hesitation. He allows no natural yearning for the offspring of his own mind to draw him aside from the higher duty of loyalty to his sovereign, Truth: to her he not only gives his affections and his wishes, but strenuous labour and scrupulous minuteness of attention.

SECT. III. TESTS OF HYPOTHESES

9.

A maxim which may be useful to recollect is this,—that *hypotheses may often be of service to science, when they involve a certain portion of incompleteness, and even errour.* The object of such inventions is to bind facts which without them are loose and detached; and if they do this, they may lead the way to a perception of the true rule by which the phenomena are associated together, even if they themselves somewhat misstate the matter. The imagined arrangement enables us to contemplate, as a whole, a collection of special cases which perplex and overload our minds when they are considered in succession; and if our scheme has so much of truth in it as to conjoin what is really connected, we may afterwards duly correct or limit the mechanism of this connexion. If our hypothesis renders a reason for the agreement of cases really similar, we may afterwards find this reason to be false, but we shall be able to translate it into the language of truth.

A conspicuous example of such an hypothesis,—one which was of the highest value to science, though very incomplete, and as a reputation of nature altogether false,—is seen in the *Doctrine of epicycles* by which the ancient astronomers explained the motions of the sun, moon, and planets. This doctrine connected the places and velocities of these bodies at particular times in a manner which was, in its general features, agreeable to nature. Yet this doctrine was erroneous in its assertion of the *circular* nature of all the celestial motions, and in making the heavenly bodies revolve *round the earth.* It was, however, of immense value to the progress of astronomical science; for it enabled men to express and reason upon many important truths which they discovered respecting the motion of the stars, up to the time of Kepler. Indeed we can hardly imagine that astronomy could, in its outset, have made so great a progress under any other form, as it did in consequence of being cultivated in this shape of the incomplete and false *epicyclical hypothesis.*

That imperfect and false hypotheses, though they may thus explain *some* phenomena, and may be useful in the progress of science, cannot explain *all* phenomena;—and that we are never to rest in our labours or acquiesce in our results, till we have found some view of the subject which *is* consistent with *all* the observed facts;—will of course be understood. We shall afterward have to speak of the other steps of such a progress.

10.

Thus the hypothesis which we accept ought to explain phenomena which we have observed. But they ought to do more than this: our hypotheses ought to *foretell* phenomena which have not yet been observed; at least all phenomena of the same kind as those which the hypothesis was invented to explain. For our assent to the hypothesis implies that it is held to be true of all particular instances. That these cases belong to past or future times, that they have or have not already occurred, makes no difference in the applicability of the rule to them. Because the rule prevails, it includes all cases; and will determine them all, if we can only calculate its real consequences. Hence it will predict the results of new combinations, as well as explain the appearances which have occurred in old ones. And that it does this with certainty and correctness, is one mode in which the hypothesis is to be verified as right and useful.

The scientific doctrines which have at various periods been established have been verified in this manner. For example, the *Epicyclical Theory* of the heavens was confirmed by its *predicting* truly eclipses of the sun and moon, configurations of the planets, and other celestial phenomena; and by its leading to the construction of Tables by which the places of heavenly bodies were given at every moment of time. The truth and accuracy of these predictions were a proof that the hypothesis was valuable, and, at least to a great extent, true; although, as was afterwards found, it involved a false representation of the structure of the heavens. In like manner, the discovery of the *Laws of Refraction* enabled mathematicians to *predict*, by calculation, what would be the effect of any new form or combination of transparent lenses. Newton's hypothesis of *Fits of Easy Transmission and Easy Reflection* in the particles of light, although not confirmed by other kinds of facts, involved a true statement of the law of the phenomena which it was framed to include, and served to *predict* the forms and colours of thin plates for a wide range of given cases. The hypothesis that Light operates by *Undulations* and *Interferences* afforded the means of *predicting* results under a still larger extent of conditions. In like manner in the progress of chemical knowledge, the doctrine of *Phlogiston* supplied the means of *foreseeing* the consequence of many combinations of elements, even before they were tried; but the *Oxygen Theory,* besides affording predictions, at least equally exact, with regard to the general results of chemical operations, included all the facts concerning the relation of weight of the elements and their compounds, and enabled chemists to *foresee* such facts in untried cases. And the Theory of *Electromagnetic Forces,* as soon as it was rightly understood, enabled those who had mastered it to *predict* motions such as had not been before observed, which were accordingly found to take place.

Men cannot help believing that the laws laid down by discoverers must be in great measure identical with the real laws of nature, when the discoverers thus determine effects beforehand in the same manner in which nature herself determines them when the occasion occurs. Those who can do this, must, to a considerable extent, have determined nature's secret;—must have fixed upon the conditions to which she attends, and must have seized the rules by which she applies them. Such a coincidence of untried facts with speculative assertions cannot be the work of chance, but implies some large portion of truth in the principles on which the reasoning is founded. To trace order and law in that which has been observed, may be considered as interpreting what nature has written down for us, and will commonly prove that we understand her alphabet. But to predict what has not been observed, is to attempt ourselves to use the legislative phrases of nature; and when she responds plainly and precisely to that which we thus utter, we cannot but suppose that we have in great measure made ourselves master of the meaning and structure of her language. The prediction of results, even of the same kind as those which have been observed, in new cases, is a proof of real success in our inductive process.

NOTES

1. Aristotle *Analytica Priora* 68b15–20.
2. The syllogism here alluded to would be this:
    Mercury, Venus, Mars, describe ellipses about the Sun;
    All Planets do what Mercury, Venus, Mars do;
    Therefore all Planets describe ellipses about the Sun.

Rudolf Carnap

"Theoretical Procedures in Science"

The activities of a scientist are in part practical: he arranges experiments and makes observations. Another part of his work is theoretical: he formulates the results of his observations in sentences, compares the results with those of other observers, tries to explain them by a theory, endeavors to confirm a theory proposed by himself or somebody else, makes predictions with the help of a theory, etc. In these theoretical activities, deduction plays an important part; this includes calculation which is a special form of deduction applied to numerical expressions. Let us consider, as an example, some theoretical activities of an astronomer. He describes his observations concerning a certain planet in a report, O_1. Further, he takes into consideration a theory T concerning the movements of the planets. (Strictly speaking, T would have to include, for the application to be discussed, laws of some other branches of physics, e.g., concerning the astronomical instruments used, refraction of light in the atmosphere, etc.) From O_1 and T, the astronomer deduces a prediction, P; he calculates the apparent position of the planet for the next night. At that time he will make a new observation and formulate it in a report O_2. Then he will compare the prediction P with O_2 and thereby find it either confirmed or not. If T was a new theory and the purpose of the procedure described was to test T, then the astronomer will take the confirmation of P by O_2 as a partial confirmation for T; he will apply the same procedure again and again and thereby obtain either an increasing degree of confirmation for T or else a disconfirmation. The same deduction of P from O_1 and T is made in the case where T is already scientifically acknowledged on the basis of previous evidence, and the present purpose is to obtain a prediction of what will happen tomorrow. There is a third situation in which a deduction of this kind may be made. Suppose we have made both the observation described in O_1 and in O_2; we are surprised by the results of the observation described in O_2 and therefore want an explanation for it. This explanation is given by the theory T; more precisely, by deducing P from O_1 and T and then showing that O_2 is in accordance with P ("What we have observed is exactly what we had to expect").

From Rudolf Carnap, "Foundations of Logic and Mathematics," in *International Encyclopedia of Unified Science*, ed. Otto Neurath, Rudolf Carnap, and Charles W. Morris (Chicago: University of Chicago Press, 1938), 1:143–44.

R. B. Braithwaite

Scientific Explanation

WHAT IS SCIENCE?

[T]he word "science" will be taken to include all the natural sciences, physical and biological, and also such parts of psychology and of the social sciences (anthropology, sociology, economics) as are concerned with an empirical subject-matter. It will exclude all philosophy which is not 'general science,' all history which is concerned merely with the occurrence of particular historical events, and the disciplines of pure mathematics and symbolic logic which are not (except, perhaps in a very peculiar sense) about empirical facts at all. This sense of the word "science" corresponds pretty closely with the most frequent modern use of the word (whose first public use was perhaps in the title of the British Association for the Advancement of Science, founded in 1831); it is synonymous with "natural science" if man is included within nature.

The function of a science, in this sense of the word, is to establish general laws covering the behavior of the empirical events or objects with which the science in question is concerned, and thereby to enable us to connect together our knowledge of the separately known events, and to make reliable predictions of events as yet unknown. This function of establishing general laws is common to all the natural sciences; it is characteristic also of those parts of psychology and the social sciences which would ordinarily be called scientific as opposed to philosophical. If the science is in a highly developed stage, as in physics, the laws which have been established will form a hierarchy in which many special laws appear as logical consequences of a small number of highly general laws expressed in a very sophisticated manner; if the science is in an early stage of development—what is sometimes called its 'natural history' stage—the laws may be merely generalizations involved in classifying things into various classes. But to classify a whale as a mammal is to assert the generalization that all infant whales are provided with milk by their mothers, and this proposition is a general law, although of limited scope. It enables us to predict that the next whale we meet will be a mammal, and it singles out an important feature in which whales differ from fishes.

From R. B. Braithwaite, *Scientific Explanation* (Cambridge: Cambridge University Press, 1953), 1–2, 9–15. Reprinted with the permission of Cambridge University Press.

To emphasize the establishment of general laws as the essential function of a science is not to overlook the fact that in many sciences the question to which the scientists attaches the most importance are historical questions about the causes of particular events rather than questions directly about general laws. Biologists ask for the origin of life upon the earth, astronomers for the origin of the solar system. But the statement that some particular event is the effect of a set of circumstances involves the assertion of a general law; to ask for the cause of an event is always to ask for a general law which applies to the particular event. Though we may be more interested in the application than in the law itself, yet we need to establish the law in order to know what law it is which we have to apply.

The fundamental concept for science is thus that of scientific law, and the fundamental aim of a science is the establishment of such laws. In order to understand the way in which a science works, and the way in which it provides explanations of the facts which it investigates, it is necessary to investigate the nature of scientific laws, and what it is to establish them.

It is self-contradictory to speak of a 'false scientific law.' Since we shall be concerned not with the truth of actual scientific laws, but with the nature of propositions which, if they are true, are scientific laws, we shall such propositions, "scientific hypotheses." A scientific hypothesis is a general proposition about all the things of a certain sort. It is an empirical proposition in the sense that it is testable by experience; experience is relevant to the question as to whether or not the hypothesis is true, i.e., as to whether or not it is a scientific law.

SCIENTIFIC LAWS

We can now turn to the general laws the establishment of which, on the basis of experience, is the function of a science. What is a scientific law?

The one thing upon which everyone agrees is that it always includes a generalization, i.e., a proposition asserting a universal connexion between properties. It always includes a proposition stating that every event or thing of a certain sort either has a certain property or stands in certain relations to other events, or things, having certain properties. The generalization may assert a concomitance of properties in the same thing or event, that everything having the property A also has the property B; e.g. that every specimen of sugar is soluble in water. Or it may assert that, of every two events or things of which the first has the property A and stands in relation R to the second, the second has the property B, e.g. that in every case of every pair of free billiard balls of which the first is moving and strikes the second, the second will move. Or it may make more complicated but similar assertions about three or four or more things. The relationship between the things may be a relationship holding between simultaneous events in the things, or it may hold between events in the same thing or in two or more things which are not simultaneous ... All these types of generalization may be brought under concomitance generalizations—that everything which is A is B—by allowing A and B to be sufficiently complex properties. So if we confine our attention for the present to concomitances, our remarks will apply also to generalizations which are of more complicated type ...

[A]n adequate theory of science to-day must explain how we come to make use of sophisticated generalizations (such as that about proton-electron constitution of the hydrogen atom) which we certainly have not derived by simple enumeration of instances. To explain this the generalization must be considered, not in isolation, but in reference to the place which it occupies within a scientific system; and induction must be considered, not primarily as induction by simple enumeration, but as the method by which we establish hypotheses within scientific systems. The combination of the constant conjunction view that scientific laws are only generalizations with a doctrine of the function of such generalizations within scientific systems puts the constant-conjunction view in a new light. It answers the complaint that the constant-conjunction view underestimates the place of reason in science; by stressing the importance of logical relationships between generalizations at different levels within the system, it goes some way to satisfy those who hanker after logically necessary scientific laws ...

Scientific hypotheses, which, if true, are scientific laws, will then, for the purpose of my exposition, be taken as equivalent to generalizations of unrestricted range in space and time of greater or less degrees of complexity and generality.

THE STRUCTURE OF A SCIENTIFIC SYSTEM

A scientific system consists of a set of hypotheses which form a deductive system; that is, which is arranged in such a way that from some of the hypotheses as premises all the other hypotheses logically follow. The propositions in a deductive system may be considered as being arranged in an order of levels, the hypotheses at the highest level being those which occur only as premises in the system, those at the lowest level being those which occur only as conclusions in the system, and those at intermediate levels being those which occur as conclusions of deductions from higher-level hypotheses and which serve as premises for deductions to lower-level hypotheses.

Let us consider as an example a fairly simple deductive system with hypotheses on three levels. The example has been selected principally because it illustrates excellently the points that need to be made, and partly because the construction and the establishment of a similar system by Galileo marks a turning-point in the history of science.

The system has one highest-level hypothesis:

I. Every body near the earth freely falling towards the Earth falls with an acceleration of 32 feet per second per second.

From this hypothesis there follows, by simple principles of the integral calculus,[1] the hypothesis:

II. Every body starting from rest and freely falling towards the Earth falls $16t^2$ feet in t seconds, whatever number t may be.

From II there follows in accordance with the logical principle (the *applicative principle*) permitting the application of a generalization to its instances, the finite set of hypotheses:

IIIa. Every body starting from rest and freely falling for 1 second towards the Earth falls a distance of 16 feet.
IIIb. Every body starting from rest and freely falling for 2 seconds towards the Earth falls a distance of 64 feet.

And so on.

In this deductive system the hypotheses at the second and third levels (II, IIIa, IIIb, etc.) follow from the one highest-level hypothesis (I); those at the third level (IIIa, IIIb, etc.) also follow from the one at the second level (II).

The hypotheses in this deductive system are empirical general propositions with diminishing generality. The empirical testing of the deductive system is effected by testing the lowest-level hypotheses in the system. The confirmation or refutation of these is the criterion by which the truth of all the hypotheses in the system is tested. The establishment of a system as a set of true propositions depends upon the establishment of its lowest-level hypotheses.

The lowest-level hypothesis IIIa is tested by applying it to a particular case. A body is allowed to fall freely for 1 second and the distance it falls is measured.[2] If it is found that it falls 16 feet, the hypothesis is confirmed; if it is found that it falls more, or less, than 16 feet, the hypothesis is refuted.

It is convenient to treat the logic of this procedure as consisting of two steps. A case is either observed, or experimentally produced, of a body falling for 1 second. The following proposition is then empirically known:

e_1. This body freely falls for 1 second towards the Earth, starting from rest.

The general hypothesis IIIa is then applied to this case by first deducing from IIIa the proposition:

IIIa'. It is only the case that this body, starting from rest, freely falls for 1 second towards the Earth if it falls a distance of 16 feet.

From this application of the general hypothesis, together with the proposition e_1, there is deduced:

f_1. This body falls 16 feet.

The testing of a scientific hypothesis thus consists in deducing from it a proposition of the form «e_1 only if f_1». Then there follows from the conjecture of e_1 with this proposition the third proposition f_1, whose truth or falsity is observed.

If f_1, the logical consequence of e_1 and IIIa', is observed to be true, the hypothesis IIIa is ordinarily said to be confirmed. The piece of evidence f_1, conjoined with e_1 (which conjunction will be called an *instance* of the hypothesis), is said to support the hypothesis. But it is clear that this one piece of evidence is insufficient to prove the hypothesis. It would only do so if the hypothesis were a logical consequence of the conjunction of f_1 with e_1. This, of course, is not the case. It is perfectly possible for the hypothesis to hold in this one instance, but to be false in some other instance, and consequently false as a general proposition. And, indeed, this is the case however many times the hypothesis is confirmed. However many conjunctions f_1 with e_1, f_2 with e_2, etc., have been examined and found to confirm the hypothesis, there will still be unexamined cases in which the hypothesis might be false without contradicting any of the observed facts.[3] Thus the empirical evidence of its instances never proves the hypothesis: in suitable cases we may say that it *establishes* the hypothesis, meaning by this that the evidence makes it reasonable to accept the hypothesis; but it never *proves* the hypothesis in the sense that the hypothesis is a logical consequence of the evidence.

The situation is different if f_1 is observed to be false. For the conjunction of not-f_1 with e_1 is logically incompatible with the hypothesis being true; the falsity of the hypothesis is a logical consequence of the conjunction of not-f_1 with e_1. Calling this conjunction a *contrary instance* of the hypothesis, we may say that a hypothesis is proved to be false, or refuted, by one known contrary instance.

This asymmetry of confirmation and refutation is a consequence of the fact that all the hypotheses of a science are general propositions of the form «Every A is B». Propositions of the form «Some A's are B's» (existential propositions),[4] which are the contradictories of general propositions, have the reverse asymmetry; they can be proved by one instance, but no number of contrary instances will suffice to disprove them.

It has been said that there is no greater tragedy than the murder of a beautiful scientific hypothesis by one discordant instance. As will be seen, it is usually possible to save any particular higher-level hypothesis from this fate by choosing instead to sacrifice some other higher-level hypothesis which is essential to the deduction. But the fact that in principle a scientific hypothesis (or a conjunction of scientific hypotheses, if more than one are required for the deduction of observable facts) can be conclusively disproved by observation, although it can never be conclusively proved, sharply distinguishes the question of the refutation of a scientific theory from that of its establishment. The former question is a simple matter of deductive logic, if the system of hypotheses is taken as a whole; the latter question involves the justification of inductive inference . . .

NOTES

1. Hypothesis I can be expressed by the differential equation $d^2s/d^2t = 32$, whose solution, under the conditions that $s = 0$ and $ds/dt = 0$ when $t = 0$, is $s = 16t^2$.

2. The system tested by Galileo was, in fact, more complicated than our example. Galileo was unable to measure the times of fall of bodies accurately enough to test IIIa; what he tested empirically was a more elaborate system in which the lowest-level hypotheses were propositions about descents of bodies rolling down grooves in inclined planes.

3. Unless, of course, the hypothesis has only a limited number of instances, and not only have all these instances been examined, but it is known that there are no unexamined instances. Generalizations with only a limited number of instances, which can be proved from a knowledge these instances by what logicians have called *perfect induction,* present no logical problem; since they are of little interest in science, they will not be considered further. All scientific hypotheses will be taken to be generalizations with an unlimited number of instances.

4. In the terminology of traditional logic my existential propositions would be called "particular propositions" and my general propositions "universal propositions," the term "general proposition" being used to cover both types.

PARADOXES OF EVIDENCE

Hypothetico-deductivism is a very persuasive system that seems to account well for much of the science we observe. But central to it is the notion of evidence, a concept that is inherently inductive. This seems to be unproblematic . . . until we start thinking hard about it. A number of philosophers have realized that the notion of evidence inherent in this sort of view of science runs into some odd problems.

In the eighteenth century, David Hume formulated the first of these problems in response to Newton's inductivism. David Hume was one of the great troublemakers in the history of philosophy. There was not an intellectual applecart in the entire marketplace of ideas he did not try to tip over. To those who sought to base their worldview on a religious foundation, he offered arguments about the unreasonableness of belief in miracles. To those who tried, like Aristotle and Descartes, to use deductive logic and mathematics as a template for all knowledge, he argued that the tools were not strong enough to cover some of the most important bits of knowledge we have. And for those who sought to use the Baconian/Newtonian approach, he developed his most well-known result, the problem of induction. For an English-speaking philosopher to undermine the Newtonian approach was, indeed, a radical act.

Hume argued that if science, the pinnacle of human knowledge, is to be based upon inductive reasoning as Bacon and Newton contend, then surely we must have a firm justification for belief in the conclusions of such arguments. But what would the justification for such beliefs rest on?

Consider a discussion between two people:

A: What do you believe?
B: I believe the sun will rise tomorrow.
A: Why do you believe that?
B: Because it always has every other morning.
A: That is an inductive argument.
B: Why, it sure is.
A: Why should I believe that the conclusions of inductive arguments are probably true?
B: Well, they have usually been so in the past.
A: So, the reason we should believe that inductive inferences will work in the future is that they have tended to work in the past?
B: Yes.
A: But that is an inductive argument.

B: Why, it sure is.

A: Why should I believe that the conclusions of inductive arguments are probably true?

B: Well, they have usually been so in the past.

A: So, the reason we should believe that inductive inferences will work in the future is that they have tended to work in the past?

B: Yes.

A: But that is an inductive argument . . .

This circle is called Hume's problem of induction.

There seem to be two ways to justify any belief: deductively and inductively. We cannot justify induction deductively because inductive conclusions outrun their premises and deductive conclusions do not, hence deduction is not strong enough to do the job. But it is also impossible and to try to justify induction inductively lest we walk into the circular argument displayed above.

If both deduction and induction are unable to give us a good reason to be confident about the basis for rational belief in inductive conclusions, then science is in a heaping helping of trouble. Those beliefs we feel most confident about are left without any philosophical grounding at all. If we do not have reason to believe in science, what do we have reason to believe in? How can we save science?

If that wasn't problem enough, Nelson Goodman came along with his new riddle of induction, the so-called grue paradox. Almost two centuries after Hume took aim at those who were so confident about the Newtonian approach to science, Goodman laid down a second layer of concern for any system that employs any sort of inductive inference. He argued that even if we somehow solved Hume's problem of induction, we still could not have good reason to believe inductively derived conclusions, because any finite set of data will have an infinite number of mutually exclusive hypotheses that can be inductively inferred from them. Induction takes us from a finite set of data to an enlarged conclusion, but there will always be multiple ways that one could extend a given set of data. Hume's problem challenges the leap itself, but Goodman points out that even if the leap is legitimate, there will always be multiple directions in which to leap. There seems to be no good reason to specify any given direction as preferred, that is to say, given a finite set of data, there will always be an infinite number of inductive conclusions that could be extrapolated from them and therefore induction does not give us reason to believe a specific one of them.

If we do something as trivial as change our language, then we would also change what seems like the natural inductive conclusion. But this then means that the scientific results would change depending upon something as trivial and conventional as what language we choose to speak. Science is supposed to be bigger than that, something objective, describing the real world beyond our choices. If science is supposed to give us good reason to believe its predictions, then rationality ends up being determined by something as arbitrary as a choice of language. Surely, the logical machinery serving as the foundation of the scientific method should be more robust than that.

C. G. Hempel added to our worries with his "raven paradox." Hempel began by setting out two simple conditions that seem incredibly obvious:

1. Nicod's criterion: a universal generalization is confirmed by its instances
2. The equivalence condition: if sentences S_1 and S_2 are logically equivalent, then whatever is evidence for S_1 must also be evidence for S_2

If I contend that all bachelors are slobs, then observing Bob, who is both single and slovenly, is to add a bit of evidence to my hypothesis. Surely, Nicod's criterion is acceptable. Similarly, any evidence that all bachelors are slobs will also have to be evidence that all unmarried men are slobs, since "bachelors" and "unmarried men" are synonymous, so any identical statements about them will be logically equivalent. That either of these conditions would be doubtful seems absurd.

But then Hempel pointed out an uncomfortable fact. The statements "all ravens are black" and "all non-black things are non-ravens" are logically equivalent. Hence, by the equivalence condition any evidence for one must also be evidence for the other. By Nicod's criterion, instances of a generalization count as evidence for that generalization, so a red herring, being a non-black thing that is also a non-raven would be an instance of the second sentence and therefore evidence that all ravens are black. Indeed, now every single thing you see that is not black and not a raven—white chalk, blue sky, green grass—would count as evidence that all ravens are black. Yet, surely all this "evidence" is not really good reason to believe something about ravens. Something is wrong. As Goodman famously remarked, "The prospect of being able to investigate ornithological theories without going out in the rain is so attractive that we know there must be a catch in it."[1] The question, of course, is where exactly *is* the catch?

The basis for all scientific reasoning in each of the views beyond deductivism has been evidence. Scientists use evidence to arrive at their results. Inferences from observable evidence is the basis of our confidence in scientific theories. If we are to keep this crucial the notion of evidence, however, we must wrestle with the puzzles of Hume, Goodman, and Hempel.

NOTES

1. Nelson Goodman, *Fact, Fiction, and Forecast* (Cambridge, MA: Harvard University Press, 1979), 70.

David Hume

· ·

Enquiry
Sceptical Doubts concerning the Operations
of the Understanding

PART I.

All the objects of human reason or enquiry may naturally be divided into two
kinds, to wit, *Relations of Ideas,* and *Matters of Fact*. Of the first kind are the sciences of Geometry, Algebra, and Arithmetic; and in short, every affirmation,
which is either intuitively or demonstratively certain. *That the square of the
hypotenuse is equal to the square of the other two sides,* is a proposition, which
expresses a relation between these figures. *That three times five is equal to the
half of thirty,* expresses a relation between these numbers. Propositions of
this kind are discoverable by the mere operation of thought, without dependence on what is any where existent in the universe. Though there never were
a circle or triangle in nature, the truths demonstrated by EUCLID, would for
ever retain their certainty and evidence. *only in our mind*

Matters of fact, which are the second object of human reason are not ascertained in the same manner; nor is our evidence of their truth, however
great, of a like nature with the foregoing. The contrary of every matter of fact
is still possible; because it can never imply a contradiction, and is conceived
by the mind with the same facility and distinctness, as if ever so confirmable
to reality. *That the sun will not rise to-morrow* is no less intelligible a proposition and implies no more contradiction, than the affirmation, *that it will rise.*
We should in vain, therefore, attempt to demonstrate its falsehood. Were it
demonstratively false, it would imply a contradiction, and could never be distinctly conceived by the mind.

It may, therefore, be a subject worthy of curiosity, to enquire what is the
nature of that evidence which assures us of any real existence and matter of
fact, beyond the present testimony of our senses, or the records of our memory. This part of philosophy, it is observable, has been little cultivated, either

From David Hume, *Philosophical Essays Concerning Human Understanding* (London: A. Millar,
1748).

by the ancients or moderns; and therefore our doubts and errors, in the prosecution of so important an enquiry, may be the more excusable; while we march through such difficult paths, without any guide or direction. They may even prove useful, by exciting curiosity, and destroying that implicit faith and security, which is the bane of all reasoning and free enquiry. The discovery of defects in the common philosophy, if any such there be, will not, I presume, be a discouragement, but rather an incitement, as is usual, to attempt something more full and satisfactory, than has yet been proposed to the public.

All reasonings concerning maters of fact seem to be founded on the relation of *Cause and Effect*. By means of that relation alone we can go beyond the evidence of our memory and senses. If you were to ask a man, why he believes any matter of fact, which is absent; for instance, that his friend is in the country, or in FRANCE; he would give you a reason; and this reason would be some other fact; as a letter received from him, or the knowledge of his former resolutions and promises. A man, finding a watch or any other machine in a desert island, would conclude, that there had once been men on the island. All our reasonings concerning fact are of the same nature. And here it is constantly supposed that there is a connexion between the present fact and that which is inferred from it. Were nothing to bind them together, the inference would be entirely precarious. The hearing of an articulate voice and rational discourse in the dark assures us of the presence of some person: Why? Because these are the effects of the human make and fabric, and closely connected with it. If we anatomize all the other reasonings of this nature, we shall find, that they are founded on the relation of cause and effect, and that this relation is either near or remote, direct or collateral. Heat and light are collateral effects of fire, and the one effect may justly be inferred from the other.

If we would satisfy ourselves, therefore, concerning the nature of that evidence, which assures us of matters of fact, we must enquire how we arrive at knowledge of cause and effect.

I shall venture to affirm, as a general proposition, which admits of no exception, that the knowledge of this relation is not, in any instance, attained by reasonings *a priori*; but arises entirely from experience, when we find, that any particular objects are constantly conjoined with each other. Let an object be presented to a man of ever so strong natural reason and abilities; if that object be entirely new to him, he will not be able, by the most accurate examination of its sensible qualities, to discover any of its causes or effects.

ADAM, though his rational faculties be supposed, at the very first, entirely perfect, could not have inferred from the fluidity, and transparency of water, that it would suffocate him, or from the light and warmth of fire, that it would consume him. No object ever discovers, by the qualities which appear to the senses, either the causes which produced it, or the effects which will arise from it; nor can our reason, unassisted by experience, ever draw any inference concerning real existence and matter of fact.

This proposition, *that causes and effects are discoverable, not by reason, but by experience,* will readily be admitted with regard to such objects, as we remember to have once been altogether unknown to us; since we must be conscious of the utter inability, which we then lay under, of foretelling, what would arise from them. Present two smooth pieces of marble to a man, who has no tincture of natural philosophy; he will never discover, that they will adhere together, in such a manner as to require great force to separate them in a direct line, while they make so small a resistance to a lateral pressure. Such events, as bear little analogy to the common course of nature, are also readily confessed to be known only by experience; nor does any man imagine that the explosion of gunpowder, or the attraction of a loadstone, could ever be discovered by arguments *a priori.* In a like manner, when the effect is supposed to depend upon an intricate machinery or secret structure of parts, we make no difficulty in attributing all our knowledge of it to experience. Who will assert, that he can give the ultimate reason, why milk or bread is proper nourishment for a man, not for a lion or a tiger?

But the same truth may not appear, at first sight, to have the same evidence with regard to events, which have become familiar to us from our first appearance in the world, which bear a close analogy to the whole course of nature, and which are supposed to depend on the simple qualities of objects, without any secret structure of parts. We are apt to imagine, that we could discover these effects by the mere operation of our reason, without experience. We fancy, that were we brought, on a sudden, into this world, we could at first have inferred, that one Billiard-ball would communicate motion to another upon impulse; and that we needed no to have waited for the event, in order to pronounce with certainty concerning it. Such is the influence of custom, that, where it is strongest, it not only covers our natural ignorance, but even conceals itself, and seems not to take place, merely because it is found in the highest degree.

But to convince us, that all the laws of nature, and all the operations of bodies without exception, are known only by experience, the following reflections may, perhaps, suffice. Were any object presented to us, and were we required to pronounce concerning the effect, which will result from it, without consulting past observation; after what manner, I beseech you, must the mind proceed in this operation? It must invent or imagine some event, which it ascribes to the object as its effect; and it is plain that this invention must be entirely arbitrary. The mind can never possibly find the effect in supposed cause, by the most accurate scrutiny and examination. For the effect is totally different from the cause, and consequently can never be discovered in it. Motion in the second Billiard-ball is a quite distinct event from motion in the first; nor is there any thing in the one to suggest the smallest hint of the other. A stone or piece of metal raised in the air, and left without any other support, immediately falls: But to consider the matter *a priori,* is there any thing we discover in this situation, which can beget the idea of a downward, rather than an upward, or any other motion, in the stone or metal?

And as the first imagination or invention of a particular effect, in all natural operations, is arbitrary, where we consult not experience; so must we also esteem the supposed tie or connexion between the cause and effect, which binds them together, and renders it impossible, that any other defect could result from the operation of that cause. When I see, for instance, a Billiard-ball moving in a straight line towards another; even suppose motion in the second ball should by accident be suggested to me, as the result of their contact or impulse; may I not conceive, that a hundred different events might as well follow from that cause? May not both these balls return in a straight line, or leap off from the second in any line or direction? All these suppositions are consistent and conceivable. Why then should we give preference to one, which is no more consistent or conceivable than the rest? All our reasonings *a priori* will never be able to show us any foundation for this preference.

In a word, then, every effect is a distinct event from its cause. It could not, therefore, be discovered in the cause, and the first invention or conception of it, *a priori,* must be entirely arbitrary. And even after it is suggested, the conjunction of it with the cause must appear equally arbitrary; since there are always many other effects, which, to reason, must seem fully as consistent and natural. In vain, therefore, should we pretend to determine any single event, or infer any cause or effect, without the assistance of observation and experience.

Hence, we may discover the reason, why no philosopher, who is rational and modest, has ever pretended to assign the ultimate cause of any natural operation, or to show distinctly the action that power, which produces any single effect in the universe. It is confessed, that the utmost effort of human reason is, to reduce the principles, productive of natural phenomena, to a greater simplicity, and to resolve the many particular effects into a few general causes, by means of reasoning from analogy, experience, and observation. But as to the causes of these general causes, we should in vain attempt their discovery; nor shall we ever be able to satisfy ourselves, by any particular explication of them. These ultimate springs and principles are totally shut up from human curiosity and enquiry. Elasticity, gravity, cohesion of parts, communication of motion by impulse; these are probably the ultimate causes and principles which we shall ever discover in nature; and we may esteem ourselves sufficiently happy, if, by accurate enquiry and reasoning, we can trace up the particular phenomena to, or near to, these general principles. The most perfect philosophy of the natural kind only staves off our ignorance a little longer: As perhaps the most perfect philosophy of the moral or metaphysical kind serves only to discover larger portions of it. Thus the observation of human blindness and weakness is the result of all philosophy, and meets us, at every turn, in spite of our endeavors to elude or avoid it.

Nor is geometry, when taken into the assistance of natural philosophy, ever able to remedy this effect, or lead us into the knowledge of ultimate causes, by all that accuracy of reasoning, for which it is so justly celebrated. Every part of mixed mathematics proceeds upon the supposition that certain laws are established by nature in her operations; and abstract reasonings are employed, either to assist experience in the discovery of these laws, or to determine their influence in particular instances, where it depends upon any precise degree of distance and quantity. Thus it is a law of motion, discovered by experience, that the moment or force of any body in motion is in the compound ratio or proportion of its solid contents and its velocity; and consequently, that a small force may remove the greatest obstacle or raise the greatest weight, if, by any contrivance or machinery, we can increase the velocity of that force, so as to make it an overmatch for its antagonist. Geometry assists us in the application of this law, by giving us the just dimensions of all the parts and figures, which can enter into any species of machine; but still the discovery of the law itself is owing merely to experience, and all the abstract reasonings in the world could never lead us one step towards the knowledge of it. When we reason *a priori*, and consider merely any operation or cause, as it appears to the mind, independent of all observation, it never could suggest to us the notion of any distinct object, such as is effect; much less, show us the inseparable and inviolable connexion between them. A man must be very sagacious, who could discover by reason, that crystal is the effect of heat, and ice of cold, without being previously acquainted with the operation of the qualities.

PART II.

But we have not, yet, attained any tolerable satisfaction with regard to the question first proposed. Each solution still gives rise to a new question as difficult as the foregoing, and leads us on to farther enquiries. When it is asked, *What is the nature of all our reasonings concerning matter of fact?* the proper answer seems to be, that they are founded on the relation of cause and effect. When again it is asked, *What is the foundation of all our reasonings and conclusions concerning that relation?* It may be replied in one word, EXPERIENCE. But if we still carry on our sifting humour, and ask, *What is the foundation of all conclusions from experience?* This implies a new question. Philosophers, that give themselves airs of superior wisdom and sufficiency, have a hard task, when they encounter persons of inquisitive dispositions, who push them from every corner, to which they retreat, and who are sure at last to bring them to some dangerous dilemma. The best expedient to prevent this confusion, is to be modest in our pretensions; and even to discover the difficulty ourselves before it is objected to by us. By this means, we may make a kind of merit of our very ignorance.

I shall content myself, in this section, with an easy task, and shall pretend only to give a negative answer to the question here proposed. I say, then, that even after we have experience of the operations of cause and effect, our conclusions from that experience are *not* founded on reasoning, or any process of the understanding. This answer we must endeavor, both to explain and defend.

It must certainly be allowed, that nature has kept us at a great distance from all her secrets, and has afforded us only the knowledge of a few superficial qualities of objects; while she conceals from us those powers and principles, on which the influence of these objects entirely depends. Our senses inform us of the colour, weight, and consistency of bread; but neither sense nor reason can ever inform us of those qualities, which fit it for the nourishment and support of a human body. Sight or feeling conveys an idea of the actual motion of bodies; but as to that wonderful force or power, which would carry on a moving body for ever in a continued change of place, and which bodies never lose but by communicating it to others; of this we cannot form the most distant conception. But notwithstanding this ignorance of natural powers and principles, we always presume, when we see like sensible qualities, that they have like secret powers, and expect, that effects, similar to those which we have experienced, will follow from them. If a body of like colour and consistence with bread, which we have formerly eaten, be presented to us, we make no scruple of repeating the experiment, and foresee, with certainty, like nourishment and support. Now this is a process of the mind or thought, of which, I would willingly know the foundation. It is allowed on all hands, that there is no known connexion between the sensible qualities and the secret powers; and consequently, that the mind is not led to form such a conclusion concerning their constant and regular conjunction, by any thing which it knows of their nature. As to past *Experience*, it can be allowed to give *direct* and *certain* information of those precise objects only, and that precise period of time, which fell under its cognizance: But why this experience should be extended to future times, and to other objects, which for aught we know, may be only in appearance similar; this is the main question on which I would insist. The bread, which I formerly eat, nourished me; that is, a body of such sensible qualities, was, at that time, endued with such secret powers: But does it follow, that other bread must also nourish me at another time, and that like sensible qualities must always be attended with like secret powers? The consequence seems nowise necessary. At least, it must be acknowledged, that there is here a consequence drawn by the mind; that there is a certain step taken; a process of thought, and an inference, which wants to be explained. These two propositions are far from being the same, *I have found that such an object has always been attended with such an effect, and I foresee, that other objects, which are, in appearance, similar, will be attended with similar effects.* I shall allow, if you please, that the one proposition may justly be inferred from the other: I know in fact, that it always is inferred. But if you insist, that the inference is made by a chain of reasoning, I desire you to produce that reasoning. The connexion between these propositions is not intuitive. There is required a medium, which may enable the mind to draw such an inference, if indeed it be drawn by reasoning and argument. What that medium is, I must confess, passes my comprehension; and it is incumbent on those to produce it, who assert, that it really exists, and is the origin of all our conclusions concerning matter of fact.

This negative argument must certainly, in process of time, become altogether convincing, if many penetrating and able philosophers shall turn their enquiries this way; and no one be ever able to discover any connecting proposition or intermediate step, which supports the understanding in this conclusion. But as the question is yet new, every reader may not trust so far to his own penetration, as to conclude, because an argument escapes his enquiry, that therefore it does not really exist. For this reason it may be requisite to venture upon a more difficult task; and enumerating all the branches of human knowledge, endeavor to show, that none of them can afford such an argument.

All reasonings may be divided into two kinds, namely demonstrative reasoning, or that concerning the relations of ideas, and moral reasoning, or that concerning matter of fact and existence. That there are no demonstrative arguments in the case, seems evident; since it implies no contradiction, that the course of nature may change, and that an object, seemingly like those which we have experienced, may be attended with different or contrary effects. May I not clearly and distinctly conceive, that a body, falling from the clouds, and which, in all other respects, resembles snow, has yet the taste of salt or the feeling of fire? Is there any more intelligible proposition than to affirm, that all the trees will flourish in DECEMBER and JANUARY, and decay in MAY and JUNE? Now whatever is intelligible, and can be distinctly conceived, implies no contradiction, and can never be proved false by any demonstrative argument or abstract reasoning *a priori*.

If we be, therefore, engaged by arguments to put trust in past experience, and make it the standard of our future judgment, these arguments must be probable only, or such as regard matter of fact and real existence, according to the division above mentioned. But that there is no argument of this kind, must appear, if our explication of that species of reasoning be admitted as solid and satisfactory. We have said, that all arguments concerning existence are founded on the relation of cause and effect; that our knowledge of that relation is derived entirely from experience; and that all our experimental conclusions proceed upon the supposition, that the future will be conformable to the past. To endeavor, therefore, the proof of this last proposition by probable arguments, or arguments regarding existence, must be evidently going in a circle, and taking that for granted, which is the very point in question.

In reality, all arguments from experience are founded on the similarity, which we discover among natural objects, and by which we are induced to expect effects similar to those, which we have found to follow from such objects. And though none but a fool or a madman will ever pretend to dispute the authority of experience, or to reject that great guide of human life; it may surely be allowed a philosopher to have such curiosity at least, as to examine the principle of human nature, which gives this mighty authority to experience, and makes us draw advantage from that similarity, which nature has placed among different objects. From causes, which appear *similar,* we expect similar effects. This is the sum of all our experimental conclusions. Now it seems evident, that, if this conclusion were formed by reason, it would be as perfect at first, and upon one instance, as after ever so long a course of experience. But the case is far otherwise. Nothing so like as eggs; yet no one, on account of this appearing similarity, expects the same taste and relish in all of them. It is only after a long course of uniform experiments in any kind, that we attain a firm reliance and security with regard to a particular event. Now where is that process of reasoning, which, from one instance, draws a conclusion, so different from that which it infers from a hundred instances, that are nowise different from that single one? This question I propose as much for the sake of information, as with an intention of raising difficulties. But I keep my mind still open to instruction; if any one will vouchsafe to bestow it on me.

Should it be said, that, from a number of uniform experiments, we *infer* a connexion between the sensible qualities and the secret powers; this, I must confess, seems the same difficulty, couched in different terms. The question still recurs, on what process of argument this *inference* is founded? Where is the medium, the interposing ideas, which join propositions so very wide of each other? It is confessed, that the colour, consistence, and other sensible qualities of bread appear not, of themselves, to have any connexion with the secret powers of nourishment and support. For otherwise we could infer these secret powers from the first appearance of these sensible qualities, without the aid of experience; contrary to the sentiment of all philosophers, and contrary to plain matter of fact. Here then is our natural state of ignorance with regard to the powers and influence of all objects. How is this remedied by experience? It only shows us a number of uniform effects, resulting from certain objects, and teaches us, that those particular objects, at that particular time, were endowed with such powers and forces. When a new object, endowed with similar sensible qualities, is produced, we expect similar powers and forces, and look for a like effect. From a body of like colour and consistence with bread, we expect like nourishment and support. But this surely is a step or progress of the mind, which wants to be explained. When a man says, *I have found, in all past instances, such sensible qualities conjoined with such secret powers:* And when he says, *similar sensible qualities will always be conjoined with similar secret powers;* he is not guilty of a tautology, nor are these propositions in any respect the same. You say that the one proposition is an inference from the other. But you must confess that the inference is not intuitive; neither is it demonstrative: Of what nature is it then? To say it is experimental, is begging the question. For all inferences from experience suppose, as their foundation, that the future will resemble the past, and that similar powers will be conjoined with similar sensible qualities. If there be any suspicion, that the course of nature may change, and that the past may be no rule for the future, all experience becomes useless, and can give rise to no inference or conclusion. It is impossible, therefore, that any arguments from experience can prove this resemblance of the past to the future; since all these arguments are founded on the supposition of that resemblance. Let the course of things be allowed hitherto ever so regular; that alone without some new argument or inference, proves not, that, for the future, it will continue so. In vain do you pretend to have learned the nature of bodies from your past experience. Their secret nature, and consequently, all their effects and influence, may change, without any change in their sensible qualities. This happens sometimes, and with regard to some objects: Why may it not happen always, and with regard to all objects? What logic, what process of argument secures you against this supposition? My practice, you say, refutes my doubts. But you mistake the purport of my question. As an agent, I am quite satisfied in the point; but as a philosopher, who has some share of curiosity, I will not say scepticism, I want to learn the foundation of this inference. No reading, no enquiry has yet been able to remove my difficulty, or give me satisfaction in a matter of such importance. Can I do better than propose the difficulty to the public, even though, perhaps, I have small hopes of obtaining a solution? We shall at least, by this means, be sensible of our ignorance, if we do not augment our knowledge.

I must confess, that a man is guilty of unpardonable ignorance, who concludes, because an argument has escaped his investigation, that therefore it does not really exist. I must also confess, that, though all the learned, for several ages, should have employed themselves in fruitless search upon any subject, it may still, perhaps, be rash to conclude positively, that the subject must, therefore, pass all human comprehension. Even though we examine all the sources of our knowledge, and conclude them unfit for such a subject, there still may remain a suspicion, that the enumeration is not complete, or the examination not accurate. But with regard to the present subject, there are some considerations, which seem to remove all this accusation of arrogance or suspicion of mistake.

It is certain, that the most ignorant and stupid peasants, nay infants, nay even brute beasts, improve by experience, and learn the qualities of natural objects, by observing the effects, which result from them. When a child has felt the sensation of pain from touching the flame of a candle, he will be careful not to put his hand near any candle; but will expect a similar effect from a cause, which is similar in its sensible qualities and appearance. If you assert, therefore, that the understanding of the child is led into this conclusion by any process of argument or ratiocination, I may justly require you to produce that argument; nor have you any pretense to refuse so equitable a demand. You cannot say, that the argument is abstruse, and may possibly escape your enquiry; since you confess, that it is obvious to the capacity of a mere infant. If you hesitate, therefore, a moment, or if, after reflection, you produce any intricate or profound argument, you, in a manner, give up the question, and confess, that it is not reasoning which engages us to suppose the past resembling the future, and to expect similar effects from causes, which are, to appearances, similar. This is the proposition which I intended to enforce in the present section. If I be right, I pretend not to have made any mighty discovery. And if I be wrong, I must acknowledge myself to be indeed a very backward scholar; since I cannot now discover an argument, which, it seems, was perfectly familiar to me, long before I was out of my cradle.

Nelson Goodman

. . . .

Fact, Fiction, and Forecast
The New Riddle of Induction

Confirmation of a hypothesis by an instance depends rather heavily upon features of the hypothesis other than its syntactical form. That a given piece of
copper conducts electricity increases the credibility of statements asserting
that other pieces of copper conduct electricity, and thus confirms the hypothesis that all copper conducts electricity. But the fact that a given man now in
this room is a third son does not increase the credibility of statements asserting that other men now in this room are third sons, and so does not confirm
the hypothesis that all men now in this room are third sons. Yet in both cases
our hypothesis is a generalization of the evidence statement. The difference is
that in the former case the hypothesis is a *lawlike* statement; while in the latter case, the hypothesis is a merely contingent or accidental generality. Only a
statement that is *lawlike*—regardless of its truth or falsity or its scientific importance—is capable of receiving confirmation from an instance of it; accidental statements are not. Plainly, then, we must look for a way of distinguishing lawlike from accidental statements.

So long as what seems to be needed is merely a way of excluding a few
odd and unwanted cases that are inadvertently admitted by our definition
of confirmation, the problem may not seem very hard or very pressing. We
fully expect that minor defects will be found in our definition and that the
necessary refinements will have to be worked out patiently one after another.
But some further examples will show that our present difficulty is of a much
graver kind.

Suppose that all emeralds examined before a certain time *t* are green. At
time *t*, then, our observations support the hypothesis that all emeralds are
green; and this is in accord with our definition of confirmation. Our evidence
statements assert that emerald *a* is green, that emerald *b* is green, and so on;
each confirms the general hypothesis that emeralds are green. So far, so good.

Now let me introduce another predicate less familiar than "green." It is the

From Nelson Goodman, *Fact, Fiction, and Forecast* (Cambridge, MA: Harvard University Press,
1979), 72–81. Reprinted with the permission of Harvard University Press.

predicate "grue" and it applies to all things examined before *t* just in case they are green but to other things just in case they are blue. Then at time *t* we have, for each evidence statement asserting that a given emerald is green, a parallel evidence statement asserting that that emerald is grue. And the statements that emerald *a* is grue, that emerald *b* is grue, and so on, will each confirm the general hypothesis that all emeralds are grue. Thus according to our definition, the prediction that all emeralds subsequently examined will be green and the prediction that all will be grue are alike confirmed by evidence statements describing the same observations. But if an emerald subsequently examined is grue, it is blue and hence not green. Thus although we are well aware which of the two incompatible predictions is genuinely confirmed, they are equally well confirmed according to our present definition. Moreover, it is clear that if we simply choose an appropriate predicate, then on the basis of these same observations we shall have equal confirmation, by our definition, for any prediction whatever about other emeralds—or indeed about anything else.[1] As in our earlier example, only the prediction subsumed under lawlike hypotheses are genuinely confirmed, but we have no criterion as yet for determining lawlikeness. And now we see that without some such criterion, our definition not merely includes a few unwarranted cases, but is so completely ineffectual that it virtually excludes nothing. We are left once again with the intolerable result that anything confirms anything. This difficulty cannot be set aside as an annoying detail to be taken care of in due course. It has to be met before our definition will work at all.

Nevertheless, the difficulty is often slighted because on the surface there seem to be easy ways of dealing with it . . . The most popular way of attacking the problem takes its cue from the fact that accidental hypotheses seem typically to involve some spatial or temporal restriction, or reference to some particular individual. They seem to concern the people in some particular room, or the objects on some particular person's desk; while lawlike hypotheses characteristically concern all ravens or all pieces of copper whatsoever. Complete generality is thus very often supposed to be a sufficient condition of lawlikeness; but to define this complete generality is by no means easy. Merely to require that the hypothesis contain no term naming, describing, or indicating a particular thing or location will obviously not be enough. The troublesome hypothesis that all emeralds are grue contains no such term; and where such a term does occur, as in hypotheses about men in *this* room, it can be suppressed in favor of some predicate (short or long, old or new) that contains no such term but applies only to exactly the same things. One might think, then, of excluding not only hypotheses that actually contain terms for specific individuals but also hypotheses that are equivalent to others that do contain such terms. But, as we have just seen, to exclude only hypotheses of which *all* equivalents contain such terms is to exclude nothing. On the other hand, to exclude all hypotheses that have some equivalent containing such a term is to exclude everything; for even the hypothesis "All grass is green" has an equivalent "All grass in London or elsewhere is green."

The next step, therefore, has been to consider ruling out predicates of certain kinds. A syntactically universal hypothesis is lawlike, the proposal runs, if its predicates are 'purely qualitative' or 'non-positional.' This will obviously accomplish nothing if a purely qualitative predicate is then conceived either as one that is equivalent to some expression free of terms for specific individuals, or as one that is equivalent to no expression that contains such a term; for this only raises again the difficulties just pointed out. The claim appears to be rather that at least in the case of a simple enough predicate we can readily determine by direct inspection of its meaning whether or not it is purely qualitative. But even aside from obscurities in the notion of 'the meaning' of a predicate, this claim seems to me wrong. I simply do not know how to tell whether a predicate is qualitative or positional, except perhaps by completely begging the question at issue and asking whether the predicate is 'well-behaved'—that is, whether simple syntactically universal hypotheses applying it are lawlike.

This statement will not go unprotested. "Consider," it will be argued, "the predicates 'blue' and 'green' and the predicate 'grue' introduced earlier, and also the predicate 'bleen' that applies to emeralds examined before time *t* just in case they are blue and to other emeralds just in case they are green. "Surely it is clear," the argument runs, "that the first two are purely qualitative and the second two are not; for the meaning of each of the latter two plainly involves reference to a specific temporal position." To this I reply that indeed I do recognize the first two as well-behaved predicates admissible in lawlike hypotheses, and the second two as ill-behaved predicates. But the argument that the former but not the latter are purely qualitative seems to be quite unsound. True enough, if we start with "blue" and "green," then "grue" and "bleen" will be explained in terms of "blue" and "green" and a temporal term. But equally truly, if we start with "grue" and "bleen," then "blue" and "green" will be explained in terms of "grue" and "bleen" and a temporal term; "green," for example, applies to emeralds examined before time *t* just in case they are grue, and to other emeralds just in case they are bleen. Thus qualitativeness is an entirely relative matter and does not by itself establish any dichotomy of predicates. This relativity seems to be completely overlooked by those who contend that the qualitative character of a predicate is a criterion for its good behavior.

Of course, one may ask why we need worry about such unfamiliar predicates as "grue" or about accidental hypotheses in general, since we are unlikely to use them in making predictions. If our definitions work for such hypotheses as are normally employed, isn't that all we need? In a sense, yes; but only in the sense that we need no definition, no theory of induction, and no philosophy of knowledge at all. We get along well enough without them in daily life and in scientific research. But if we seek a theory at all, we cannot excuse gross anomalies resulting from a proposed theory by pleading that we can avoid them in practice. The odd cases we have been considering are clinically pure cases that, though seldom encountered in practice, nevertheless display to best advantage the symptoms of a widespread and destructive malady.

We have so far neither any answer nor any promising clue to an answer to the question what distinguishes lawlike or confirmable hypotheses from accidental or non-confirmable ones; and what may at first have seemed a minor technical difficulty has taken on the stature of a major obstacle to the development of a satisfactory theory of confirmation. It is this problem that I call the new riddle of induction.

NOTES

1. For instance, we shall have equal confirmation, by our present definition, for the prediction that roses subsequently examined will be blue. Let "emerose" apply just to emeralds examined before time *t*, and to roses examined later. Then all emeroses so far examined are grue, and this confirms the hypothesis that all emeroses are grue and hence that roses subsequently examined will be blue. The problem raised by such antecedents has been little noticed, but is no easier that that raised by similarly perverse consequents.

Carl G. Hempel

· ·

"Studies in the Logic of Confirmation"

1. OBJECTIVE OF THE STUDY[1]

The defining characteristic of an empirical statement is its capability of being tested by a confrontation with experimental findings, i.e. with the results of suitable experiments or focused observations. This feature distinguishes statements which have empirical content both from the statements of the formal sciences, logic and mathematics, which require no experimental test for their validation, and from the formulations of transempirical metaphysics, which admit of none.

The testability here referred to has to be understood in the comprehensive sense of "testability in principle" or "theoretical testability"; many empirical statements, for practical reasons, cannot actually be tested now. To call a statement of this kind testable in principle means that it is possible to state just what experimental findings, if they were actually obtained, would constitute favourable evidence for it, and what findings or "data," as we shall say for brevity, would constitute unfavourable evidence; in other words, a statement is called testable in principle if it is possible to describe the kind of data which would confirm or disconfirm it.

The concepts of confirmation and of disconfirmation as here understood are clearly more comprehensive than those of conclusive verification and falsification. Thus, e.g., no finite amount of experiential evidence can conclusively verify a hypothesis expressing a general law such as the law of gravitation, which covers an infinity of potential instances, many of which belong either to the as yet inaccessible future or to the irretrievable past; but a finite set of relevant data may well be "in accord with" the hypothesis and thus constitute confirming evidence for it.

While, in the practice of scientific research, judgments as to the confirming or disconfirming character of experimental data obtained in the test of a hypothesis are often made without hesitation and with a wide consensus of

From Carl G. Hempel, *Aspects of Scientific Explanation* (New York: Free Press, 1965), 10–11, 18–23.

opinion, it can hardly be said that these judgments are based on an explicit theory providing general criteria of confirmation and disconfirmation. In this respect, the situation is comparable to the manner in which deductive inferences are carried out in the practice of scientific research: this, too, is often done without reference to an explicitly stated system of rules of logical inference. But while criteria of valid deduction can be and have been supplied by formal logic, no satisfactory theory providing general criteria of confirmation and disconfirmation appears so far. . . .

3. NICOD'S CRITERION OF CONFIRMATION AND ITS SHORTCOMINGS

We consider first a conception of confirmation which underlies many recent studies of induction and of scientific method. A very explicit statement of this conception has been given by Jean Nicod in the following passage: "Consider the formula or the law *A entails B*. How can a particular proposition, or more briefly, a fact, affect its probability? If this fact consists of the presence of B in the case of A, it is favorable to the law 'A entails B'; on the contrary, if it consists of the absence of B in a case of A, it is unfavorable to the law. It is conceivable that we have here the only two direct modes in which a fact can influence the probability of a law. . . . Thus, the entire influence of particular truths or facts on the probability of universal propositions or laws would operate by means of these two elementary relations which we shall call *confirmation* and *invalidation*."[2] Note that the applicability of this criterion is restricted to hypotheses of the form 'A entails B'. Any hypothesis H of this kind may be expressed in the notation of symbolic logic[3] by means of a universal conditional sentence, such as, in the simplest case,

$$(x)[P(x) \supset Q(x)]$$

i.e. 'For any object x: if x is a P, then x is a Q,' or also 'Occurrence of the quality P entails occurrence of the quality Q.' According to the above criterion this hypothesis is confirmed by an object a if a is P, but not Q.[4] In other words, an object confirms a universal conditional hypothesis if and only if it satisfies both the antecedent (here: 'P(x)') and the consequent (here: 'Q(x)') of the condition; it disconfirms the hypothesis if and only if it satisfies the antecedent, but not the consequent of the conditional; and (we add this to Nicod's statement) it is neutral, or irrelevant, with respect to the hypothesis if it does not satisfy the antecedent.

This criterion can readily be extended so as to be applicable also to universal conditionals containing more than one quantifier, such as 'Twins always resemble each other,' or in symbolic notation, '$(x)(y)(Twins(x,y) \supset Rsbl(x,y))$.' In these cases, a confirming instance consists of an ordered couple, or triple, etc., of objects satisfying the antecedent and the consequent of the conditional. (In the case of the last illustration, any two persons who are twins and resemble each other would confirm the hypothesis; twins who do not resemble each other would disconfirm it; and any two persons not twins—no matter whether they resemble each other or not—would constitute irrelevant evidence.)

We shall refer to this criterion as Nicod's criterion.[5] It states explicitly what is perhaps the most common tacit interpretation of the concept of confirmation. While seemingly quite adequate, it suffers from serious shortcomings, as will now be shown.

(*a*) First, the applicability of this criterion is restricted to hypotheses of universal conditional form; it provides no standards of confirmation for existential hypotheses (such as 'There exists organic life on other stars' or 'Poliomyelitis is caused by some virus') or for hypotheses whose explicit formulation calls for the use of both universal and existential quantifiers (such as 'Every human being dies some finite number of years after his birth' or the psychological hypothesis 'You can fool all of the people some of the time and some of the people all of the time, but you cannot fool all of the people all of the time,' which may be symbolized by '$(x)(Et)Fl(x,t)\cdot(Ex)(t)Fl(x,t)\cdot \sim (x)(t)Fl(x,t)$' (where '$Fl(x,t)$' stands for 'You can fool person x at time t'). We note, therefore, the desideratum of establishing a criterion of confirmation which is applicable to hypotheses of *any* form.[6]

(*b*) We now turn to a second shortcoming of Nicod's criterion. Consider the two sentences

$$S_1: \text{'}(x)[Raven(x) \supset Black(x)]\text{'};$$
$$S_2: \text{'}(x)[\sim Black(x) \supset \sim Raven(x)]\text{'}$$

(i.e. 'All ravens are black' and 'Whatever is not black is not a raven'), and let *a, b, c, d* be four objects such that *a* is a raven and black, *b* is a raven but not black, *c* is not a raven but black, and *d* neither a raven nor black. Then according to Nicod's criterion, *a* would confirm S_1, but be neutral with respect to S_2; *b* would disconfirm both S_1 and S_2; *c* would be neutral with respect to both S_1 and S_2; and *d* would confirm S_2, but be neutral with respect to S_1.

But S_1 and S_2 are logically equivalent; they have the same content, they are different formulations of the same hypothesis. And yet, by Nicod's criterion, either of the objects *a* and *d* would be confirming for one of the two sentences, but neutral with respect to the other. This means that Nicod's criterion makes confirmation depend not only on the content of the hypothesis, but also on its formulation.[7]

One remarkable consequence of this situation is that every hypothesis to which the criterion is applicable—i.e. every universal conditional—can be stated in a form for which there cannot possibly exist any confirming instances. Thus, e.g. the sentence

$$(x)[(Raven(x)\cdot \sim Black(x)) \supset (Raven(x)\cdot \sim Raven(x))]$$

is readily recognized as equivalent to both S_1 and S_2 above; yet no object whatever can confirm this sentence, i.e. satisfy both its antecedent and its consequent; for the consequent is contradictory. An analogous transformation is, of course, applicable to any other sentence of universal conditional form.

4. THE EQUIVALENCE CONDITION

The results just obtained call attention to the following condition which an adequately defined concept of confirmation should satisfy, and in light of which Nicod's criterion has to be rejected as inadequate:

Equivalence condition: Whatever confirms (disconfirms) one of two equivalent sentences, also confirms (disconfirms) the other.

Fulfillment of this condition makes the confirmation of a hypothesis independent of the way in which it is formulated; and no doubt it will be conceded that this is a necessary condition for the adequacy of any proposed criterion of confirmation. Otherwise, the question as to whether certain data confirm a given hypothesis would have to be answered by saying: "That depends on which of the different equivalent formulations of the hypothesis is considered"—which appears absurd. Furthermore—and this is a more important point than an appeal to a feeling of absurdity—an adequate definition of confirmation will have to do justice to the way in which empirical hypotheses function in theoretical scientific contexts such as explanations and predictions; but when hypotheses are used for purposes of explanation or prediction,[8] they serve as premises in a deductive argument whose conclusion is a description of the event to be explained or predicted. The deduction is governed by the principles of formal logic, and according to the latter, a deduction which is valid will remain so if some or all of the premises are replaced by different but equivalent statements; and indeed, a scientist will feel free, in any theoretical reasoning involving certain hypotheses, to use the latter in whichever of their equivalent formulations are most convenient for the development of his conclusions. But if we adopted a concept of confirmation which did not satisfy the equivalence condition, then it would be possible, and indeed necessary, to argue in certain cases that it was sound scientific procedure to base a prediction on a given hypothesis if formulated in a sentence S_1, because a good deal of confirming evidence had been found for S_1; but that it was altogether inadmissible to base the prediction (say, for convenience of deduction) on an equivalent formulation S_2, because no confirming evidence for S_2 was available. Thus, the equivalence condition has to be regarded as a necessary condition for the adequacy of any definition of confirmation.

5. PARADOXES OF CONFIRMATION

Perhaps we seem to have been labouring the obvious in stressing the necessity of satisfying the equivalence condition. This impression is likely to vanish upon consideration of certain consequences which derive from a combination of the equivalence condition with a most natural and plausible assumption concerning a sufficient condition of confirmation.

The essence of the criticism we have leveled so far against Nicod's criterion is that it certainly cannot serve as a necessary condition of confirmation; thus, in the illustration given in the beginning of section 3, object a confirms S_1 and should therefore also be considered as confirming S_2, while according to Nicod's criterion it is not. Satisfaction of the latter is therefore not a necessary condition for confirming evidence.

On the other hand, Nicod's criterion might still be considered as stating a particularly obvious and important sufficient condition of confirmation. And indeed, if we restrict ourselves to universal conditional hypotheses in one variable[9]—such as S_1 and S_2 in the above illustration—then it seems perfectly reasonable to qualify an object as confirming such a hypothesis if it satisfies both its antecedent and its consequent. The plausibility of this view will be further corroborated in the course of our subsequent analyses.

Thus, we shall agree that if a is both a raven and black, then a certainly confirms S_1: '$(x)(Raven(x) \supset Black(x))$' and if d is neither black nor a raven, d certainly confirms S_2: '$(x)[\sim Black(x) \supset \sim Raven(x)]$.'

Let us now combine this simple stipulation with the equivalence condition. Since S_1 and S_2 are equivalent, d is confirming for also for S_1; and thus, we have to recognize as confirming for S_1 any object which is neither black nor a raven. Consequently, any red pencil, any green leaf, any yellow cow, etc., becomes confirming evidence for the hypothesis that all ravens are black. This surprising consequence of two very adequate assumptions (the equivalence condition and the above sufficient condition of confirmation) can be further expanded: The sentence S_1 can readily be shown to be equivalence to S_3: '$(x)[(Raven(x) \lor \sim Raven(x)) \supset (\sim Raven(x) \lor Black(x))]$,' i.e. 'Anything which is or is not a raven is either no raven or black.' According to the above sufficient condition, S_3 is certainly confirmed by any object, say e, such that (1) e is or is not a raven and, in addition to (2) e is not a raven or is also black. Since (1) is analytic, these conditions reduce to (2). By virtue of the equivalence condition, we have therefore to consider as confirming for S_1 any object which is either no raven at all or also black (in other words: any object which is no raven at all, or a black raven).

Of the four objects characterized in section 3, a, c, and d would therefore constitute confirming evidence for S_1, while b would be disconfirming for S_1. This implies that any non-raven represents confirming evidence for the hypothesis that all ravens are black.[10]

We shall refer to these implications of the equivalence conditions and of the above sufficient condition of confirmation as the *paradoxes of confirmation*.

NOTES

1. The present analysis of confirmation was to a large extent suggested and stimulated by a cooperative study of certain more general problems which were raised by Dr. Paul Oppenheim, and which I have been investigating with him for several years. These problems concern the form and the function of scientific laws and the comparative methodology of different branches of empirical science.

In my study of the logical aspects of confirmation, I have benefited greatly from discussions with Professor R. Carnap, Professor A. Tarski, and particularly Dr. Nelson Goodman, to whom I am indebted for several valuable suggestions which will be indicated subsequently.

A detailed exposition of the more technical aspects of the analysis of confirmation presented in this essay is included in my article "A Purely Syntactical Definition of Confirmation," *The Journal of Symbolic Logic*, vol. 8 (1943).

2. Jean Nicod, *Foundation of Geometry and Induction* (trans. by P. P. Wiener), London, 1930, 219; cf. also R. M. Eaton's discussion of "Confirmation and Infirmation," which is based on Nicod's views; it is included in ch. III of his *General Logic* (New York, 1931).

3. In this essay, only the most elementary devices of this notation are used; the symbolism is essentially that of *Principia Mathematica*, except that parentheses are used instead of dots, and that existential quantification is symbolized by '(E)' instead of by the inverted 'E.'

4. More precisely we would have to say, in Nicod's parlance that the hypothesis is confirmed by the *proposition* that a is both P and Q, and is disconfirmed by the *proposition* that a is P but not Q.

5. This term is chosen for convenience, and in view of the above explicit formulation given by Nicod; it is not, of course, intended to imply to that this conception of confirmation originated with Nicod.

6. For a rigorous formulation of the problem, it is necessary first to lay down assumptions as to the means of expression and the logical structure of the language in which the hypotheses are supposed to be formulated; the desideratum then calls for a definition of confirmation applicable to any hypothesis which can be expressed in the given language. Generally speaking, the problem becomes increasingly difficult with increasing richness and complexity of the assumed language of science.

7. This difficulty was pointed out, in substance, in my article "Le Problème de la vérité," *Theoria* (Göteborg), 3 (1937), esp. p. 222.

8. For a more detailed account of the logical structure of scientific explanation and prediction, cf. C. G. Hempel, "The Function of General Laws in History," *Journal of Philosophy*, 39 (1942), esp. sections 2, 3, 4. The characterization, given in that paper as well as in the above text, of explanations and predictions as arguments of a deductive logical structure, embodies an oversimplification: . . . explanations and predictions often involve "quasi-inductive" steps besides deductive ones. This point, however, does not affect the validity of the above argument.

9. This restriction is essential: in its general form which applies to universal conditions in any number of variable, Nicod's criterion cannot even be construed as expressing a sufficient condition of confirmation. This is shown by the following rather surprising example: Consider the hypothesis:

$$S_1 : (x)(y)[\sim(R(x,y) \cdot R(y,x)) \supset (R(x,y) \cdot \sim R(y,x))].$$

Let a, b be two objects such that $R(a,b)$ and $\sim R(b,a)$. Then clearly, the couple (a,b) satisfies both the antecedent and the consequent of the universal conditional S_1; hence, if Nicod's criterion in its general form is accepted as stating a sufficient condition of confirmation, (a,b) constitutes confirming evidence for S_1. But S_1 can be shown to be equivalent to

$$S_2 : (x)(y)R(x,y)$$

Now, by hypothesis, we have $\sim R(b,a)$; and this flatly contradicts S_2 and thus S_1. Thus, the couple (a,b) although satisfying both the antecedent and the consequent of the universal conditional S_1, actually constitutes disconfirming evidence of the strongest kind (conclusively disconfirming evidence, as we shall say later) for that sentence. This illustration reveals a striking and—hitherto unnoticed weakness of that conception of confirmation which underlies Nicod's criterion. In order to realize the bearing of our illustration upon Nicod's original formulation, let A and B be $\sim (R(x,y) \cdot R(y,x))$ and $R(x,y) \cdot \sim R(y,x)$, respectively. Then S_1 asserts that A entails B, and the couple (a,b) is a case of the presence of B in the presence of A; this should, according to Nicod, be favourable to S_1.

10. The following further "paradoxical" consequence of our two conditions might be noted: any hypothesis of universal conditional form can be equivalently rewritten as another hypothesis of the same form which, even if true, can have no confirming instances in Nicod's sense at all, since the proposition that a given object satisfies the antecedent and the consequent of the second hypothesis is self-contradictory. For example, '$(x)[P(x) \supset Q(x)]$' is equivalent to the sentence '$(x)[P(x) \cdot \sim Q(x)) \supset (P(x) \cdot \sim P(x))]$,' whose consequent is true of nothing.

Responses to the Paradoxes of Evidence

Few topics affect students like the paradoxes of evidence. The arguments from Hume, Goodman, and Hempel seem so clean and tight that they appear to constitute an ironclad attack on even the possibility of doing science. They are indeed important arguments in the history of the philosophy of science, and they require tangling with. But one should not give up hope that there is good reason to believe in science as a rational enterprise upon confronting these very powerful puzzles. Indeed, there is a long history of very smart thinkers trying to wrestle with them. So let me sketch approaches that some of them have taken in trying to defuse these problems.

One approach is acceptance. This is what Hume thought was necessary. We just acknowledge that we cannot help but make inductive inferences, even if there is no justification for them. It is a habit of the mind, he contended, a matter of custom that we cannot avoid. Immanuel Kant picked up on Hume's idea that the source of cause-and-effect reasoning is the structure of the mind and took it in a slightly different direction.

Kant argued that perception doesn't just happen to our minds, but rather is the result of a combination of the input from the sense organs joined together with concepts from our minds.[1] Think of a voltmeter or collecting data about circuits. We have to plug the meter into the circuit so that it gets a sense of the system, of course, but it also needs to have a certain internal wiring if it is going to provide us with data. In the same sort of way, our minds are not just blank slates drawn upon by perception; they have an active role in creating these perceptions, and cause and effect is part of our internal wiring. Causation is something we impress onto the world, not something we get from it.

Others have taken different approaches. Hans Reichenbach was one of many who thought that Hume's problem came from a naive a picture of induction.[2] He sought to transform inductive reasoning into rigorous statistical reasoning. If we equate degrees of belief with numbers between zero and one, with zero being complete disbelief and one being absolute deductive certainty, then we can talk about inductive inferences in terms of the mathematical theory of probability, meaning that we just need to establish a firm foundation for our understanding of statistics and this requires something more akin to deduction which does not seem to have problems of the sort Hume sets out.

Goodman's grue paradox gave rise to a complete industry of its own. W. V. O. Quine argued that Goodman was wrong to hold "grue" to be a "natural kind term," as a word that denotes an actual division in nature.[3] Green is world-referring in a way that grue is not, Quine argued, because things that are grue before the appointed time and bleen after are more similar, more part of a species, than things that are grue before and after. Therefore, the notion of green as a natural kind term is projectable into the future in a way that grue is not.

Others like Gilbert Harman argued that it is not the nature of the terms, but rather simplicity concerns that lead us to asymmetrically prefer projections of green into the future instead of grue.[4] Some version of Occam's razor, these lines go, lead us to prefer green claims over grue claims. Like Reichenbach with Hume's problem, Harman worked out his notion of simplicity in terms of theorems of probability.

Similarly, Hempel's raven paradox has also attracted much attention from some of the most well-known thinkers throughout the twentieth century. Some follow Hempel's own resolution in which there is no real problem with red herrings being evidence that all ravens are black if we realize that it is incredibly weak evidence — so weak that it makes no real difference in our degree of belief. Others like Peter Achinstein argued that there is a problem with Hempel's reasoning.[5] Achinstein contended that Nicod's criterion is too generous in taking all instances to be evidence for their generalizations. Whether an instance O of a generalization H is supporting evidence depends at least in part upon how one goes about getting O. There is a reason we demand random samples and double-blind studies. How we get our data determines whether the data is relevant to scientific inferences and so we need to look at how we collected our non-black non-ravens before we declare them evidence at all because they may have come from biased selection procedures.

These are just quick sketches of a few of the arguments that very important philosophers of science have put forward in wrestling with the paradoxes of evidence. Are any of them ultimately effective? Do the paradoxes get dissolved? Like so much in philosophy, this is a matter of very lively, very smart debate. Students who are interested should be encouraged to read further on all of these questions. These are problems to be reckoned with and if you found yourself disturbed by them, well . . . you should be. But at the same time, take heart that there may be routes around these problems or at least hints offered by people who have thought very hard and deeply about them.

NOTES

1. See Immanuel Kant, "The Transcendental Analytic," in *Critique of Pure Reason*, 21–43 (Buffalo, NY: Prometheus, 1990).

2. Hans Reichenbach, *Experience and Prediction* (Chicago: University of Chicago Press, 1938), esp. chapter 5, "Probability and Induction" (297–406).

3. W. V. O. Quine, "Natural Kinds," in *Ontological Relativity and Other Essays*, 26–68 (New York: Columbia University Press, 1969).

4. Gilbert Harman, "Simplicity as a Pragmatic Criterion for Deciding What Hypotheses to Take Seriously" in *Grue!*, ed. Douglas Stalker, 153–72 (Chicago: Open Court, 1994). Stalker collects fifteen approaches to unraveling Goodman's riddle and is a must-read for anyone deeply bothered by the argument.

5. Peter Achinstein, *The Nature of Explanation.* (Oxford: Oxford University Press, 1983), esp. "The Paradox of the Ravens" (367–74).

FALSIFICATIONISM

One response to the problems of evidence is to try to solve them. Another is to try and dissolve them, showing that they only appear to be problems. A third option is to accept them as unavoidably problematic and then to try to go about reshaping our view of science in a way that accepts them. This last route is the path of Karl Popper.

The problems of evidence are all tied to the notions of confirmation, that is, the use of induction to show that scientific beliefs are probably true or reasonable to believe based on observable evidence. If, like Popper, we want to avoid the problem raised by David Hume and company, scientific inferences need to have all inductive steps eliminated from them.

Popper's main concern was not to explain the scientific method. Popper began his work in 1920s Vienna where he was exposed to several major fads among the educated class at the time. One was discussions of paranormal phenomena. Even figures as important as Rudolf Carnap and the great mathematician Kurt Gödel seriously considered whether séances and other means of contacting the deceased were scientifically testable. Another was a revision of our understanding of the human mind. Popper began his studies in psychology at a time and place—early-twentieth-century Vienna—when one could scarcely think about that subject without being bombarded with the work of Freud and his placement of neuroses in the dark and seemingly inaccessible corners of the subconscious. Austria was in the midst of a political upheaval with the population actively considering what should replace the old Austro-Hungarian empire, much talk focused on Marxism, which billed itself as a "scientific" approach to governance.

All of these appealed to science in their justifications. Popper felt uncomfortable with each of them in turn and began to consider the foundations of their support. From this he was led to try and figure out what are the essential properties which actually make a claim scientific. He was trying to answer what we call "the problem of demarcation," that is, how we go about drawing the line between science and nonscience, or science and pseudo-science. Those who came before Popper based the line on support from observation. A claim is scientific, on this view, if it is possible for there to be observational evidence in favor of it, that is, if it is empirically verifiable. But, of course, this reliance on evidence runs smack dab into Hume's problem, a problem Popper thought was insoluble.

Popper's solution was, like Aristotle and Descartes, to limit the logical machinery of science to deductive inferences only. But unlike the classical deductivists, he was not proposing deductive reasoning as a logic of discovery. Indeed, he explicitly denied that a logic of discovery exists and argued that even considering one would be to harm the progress of science that requires scientists to be free. Just like the hypothetico-deductivists, he wanted a completely open context of discovery in which scientists are free to come up with the most novel, interesting, and boldest hypotheses possible.

Indeed, it is this boldness that he took to be the dividing line between science and nonscience. The job of science is to describe the workings of the world. Ideally, we would like scientific proclamations to be true. Does that mean that true statements are therefore always scientific? Absolutely not. Suppose you turn on the television at dinnertime and watch the local news. The weatherman comes on and claims that his forecasts are guaranteed to be 100 percent accurate. It's a proposition that would surely catch your attention. How can this guy offer absolute certainty when the meteorologist on the other channel at best gives a statistic, something like a 30 percent chance of rain? You listen as he says, "Tomorrow, it will rain or it won't. Back to you, Bob."

Now is he right? Yes. He is absolutely right—it will either rain or it won't. But did he actually tell you anything about the weather tomorrow? Do you know any more than you did about whether you should bring your umbrella? No. He told you nothing. His sentence is guaranteed to be true because it has no content. A statement about the world tells you how the world is and is not. His sentence is true no matter how the world will be and therefore doesn't actually describe anything. This is what we call a tautology, a sentence that is always true no matter the state of the universe.

But if science is supposed to describe how the universe is and how the universe works, then a sentence that is true regardless of the state of the universe says nothing about the universe and therefore cannot be scientific. A sentence can be scientific, Popper argues, only if the universe could be a certain way that would show the sentence to be false, that is, if the sentence is falsifiable. For a sentence to be scientific, you need to be able to tell me what I could observe that would show this sentence is not true. Science means intellectual risk. The sort of testability that comes with going out on a limb, what Popper terms "falsifiability," is the defining characteristic of scientific claims.

Deduction therefore figures into the scientific method in two places according to Popper. First, deduction is used, as Mill argued in his notion of "ratiocination," to derive observable consequences from the more general purported laws of nature. If the hypothesis h is true, then empirical consequences $o_1, o_2, o_3, \ldots, o_n$ must follow as a matter of course. These are our falsifying instances, the possible observations that if they were false would eliminate the hypothesis from consideration. With testable predictions from the theory, we can now go to the lab, set up the appropriate experiment and see if we do, in fact, observe $o_1, o_2, o_3, \ldots, o_n$.

There are two possibilities. If we fail to observe any of the results that necessarily follow from h's supposed truth, we can now employ deductive reasoning in the form of modus tollens and reject h as false. In other words, if hypothesis h is true, then we would observe $o_1, o_2, o_3, \ldots, o_n$. We didn't observe $o_1, o_2, o_3, \ldots, o_n$. Therefore, we must conclude that h is false. This far, Popper's sketch of the logic of science is no different from that of the hypothetico-deductivists.

But what if we do observe $o_1, o_2, o_3, \ldots, o_n$? According to the H-D theorists, this gives us evidence for the hypothesis and if it is broad and strong enough evidence, it gives us good inductive reason to think that h is true. But this is just where we ran into the paradoxes of evidence.

Popper argued that the observations are good things for the hypothesis, but not because it provides probabilistic support for its likely truth. The scientific game is like professional boxing. Just because someone is now the champ does not mean they will be forever, but it becomes a matter of interest to see how many serious challenges the champ can successfully face. Each new correct prediction, like each defense of his title, brings more prestige for the champ. Popper's term for this is "corroboration." A highly corroborated hypothesis is one that has withstood the test of many challenges, many experiments that could have falsified it, but has emerged victorious.

In science, we want two things in our theories. We want theories that are highly corroborated and those that are as universal as possible. The idea in science is to find propositions that account for as many potential observations as possible. The great champs, the legendary fighters, are the ones who stood the test of time in defending their crown the longest and those who withstood the most difficult challengers. So it is with scientific theories: those we are looking for are those that make the boldest, broadest claims and therefore leave themselves open to the most risk. If they survive the risk for a long period of time while facing the most rigorous, stringent, and spirited attempts to knock them down, they are considered great. Scientists must not only propose bold theories, they must also do their utmost to knock them down, to show them to be false. The rigor of science lays in its good faith efforts to challenge its own products.

Scientists, Popper argued, are in the strange position of wanting truth, of seeking truth, of proposing theories to be the truth while all the time not only knowing that their best efforts will likely fail, but that it is their job to work as hard as possible to make them fail. With each new step, scientists must derive more difficult, clever, and intricate ways of seeing if they can undermine their own creations.

Karl Popper

The Logic of Scientific Discovery

DEDUCTIVE TESTING OF THEORIES

According to the view that will be put forward here, the method of critically testing theories, and selecting them according to the results of tests, always proceeds on the following lines. From a new idea, put up tentatively, and not yet justified in any way—an anticipation, a hypothesis, a theoretical system, or what you will—conclusions are drawn by means of logical deduction. These conclusions are then compared with one another and with other relevant statements, so as to find what logical relations (such as equivalence, derivability, compatibility, or incompatibility) exist between them.

We may if we like distinguish four different lines along which the testing of a theory could be carried out. First there is the logical comparison of the conclusions among themselves, by which the internal consistency of the system is tested. Secondly, there is the investigation of the logical form of the theory, with the object of determining whether it has the character of an empirical or scientific theory, or whether it is, for example, tautological. Thirdly, there is the comparison with other theories, chiefly with the aim of determining whether the theory would constitute a scientific advance should it survive our various tests. And finally, there is the testing of the theory by way of empirical applications of the conclusions which can be derived from it.

The purpose of this last kind of test is to find out how far the new consequences of the theory—whatever may be new in what it asserts—stand up to the demands of practice, whether raised by purely scientific experiments, or by practical technological applications. Here too the procedure of testing turns out to be deductive. With the help of other statements, previously accepted, certain singular statements—which we may call 'predictions'—are deduced from the theory; especially predictions that are easily testable or applicable. From among these statements. those are selected which are not derivable from the current theory, and more especially those which the current theory contradicts. Next we seek a decision as regards these (and other) de-

From Karl Popper, *The Logic of Scientific Discovery* (New York: Routledge, 1992), 32–33, 39–42, 84–86, 112–13, 276–79. Reprinted by permission of University of Klagenfurt / Karl Popper Library.

rived statements by comparing them with the results of practical applications and experiments. If this decision is positive, that is, if the singular conclusions turn out to be acceptable, or *verified,* then the theory has, for the time being, passed its test: we have found no reason to discard it. But if the decision is negative, or in other words, if the conclusions have been *falsified,* then their falsification also falsifies the theory from which they were logically deduced.

It should be noticed that a positive decision can only temporarily support the theory, for subsequent negative decisions may always overthrow it. So long as a theory withstands detailed and severe tests and is not superseded by another theory in the course of scientific progress, we may say that it has 'proved its mettle' or that it is '*corroborated*' by past experience.

Nothing resembling inductive logic appears in the procedure here outlined. I never assume that we can argue from the truth of singular statements to the truth of theories. I never assume that by force of 'verified' conclusions, theories can be established as 'true,' or even as merely 'probable.'

EXPERIENCE AS A METHOD

The task of formulating an acceptable definition of the idea of an 'empirical science' is not without its difficulties. Some of these arise from *the fact that there must be many theoretical systems* with a logical structure very similar to the one which at any particular time is the accepted system of empirical science. This situation is sometimes described by saying that there is a great number—presumably an infinite number—of 'logically possible worlds.' Yet the system called 'empirical science' is intended to represent only *one* world: the 'real world' or the 'world of our experience.'

In order to make this idea a little more precise, we may distinguish three requirements which our empirical theoretical system will have to satisfy. First, it must be *synthetic,* so that it may represent a noncontradictory, a *possible* world. Secondly, it must satisfy the criterion of demarcation (*cf.* sections 6 and 21), *i.e.* it must not be metaphysical, but must represent a world of possible *experience.* Thirdly, it must be a system distinguished in some way from other such systems as the one which represents *our* world of experience.

But how is the system that represents our world of experience to be distinguished? The answer is: by the fact that it has been submitted to tests, and has stood up to tests. This means that it is to be distinguished by applying to it that deductive method which it is my aim to analyse, and to describe.

'Experience,' on this view, appears as a distinctive *method* whereby one theoretical system may be distinguished from others; so that empirical science seems to be characterized not only by its logical form but, in addition, by its distinctive *method*. (This, of course, is also the view of the inductivists, who try to characterize empirical science by its use of the inductive method.)

The theory of knowledge, whose task is the analysis of the method or procedure peculiar to empirical science, may accordingly be described as a theory of the empirical method—a *theory of what is usually called 'experience.'*

FALSIFIABILITY AS A CRITERION OF DEMARCATION

The criterion of demarcation inherent in inductive logic—that is, the positivistic dogma of meaning—is equivalent to the requirement that all the statements of empirical science (or all 'meaningful' statements) must be capable of being finally decided, with respect to their truth *and* falsity; we shall say that they must be *'conclusively decidable.'* This means that their form must be such that *to verify them and to falsify them* must both be logically possible. Thus Schlick says: '. . . a genuine statement must be capable of *conclusive verification*,'[1] and Waismann says still more clearly: 'If there is no possible way to *determine whether a statement is true* then that statement has no meaning whatsoever. For the meaning of a statement is the method of its verification.'[2]

Now in my view there is no such thing as induction.[3] Thus inference to theories, from singular statements which are 'verified by experience' (whatever that may mean), is logically inadmissible. Theories are, therefore, *never* empirically verifiable. If we wish to avoid the positivist's mistake of eliminating, by our criterion of demarcation, the theoretical systems of natural science, then we must choose a criterion which allows us to admit to the domain of empirical science even statements which cannot be verified.

But I shall certainly admit a system as empirical or scientific only if it is capable of being *tested* by experience. These considerations suggest that not the *verifiability* but the *falsifiability* of a system is to be taken as a criterion of demarcation. In other words: I shall not require of a scientific system that it shall be capable of being singled out, once and for all, in a positive sense; but I shall require that its logical form shall be such that it can be singled out, by means of empirical tests, in a negative sense: *it must be possible for an empirical scientific system to be refuted by experience.*[4]

(Thus the statement, 'It will rain or not rain here tomorrow' will not be regarded as empirical, simply because it cannot be refuted; whereas the statement, 'It will rain here tomorrow' will be regarded as empirical.)

It might be said that even if the asymmetry is admitted, it is still impossible, for various reasons, that any theoretical system should ever be conclusively falsified. For it is always possible to find some way of evading falsification, for example by introducing *ad hoc* an auxiliary hypothesis, or by changing *ad hoc* a definition. It is even possible without logical inconsistency to adopt the position of simply refusing to acknowledge any falsifying experience whatsoever. Admittedly, scientists do not usually proceed in this way, but logically such procedure is possible; and this fact, it might be claimed, makes the logical value of my proposed criterion of demarcation dubious, to say the least.

I must admit the justice of this criticism; but I need not therefore withdraw my proposal to adopt falsifiability as a criterion of demarcation. For I am going to propose (in sections 20 *f.*) that the *empirical method* shall be characterized as a method that excludes precisely those ways of evading falsification which, as my imaginary critic rightly insists, are logically possible. According to my proposal, what characterizes the empirical method is its manner of exposing to falsification, in every conceivable way, the system to be tested. Its aim is not to save the lives of untenable systems but, on the contrary, to select the one which is by comparison the fittest, by exposing them all to the fiercest struggle for survival.

The proposed criterion of demarcation also leads us to a solution of Hume's problem of induction—of the problem of the validity of natural laws. The root of this problem is the apparent contradiction between what may be called 'the fundamental thesis of empiricism'—the thesis that experience alone can decide upon the truth or falsity of scientific statements—and Hume's realization of the inadmissibility of inductive arguments. This contradiction arises only if it is assumed that all empirical scientific statements must be 'conclusively decidable,' i.e. that their verification and their falsification must both in principle be possible. If we renounce this requirement and admit as empirical also statements which are decidable in one sense only—unilaterally decidable and, more especially, falsifiable—and which may be tested by systematic attempts to falsify them, the contradiction disappears: the method of falsification presupposes no inductive inference, but only the tautological transformations of deductive logic whose validity is not in dispute.

LOGICAL INVESTIGATION OF FALSIFIABILITY

Only in the case of systems which would be falsifiable if treated in accordance with our rules of empirical method is there any need to guard against conventionalist stratagems. Let us assume that we have successfully banned these stratagems by our rules: we may now ask for a *logical* characterization of such falsifiable systems. We shall attempt to characterize the falsifiability of a theory by the logical relations holding between the theory and the class of basic statements.

As a first attempt one might perhaps try calling a theory 'empirical' whenever singular statements can be deduced from it. This attempt fails, however, because in order to deduce singular statements from a theory, we always need other singular statements—the initial conditions that tell us what to substitute for the variables in the theory. As a second attempt, one might try calling a theory 'empirical' if singular statements are derivable with the help of other singular statements serving as initial conditions. But this will not do either; for even a non-empirical theory, for example a tautological one, would allow us to derive some singular statements from other singular statements. (According to the rules of logic we can for example say: From the conjunction of 'Twice two is four' and 'Here is a black raven' there follows, among other things, 'Here is a raven.') It would not even be enough to demand that from the theory together with some initial conditions we should be able to deduce *more* than we could deduce from those initial conditions alone. This demand would indeed exclude tautological theories, but it would not exclude synthetic metaphysical statements. (For example from 'Every occurrence has a cause' and 'A catastrophe is occurring here,' we can deduce 'This catastrophe has a cause.')

In this way we are led to the demand that the theory should allow us to deduce, roughly speaking, more *empirical* singular statements than we can deduce from the initial conditions alone. This means that we must base our definition upon a particular class of singular statements; and this is the purpose for which we need the basic statements. Seeing that it would not be very easy to say in detail how a complicated theoretical system helps in the deduction of singular or basic statements, I propose the following definition. A theory is to be called 'empirical' or 'falsifiable' if it divides the class of all possible basic statements unambiguously into the following two non-empty subclasses. First, the class of all those basic statements with which it is inconsistent (or which it rules out, or prohibits): we call this the class of the *potential falsifiers* of the theory; and secondly, the class of those basic statements which it does contradict (or which it ·permits'). We can put this more briefly by saying: a theory is falsifiable if the class of its potential falsifiers is not empty.

DEGREES OF TESTABILITY

A theory is falsifiable, as we saw in section 23, if there exists at least one non-empty class of homotypic basic statements which are forbidden by it; that is, if the class of its potential falsifiers is not empty. If, as in section 23, we represent the class of all possible basic statements by a circular area, and the possible events by the radii of the circle, then we can say: At least *one* radius—or perhaps better, one narrow sector whose width may represent the fact that the event is to be 'observable'—must be incompatible with the theory and ruled out by it. One might then represent the potential falsifiers of various theories by sectors of various widths. And according to the greater and lesser width of the sectors ruled out by them, theories might then be said to have more, or fewer, potential falsifiers. (The question whether this 'more' or 'fewer' could be made at all precise will be left open for the moment.) It might then be said, further, that if the class of potential falsifiers of one theory is 'larger' than that of another, there will be more opportunities for the first theory to be refuted by experience; thus compared with the second theory, the first theory may be said to be 'falsifiable in a higher degree.' This also means that the first theory *says more* about the world of experience than the second theory, for it rules out a larger class of basic statements. Although the class of permitted statements will thereby become smaller, this does not affect our argument; for we have seen that the theory does not assert anything about this class. Thus it can be said that the amount of empiric information conveyed by a theory, or its *empirical content,* increases with its degree of falsifiability.

Let us now imagine that we are given a theory, and that the sector representing the basic statements which it forbids becomes wider and wider. Ultimately the basic statements *not* forbidden by the theory will be represented by a narrow remaining sector. (If the theory is to be consistent, then some such sector must remain.) A theory like this would obviously be very easy to falsify, since it allows the empirical world only a narrow range of possibilities; for it rules out almost all conceivable, *i.e.* logically possible, events. It asserts so much about the world of experience, its empirical content is so great, that there is, as it were, little chance for it to escape falsification.

Now theoretical science aims, precisely, at obtaining theories which are easily falsifiable in this sense. It aims at restricting the range of permitted events to a minimum; and, if this can be done at all, to such a degree that any further restriction would lead to an actual empirical falsification of the theory. If we could be successful in obtaining a theory such as this, then this theory would describe 'our particular world' as precisely as a theory can; for it would single out the world of 'our experience' from the class of all logically possible worlds of experience with the greatest precision attainable by theoretical science. All the events or classes of occurrences which we actually encounter and observe, and only these, would be characterized as 'permitted.'

One may discern something like a general direction in the evolution of phys-ics—a direction from theories of a lower level of universality to theories of a higher level. This is usually called the 'inductive' direction; and it might be thought that the fact that physics advances in this 'inductive' direction could be used as an argument in favour of the inductive method.

Yet an advance in the inductive direction does not necessarily consist of a sequence of inductive inferences. Indeed we have shown that it may be ex-plained in quite different terms—in terms of degree of testability and cor-roborability. For a theory which has corroborated can only be superseded by one of a higher level of universality; that is, by a theory which is better test-able and which, in addition, *contains* the old, well corroborated theory—or at least a good approximation to it. It may be better, therefore, to describe that trend—the advance towards theories of an ever higher level of universality—as 'quasi-inductive.'

The quasi-inductive process should be envisaged as follows. Theories of some level of universality are proposed, and deductively tested; after that, the-ories of a higher level of universality are proposed, and in their turn tested with the help of those of the previous levels of universality, and so on. The methods of testing are invariably based on deductive inferences from the higher to the lower level;[5] on the other hand, the levels of universality are reached, in the order of time, by proceeding from lower to higher levels.

The question may be raised: 'Why not invent theories of the highest level of universality straight away? Why wait for this quasi-inductive evolution? Is it not perhaps because there is after all an inductive element contained in it?' I do not think so. Again and again suggestions are put forward—conjectures, or theories—of all possible levels of universality. Those theories which are on too high a level of universality, as it were (that is, too far removed from the level reached by the testable science of the day) give rise, perhaps, to a 'metaphysical system.' In this case, even if from this system statements should be deducible (or only semi-deducible, as for example in the case of Spinoza's system), which belong to the prevailing scientific system, there will be no *new* testable state-ment among them; which means that no crucial experiment can be designed to test the system in question.[6] If, on the other hand, a crucial experiment can be designed for it, then the system will contain, as a first approximation, some well corroborated theory, and at the same time also something new—and something that can be tested. Thus the system will not, of course, be 'meta-physical.' In this case, the system in question may be looked upon as a new advance in the quasi-inductive evolution of science. This explains why a link with the science of the day is as a rule established only by those theories which are proposed in an attempt to meet the current problem situation; that is, the current difficulties, contradictions, and falsifications. In proposing a solution to these difficulties, these theories may point the way to a crucial experiment.

To obtain a picture or model of this quasi-inductive evolution of science, the various ideas and hypotheses might be visualized as particles suspended in a fluid. Testable science is the precipitation of these particles at the bottom of the vessel: they settle down in layers (of universality). The thickness of the deposit grows with the number of these layers, every new layer corresponding to a theory more universal than those beneath it. As the result of this process ideas previously floating in higher metaphysical regions may sometimes be reached by the growth of science, and thus make contact with it, and settle. Examples of such ideas are atomism; the idea of a single physical 'principle' or ultimate element (from which the others derive); the theory of terrestrial motion (opposed by Bacon as fictitious); the age-old corpuscular theory of light; the fluid-theory of electricity (revived as the electron-gas hypothesis of metallic conduction). All these metaphysical concepts and ideas may have helped, even in their early forms, to bring order into man's picture of the world, and in some cases they may even have led to successful predictions. Yet an idea of this kind acquires scientific status only when it is presented in falsifiable form; that is to say, only when it has become possible to decide empirically between it and some rival theory.

Science is not a system of certain, or well-established, statements; nor is it a system which steadily advances towards a state of finality. Our science is not knowledge (*epistēmē*): it can never claim to have attained truth, or even a substitute for it, such as probability.

Yet science has more than mere biological survival value. It is not only a useful instrument. Although it can attain neither truth nor probability, the striving for knowledge and the search for truth are still the strongest motives of scientific discovery.

We do not know: we can only guess. And our guesses are guided by the unscientific, the metaphysical (though biologically explicable) faith in laws, in regularities which we can uncover—discover. Like Bacon, we might describe our own contemporary science—'the method of reasoning which men now ordinarily apply to nature'—as consisting of 'anticipations, rash and premature' and of 'prejudices.'[7]

But these marvellously imaginative and bold conjectures or 'anticipations' of ours are carefully and soberly controlled by systematic tests. Once put forward, none of our 'anticipations' are dogmatically upheld. Our method of research is not to defend them, in order to prove how right we were. On the contrary, we try to overthrow them. Using all the weapons of our logical, mathematical, and technical armoury, we try to prove that our anticipations were false—in order to put forward, in their stead, new unjustified and unjustifiable anticipations, new 'rash and premature prejudices,' as Bacon derisively called them.

The advance of science is not due to the fact that more and more perceptual experiences accumulate in the course of time. Nor is it due to the fact that we are making ever better use of our senses. Out of uninterpreted sense-experiences science cannot be distilled, no matter how industriously we gather and sort them. Bold ideas, unjustified anticipations, and speculative thought, are our only means for interpreting nature: our only organon, our only instrument, for grasping her. And we must hazard them to win our prize. Those of us who are unwilling to expose their ideas to the hazard of refutation do not take part in the scientific game.

The old scientific ideal of *epistēmē*—of absolutely certain, demonstrable knowledge—has proved to be an idol. The demand for scientific objectivity makes it inevitable that every scientific statement must remain *tentative for ever.* It may indeed be corroborated, but every corroboration is relative to other statements which, again, are tentative. Only in our subjective experiences of conviction, in our subjective faith, can we be 'absolutely certain.'

With the idol of certainty (including that of degrees of imperfect certainty or probability) there falls one of the defences of obscurantism which bar the way of scientific advance. For the worship of this idol hampers not only the boldness of our questions, but also the rigour and the integrity of our tests. The wrong view of science betrays itself in the craving to be right; for it is not his *possession* of knowledge, of irrefutable truth, that makes the man of science, but his persistent and recklessly critical *quest* for truth.

Has our attitude, then, to be one of resignation? Have we to say that science can fulfil only its biological task; that it can, at best, merely prove its mettle in practical applications which may corroborate it? Are its intellectual problems insoluble? I do not think so. Science never pursues the illusory aim of making its answers final, or even probable. Its advance is, rather, towards an infinite yet attainable aim: that of ever discovering new, deeper, and more general problems, and of subjecting our ever tentative answers to ever renewed and ever more rigorous tests.

NOTES

1. Schlick, *Naturwissenschaften* 19, 1931, p. 150.

2. Waismann, *Erkenntnis* I, 1903, p. 229.

3. I am not, of course, here considering so-called 'mathematical induction.' What I am denying is that there is such a thing as induction in the so-called 'inductive sciences': that there are either 'inductive procedures' or 'inductive inferences.'

4. Related ideas are to be found, for example, in Frank, *Die Kausalität und ihre Grenzen* (1931), ch. I, §10 (p. 15 *f.*); Dubislav, *Die Definition* (3rd edition 1931), p. 100 *f.*

5. The 'deductive inferences from the higher to the lower level' are, of course, *explanations* (in the sense of section 12); thus the hypotheses on the higher level are *explanatory* with respect to those on the lower level.

6. It should be noted that I mean by a crucial experiment one that is designed to refute a theory (if possible) and more especially one which is designed to bring about a decision between two competing theories by refuting (at least) one of them—without, of course, proving the other.

7. Bacon, *Novum Organum,* I, 26.

Case Studies

· ·

Astronomy Track

PAPER 3: GALILEO AND THE HELIOCENTRIC PICTURE OF THE SOLAR SYSTEM

THE CASE

In 1543, Nicholas Copernicus published *On the Revolutions of the Heavenly Spheres*, which set out a heliocentric system of the universe that competed with Ptolemy's geocentric picture. Like Ptolemy, he used circular orbits and Apollonius's notions of eccentric orbits and epicycles, although he was determined to get rid of Ptolemy's equants. The notion of a moving Earth—which revolved and rotated—and a stationary sun allowed for a system as predictive as Ptolemy's, but with significantly simpler calculations using fewer epicycles.

While Copernicus himself most likely thought his sun-centered picture was true of the universe, an introduction to *On the Revolutions of the Heavenly Spheres* written by Copernicus's colleague Andrew Osiander, a Lutheran minister, was appended to it at the time of publication. This introduction contended that Copernicus's system need not be understood as a competing picture of the universe to that of Ptolemy, something accepted not only by scientists but also the powerful religious institutions of the time, but rather could be considered as a mathematical tool to allow for more convenient calculation.

This did not dissuade certain thinkers, however, from considering the heliocentric picture to be literally true. Some, such as Thomas Digges, Giordano Bruno, and Galileo Galilei, posited such a system with the complete absence of epicycles and eccentricities. Galileo argued that the simplified Copernican picture was, in fact, the way the universe worked. In support of his rejection of the earth-centered model, Galileo famously presented new pieces of evidence inaccessible to the ancients.

Galileo was not the first to construct a telescope—Dutch scientists had developed them not long before—but it was Galileo who was clever enough to realize that he could use it at night to change the world. With his telescope, Galileo made several significant discoveries that supported his Copernicanism. First, he discovered that the moon had mountains and craters. These deviations from smoothness contradicted the Catholic Church's acceptance of Aristotle's physics in which all heavenly bodies were made of the more perfect element ether and therefore were of a more perfect shape than the more corruptible terrestrial bodies. Second, he discovered the moons of Jupiter. If moons orbit Jupiter, then not everything could orbit Earth. Third, he found that Venus had phases like the moon, in other words, at different times Venus would appear as a crescent that waxes and wanes, getting fuller and fuller and then thinner and thinner. However, Galileo found that we never quite get a full Venus like we get a full moon. This, he realized, must mean that while the moon orbits Earth, Venus could not. Based upon this evidence, Galileo argued that the Copernican picture was not a mere calculation device, but actually descriptive of the structure of the universe.

THE SCIENTIST

Galileo Galilei (1564–1642) was born in Pisa, where he was both a student and professor. Moving to Padua and then Florence, Galileo was able to conduct research on mechanics and astronomy that might have been too scandalous in the more conservative Pisa. He wrote up his earth-shaking astronomical observations of Earth's moon and those of Jupiter in the short book *The Starry Messenger*. The Copernican viewpoint they supported ran counter to that of the Catholic Church, and Galileo was brought before the Inquisition in 1615, where he was forced to declare his opinions false under threat of being charged with heresy. Sixteen years later, when Galileo's friend Cardinal Barberini became Pope Urban VIII, Galileo received permission from the Church to write a balanced account of the controversy between Ptolemaic Aristotleanism and Copernicanism. But Galileo's book, *Dialogue on the Two Chief Systems of the Universe*, was anything but balanced, presenting a full-throated defense of heliocentrism and an attack on geocentrism that included a scathing caricature of the pope himself. As a result, he was called back before the Inquisition and sentenced as a heretic to remain in his home and to never again write on astronomical subjects.

YOUR JOB

Galileo was a Copernican adherent before his famous telescopic observations. Find Galileo's discussion of one of these observations in *A Dialogue on Two Chief Systems of the Universe* (hint: look in the third day's conversation). Quote Galileo and explain how the observed phenomenon supports the simplified Copernican picture and not the Ptolemaic. Explain the hypothetico-deductivist and falsificationist approaches to science, noting their similarities and differences. Would an adherent of the hypothetico-deductivist or falsificationist view be more accurate in describing Galileo's approach to science? Explain how Galileo's work fits better into one or the other of these approaches.

RESEARCH HINT

- Once you have located your source, make a photocopy of it. As you read it, make notes on your copy, underlining passages to quote and picking out the author's premises and conclusions. These notes will make writing your exposition much easier, as you may forget what you had been thinking while immersed in the details of the work.

Physics Track

PAPER 3: RUTHERFORD AND THE DISCOVERY OF THE NUCLEUS

THE CASE

Maxwell argued for the existence of molecules, but said nothing about their actual atomic nature, that is, whether they were indeed uncuttable. In his work, he treated them as tiny spheres interacting in early versions only through contact; he later added in electrodynamic considerations. But they were treated as individual particles without parts.

As research continued, scientists came to realize that atoms were electrically neutral, but that under proper circumstances, charged particles could be emitted by them. It became clear that the atomic particles were made up of charged constituents. The negatively charged bits were seen to be of relatively small mass. J. J. Thomson came up with his "plum pudding" model, which was designed to account for this by embedding small negatively charged particles in a larger "soup" of positive charge.

From 1909 to 1911, Ernest Rutherford, assisted by Hans Geiger and Ernest Marsden, conducted an experiment to shed light on the composition of atoms. Rutherford was trying to figure out what the particles emitted by certain radioactive elements were and where they came from. To do this, Rutherford and his team sent a beam of electrons at a very thin foil of gold. Since the electron had a known negative charge, the angle of deflection would say something about the microconstruction and of the gold atoms in the foil.

As the junior member of team, Marsden's job was documenting the location of the electrons after interacting with the atoms in the foil. While most of the data was unremarkable, Marsden observed the occasional electron bouncing back off the foil at an angle greater than 90°, that is, it bounced back off the foil like a ball off a wall. The nonconcentrated electrical charge could not produce a result so striking, it had to be that the electron was bouncing off of something of a much greater strength. But since so few of the electrons bounced back, there must be very few of them in a given area.

Taking Marsden's observations seriously, Rutherford concluded that instead of a spreading of the atom's mass and positive charge as Thomson proposed, it must be concentrated in a small region. What the experiment showed, Rutherford argued, was that atoms have a nucleus.

Ernest Rutherford (1871–1937) was a New Zealander who studied under J. J. Thomson at Cambridge. Much less culturally refined than his fellow students at Cambridge, Rutherford was loud and brash and determined to make a major breakthrough in physics. And so he did, striding through the lab singing his unique, off-key version of "Onward Christian Soldier." His work led to the discovery of the nucleus and major advances in the understanding of radioactivity, and for this, he was awarded—to his dismay—the Nobel Prize in chemistry.

YOUR JOB

Explain how Rutherford's experiment worked and why the results were surprising. Why did this lead him to posit the existence of a nucleus? (Hint: you may want to look at Rutherford's own words on the matter in his May 1911 article in *Philosophical Magazine* entitled "The Scattering of α and β Particles by Matter and the Structure of the Atom," especially sections 1, 2, and 7.) Explain in terms of either hypothetico-deductivism or falsificationism. How would a proponent of this view argue that Rutherford was following their approach to the scientific method? Are there aspects to Rutherford's work that deviate from the proposed method?

RESEARCH HINT

- Once you have located your source, make a photocopy of it. As you read it, make notes on your copy, underlining passages to quote and picking out the author's premises and conclusions. These notes will make writing up your exposition much easier, as you may forget what you had been thinking while immersed in the details of the work.

Chemistry Track

PAPER 3: JOSEPH PRIESTLEY, PHLOGISTON, AND THE DISCOVERY OF OXYGEN

THE CASE

While air was still generally considered to be an elemental substance, several chemists of the period posited the existence of other gasses, what Boyle called "factitious airs," which they thought to differ from what they took to be elemental air. Joseph Black, for example, showed that "fixed air," what we now call carbon dioxide, is different from the air we breathe. This kicked off a period where chemists became interested in gasses.

A leader in this movement was Joseph Priestley, whose work with the pneumatic trough led to his discovery of many gasses, including ammonia, sulfur dioxide, the gaseous form of hydrochloric acid, nitrous oxide, nitric oxide, ethylene, nitrogen, and carbon monoxide. But his most important discovery was oxygen, which he isolated using mercury oxide. Similar work was done by Karl-Wilhelm Scheele in Sweden and Antoine Lavoisier in France. Indeed, it was Scheele who first isolated the gas and Lavoisier who realized it was its own element, but it was Priestley who discovered its chemical properties.

Priestley had isolated what we now recognize as both nitrogen and oxygen. Nitrogen, he thought, was the common air we breathe, oxygen was the air we breathe separated from its phlogiston, the elemental substance of heat. Common air, he argued, was therefore a combination of two elements—air and heat—which could be separated. As such, the long-held view that our air is an element is false, since we could isolate a part of it. Priestley originally gathered the gas from a calx of mercury, but when he was able to also separate it from red lead, he contended that it was, in fact, something more generally found.

Priestley observed that the gas he isolated would cause candle flames to glow brighter and burn longer. This meant that it encouraged combustion and therefore, he concluded, increased the flow of heat, or phlogiston, from that which was burning. In other words, the new gas had a deficiency of phlogiston, or, was dephlogisticated, and the more vivid flame was evidence of increased flow of phlogiston out of the candle wax and into the dephlogisticated air, seeking equilibrium. The lack of flammability of nitrogen, on the other hand, was evidence that it had a sufficient supply of phlogiston and so retarded the flow from the candle. These differences, and others he was able to determine through experimentation, allowed Priestley to describe the chemical properties of his dephlogisticated air and therefore he is widely credited with the discovery of oxygen.

THE SCIENTIST

Joseph Priestley (1733–1804) was a Presbyterian minister who had the good fortune to live next to a brewery. His interests in science had largely been in physics, having worked with Benjamin Franklin on questions related to electricity when Franklin was in Europe. But exciting advances in chemistry and his observations of the brewing process led to his interest in gasses. His discoveries brought him in contact with many of the great scientific minds of his times, but his commitment to democracy and his opposition to King George III led to an attack on his home by royal loyalists and Priestley fled England for the American colonies, where he settled in Pennsylvania.

YOUR JOB

Explain how Priestley discovered oxygen. How did he know it was a different substance from air? Explain in terms of either hypothetico-deductivism or falsificationism. Priestley repeatedly expresses surprise at the results (hint: see section 3, "Of Dephlogisticated Air, and of the Constitution of the Atmosphere," in Priestley's *Experiments and Observations on Different Kinds of Air*). How would a hypothetico-deductivism theorist or a falsificationist explain the effects on the view that the air we breathe is a simple element? Explain how an hypothetico-deductivist or falsificationist would treat Priestley's hypothesis of air as phlogisticated nitrogen. Are there aspects that do not sit well in the hypothetico-deductive or falsificationist accounts?

RESEARCH HINT

- Once you have located your source, make a photocopy of it. As you read it, make notes on your copy, underlining passages to quote and picking out the author's premises and conclusions. These notes will make writing up your exposition much easier, as you may forget what you had been thinking while immersed in the details of the work.

Genetics Track

PAPER 3: MORGAN, FRUIT FLIES, AND GENES

THE CASE

Just a few decades after Mendel, advances in microscopy allowed scientists to observe cells dividing and the creation of zygotes in sexually reproducing species. It was noted that a cell's nucleus would divide immediately before the cell itself divided, and thin strands were found in the nuclei that stained especially well with a particular new dye. Because of this ability to be stained, they were called chromosomes. It was conjectured that they were related to the inherited properties of the cells, especially when it was discovered that the sex of the offspring was determined by the chromosomes and the genes that make them up.

But not everyone was convinced. Among the skeptics was Thomas Hunt Morgan who thought that (a) Mendel's view was far too simplistic to explain the intricacies of heredity in complex organisms, especially with a number cases that the genetics camp could not account for, (b) the whole thing seemed to smell of the Aristotelian germ theory, where properties were predetermined in a germ of the organism before birth, and (c) it did not appear to fit naturally with Darwin's theory of natural selection. So Morgan set out to undermine the Mendelian theory and the contemporary theory that located the mechanism of inheritance in the chromosomes.

To do this, he would subject organisms to a series of selection pressures and see what would happen to them and their chromosomes. He needed a subject that was cheap and easy to breed, could be observed over a number of generations in a relatively short period of time, and had easily observable traits, so he selected the fruit fly

(*Drosophila melanogaster*). When he observed a mutant male fruit fly with white eyes, different from the nearly universal red, he realized he had an experiment.

He mated the white-eyed male with a red-eyed female, and all of the offspring had red eyes regardless of sex. But when he mated this first generation with itself, he found that all females had red eyes, but half of the males had white eyes, a result not only consistent with Mendel's approach, but one that could be explained in terms of genetics given the discovery of the role of chromosomes in determining sex. The gene for white eyes was recessive and located on the Y chromosome.

Morgan's mind began to change and as more experiments were performed and more sex-linked mutations were found. Characteristics, that is, statistical correlations, could be found showing that when one was present the other likely was also. From this connection, it could be shown that the genes for them were located in close proximity on the chromosome, making it likely that they would get transferred together. This led to the sort of mechanical explanation that Morgan preferred, and made him a staunch genetic advocate of the Mendelian theory with its chromosomal account of heritability.

THE SCIENTIST

Thomas Hunt Morgan (1866–1945), the great-grandson of Francis Scott Key, earned his PhD from Johns Hopkins University and eventually became a faculty member at Columbia University, where his work in the "fly room" began modern genetics and earned him a Nobel Prize, the first ever in medicine for work on genetics. He finished his career at the California Institute of Technology, where he worked in the biology department with an eye toward making it the premier institution for the study of evolution and genetics, a reputation it has maintained.

YOUR JOB

Explain why Morgan had his doubts about Mendelian genetics. Explain his work with fruit flies and how this changed his mind (hint: see Morgan's discussion of it in either *The Theory of the Gene* or *Mechanism of Mendelian Heredity*). Choose either hypothetico-deductivism or falsificationism and explain how they see science to work. How would an advocate of this view make sense of Morgan's work? Are there aspects that do not fit well with this view?

RESEARCH HINT

- Once you have located your source, make a photocopy of it. As you read it, make notes on your copy, underlining passages to quote and picking out the author's premises and conclusions. These notes will make writing up your exposition much easier, as you may forget what you had been thinking while immersed in the details of the work.

Evolutionary Biology Track

PAPER 3: LAMARCK AND ACQUIRED CHARACTERISTICS

THE CASE

The Linnaean system was not accepted by everyone. Georges-Louis Leclerc, comte de Buffon, was a naturalist and Keeper of the Royal Gardens, and he objected to Linnaeus's work on the grounds that not only were species not fixed but the concept was not entirely useful. He thought that there were internal molds to types of animals, but that particular instances were deviant versions. The zebra, for example, was a degenerated version of the horse. God provided the perfect forms, but the material instantiations of them became more and more corrupted. Individuals could be grouped according to their internal molds—say, horses or cats—but individuals degenerated, and so subsequent groupings were meaningless.

Working under Buffon was a younger scientist who had been appointed botanist at the Royal Gardens named Jean-Baptiste Lamarck. Moving from plants to lower animals, invertebrates, he too abandoned the fixity of species but not in as radical a fashion. Lamarck held that the notion of species was still a meaningful one, but that species could change over time.

Life, he argued, arose from lifeless matter by spontaneous generation. First the simplest sorts of living beings appeared. As these organisms interacted with their environment, they were altered. These alterations were then passed on to offspring, who were born with them. They would then interact with their environments and again be transformed by it, creating a different and more complex offspring. Where Buffon held that change was degeneration from a preexisting plan or form, Lamarck held that these acquired mutations led to more and more anatomically intricate beings. As such, species progressed up the ladder of complexity.

Organs that are used more would increase in size, just as human muscles that are used more grow larger. The next generation would then be born with naturally larger organs. For isolated groups exposed to different environmental challenges, this would mean that different aspects would change and the offspring would less and less resemble each other, causing branching, as we see in Linnaeus's work.

THE SCIENTIST

Jean-Baptiste Pierre Antoine de Monet, chevalier de Lamarck (1744–1829) came from a French military family and was a decorated soldier for bravery. He retired from the military because peacetime made the job less than exciting. He took a job in a bank for a while before deciding to study medicine, which led to an interest in botany. His botanical work caught the eye of Buffon, who secured him a job in the Royal Gardens. When the King Louis XVI and Queen Marie Antoinette were beheaded, the garden was reorganized, and Lamarck began his study of invertebrates. His work in the field led not only to his view of acquired mutation (a position widely rejected at the time) but also to his contributions to taxonomy in which he distinguished crustaceans, spiders, and worms from the family of insects. He eventually went blind and died in poverty. His remains were exhumed from the grave that his family was able to temporarily rent for five years after his death, and no one knows where they currently rest.

YOUR JOB

Explain Lamarck's theory of the inheritance of acquired traits. How do changes in environment create changes in individual organisms and therefore changes in a species (hint: Lamarck sets out clearly the job of vital fluids to enlarge and create new organs in the introduction to part 2 of his *Zoological Philosophy*)? Explain this in terms of hypothetico-deductivism or falsificationism. How would a hypothetico-deductivist or a falsificationist account for Lamarck's work? Are there aspects that do not fit this methodological approach well?

RESEARCH HINT

- Once you have located your source, make a photocopy of it. As you read it, make notes on your copy, underlining passages to quote and picking out the author's premises and conclusions. These notes will make writing up your exposition much easier, as you may forget what you had been thinking while immersed in the details of the work.

Geology Track

PAPER 3: CHARLES LYELL, UNIFORMITARIANISM, AND THE STEADY STATE OF THE EARTH

THE CASE

The history of geology is pregnant with theology. From the Creation to the Great Flood, many explanations tried to account for the emergence of geological features in terms of major, catastrophic events taking place over relatively short periods of time from causes that we do not see in our day and age. Hutton argued against this approach, contending that geology should only make use of the sorts of forces we observe and taking up this position and advancing it to its logical end was Charles Lyell.

Lyell's view, termed uniformitarianism, asserted that the earth was a stable system in a steady state. There would be volcanoes and earthquakes in some places, but the effects would be local and always offset. Erosion in one place would be balanced by a similar amount of the depositing of new minerals in another. Subsiding land would be balanced by elevation elsewhere. Species may go extinct from time to time, but they would then be replaced by similar organisms, and it can be seen from fossils that they were not terribly different anatomically from the species we see now. The extinctions were not to the result of major changes, just shifts in the distribution of land and water along the surface of the earth. And all of it was governed by the laws of nature that we observe operating in the world around us. We were not to explain any unusual features in terms of forces we do not currently see operative.

This view had several significant effects. First, if major features like volcanoes and mountains were to be accounted for in terms of current-day forces, then the time frame that geology was dealing with would be much, much longer than anyone had previously considered. Further, it meant that species were not immutable and present from the creation, an insight not lost on Charles Darwin, who read Lyell aboard the HMS *Beagle*. Both of these views clashed with the theological presuppositions that sat beneath much of science at that time.

THE SCIENTIST

Charles Lyell (1797–1875) trained as a lawyer, a job he hated. Using poor eyesight as an excuse to leave his legal practice, he became a geologist, although one who read the findings of other naturalists more than he went out and investigated for himself. His book, *Principles of Geology*, became a standard text and went through eleven editions with Lyell revising the earlier parts as quickly as he produced the later ones.

YOUR JOB

Explain Lyell's uniformitarianism. How is it different from catastrophism (hint: look at Book 1, chapter 5, "Causes which have Retarded the Progress of Geology," in *Principles of Geology*)? Select either hypothetico-deductivism or falsificationism and explain the approach. How would an advocate of either of those views make sense of Lyell's work? Are there aspects that do not fit in well?

RESEARCH HINT

- Once you have located your source, make a photocopy of it. As you read it, make notes on your copy, underlining passages to quote and picking out the author's premises and conclusions. These notes will make writing up your exposition much easier, as you may forget what you had been thinking while immersed in the details of the work.

Psychology Track

PAPER 3: PAVLOV, CONDITIONED REFLEXES, AND EXPERIMENTATION

THE CASE

Following in the footsteps of Weber and Fechner, psychologists and physiologists began conducting experiments that tested measurable and repeatable relationships between perception and behavior or between perception and physiological reactions. The most famous of these were I. P. Pavlov's work on the conditioned reflex.

Pavlov, a physiologist, studied digestion, using dogs as his test subjects. For part of his work, he measured the amount of saliva the dogs would generate while eating dry food. He came to realize that the dogs would begin to salivate before they ate, indeed, before they were even given food; they salivated upon hearing the familiar footsteps of their handlers who brought them the food. The correlation between a perception that itself cannot produce the physiological effect and the observation of that very effect made Pavlov interested.

He began to develop experiments to test how one could take such reflexes and intentionally develop them. He knew that putting a mild acid in a dog's mouth would cause the dog to generate significant amounts of saliva. It is a natural reaction designed to dilute the acid and wash it out of the mouth. Pavlov had the acid administered to the dogs accompanied by a tone of 500 hertz. The tone would always accompany the acid (this was termed "conditioning"), although tones of other frequencies would be generated randomly and never accompanied by the acid. Eventually, the acid was not given, but tones were played. When the 500-hertz tone sounded, the dogs salivated. But other tones, even those that were very close, say, 480 hertz, would not give rise to salivation.

Pavlov not only noted that the effect occurred but found that over time it diminished and eventually disappeared. The longer the conditioning continued, the longer the diminution would take. From the salivation of dogs, Pavlov argued that we can understand emotional reactions in humans ranging over everything from hunger, pleasure, and anger to more complex feelings like triumph and despair—even sexual attraction. The brain not only controls basic functions and allows for perception of the world around us but is also wired in such a way as to interconnect these functions in complex ways. The brain itself learns, independently of our intentional,\ rational thought processes.

THE SCIENTIST

Ivan Petrovich Pavlov (1849–1936) was a Russian physiologist who received the Nobel Prize for medicine in 1904 for his work in understanding the details of the processes involved in digestion. He also was the first to figure out other methods of observing functioning organs in living beings instead of relying on vivisection. He was one of the few elite members of Russian society to be heralded after the Bolshevik Revolution; Lenin himself signed a proclamation extolling Pavlov's work in science.

YOUR JOB

Explain Pavlov's work on operant conditioning. What sort of experiments did he use (hint: see the first section of lecture 2, "Technical Methods Employed in the Objective Investigation of the Functions of the Cerebral Hemispheres," in his collected lectures from 1924, *Conditioned Reflexes*)? Select either hypothetico-deductivism or falsification-ism and explain the view of scientific methodology. How would an advocate of either of these views account for Pavlov's work? Are there aspects that do not fit well in to these schemes?

RESEARCH HINT

- Once you have located your source, make a photocopy of it. As you read it, make notes on your copy, underlining passages to quote and picking out the author's premises and conclusions. These notes will make writing up your exposition much easier, as you may forget what you had been thinking while immersed in the details of the work.

Sociology Track

PAPER 3: WEBER, RELIGION, AND WEALTH

THE CASE

It was a well-discussed fact of Europe in the late nineteenth century that a significantly larger percentage of high-paying jobs, especially managerial jobs with authority, were held by Protestants, while Catholics tended to do the lower-paying, dirtier, more physical labor. This regularity held across national borders in pluralistic societies with generally different cultural historical narratives. Much time and space, especially in Catholic publications, was devoted to considering why this might be.

Max Weber, in his book *The Protestant Ethic and the Spirit of Capitalism,* took up this question, deeming it to be the sort of issue designed for the sociologist. In the first part of the book, Weber debunked a number of seemingly reasonable hypotheses that sought to explain the phenomenon. For example, one might suppose that the hierarchical structure of the Catholic Church, as opposed to the diffusion of power in Protestantism, might put religion in a prominent place that would leave little room for financial concerns. But Weber pointed out that while Protestantism has removed a central authority, it in fact replaced one authority with another that puts a religious spin on every aspect of life. Similarly, one might think that the difference is one of theology, in which the doctrinal beliefs stemming from the Reformation encouraged the sort of money-making interests that were reported in the Protestant community. But examinations of early Protestant doctrine shows quite the opposite. And so it goes with several possible explanations.

Weber took views typical of Americans, particularly the writings of Benjamin Franklin, as prime examples of the statement of the Protestant work ethic, wherein making money was not seen as a means to a better life but as a moral imperative that governed the very process of living itself. The Protestants made more money, but they did not seem to enjoy it; indeed, the enjoyment of the wealth they slaved so hard to achieve was seen as sinful. It was in the ethical framing of Franklin's language that Weber found the sociological clue for what he claimed to be the best explanation for the capitalist leanings of Protestants and the nonmaterial lifestyle of Catholics. As the Protestants gained political power, they framed those sorts of behaviors that benefited them as a group in moral terms; thus, the basic concepts of capitalism became internalized within the structure of society and within the consciousnesses of its members. The interesting question for the sociologist, then, was how such normative structures that determine how a person lives, acts, and feels in society came to be.

He traces the move to Luther's notion of calling. In Catholicism, one's calling takes one out of the world. The secular is to be left behind, to be transcended. But Luther elevated the secular to the status of the sacred. All work can be divine, and this change permeates the boundary that the Catholics had placed between inferior worldly work and that which was divine. Now the divine was contained within the ordinary, and this meant that even the economic could be brought under the umbrella of the religious. Hence one was not working for the earthly rewards—those were to be avoided—but rather for the sake of working, a notion to be encouraged by the nature of the social relations in an emerging capitalist society.

THE SCIENTIST

Max Weber (1864–1920) was born to a political family; his father was a prominent liberal politician and the house was abuzz with history, politics, and economics. Following in this path, Weber studied law and economics, from which he helped launch sociology as a modern field of study. His father died immediately following a terrible quarrel between the two and the unresolved fight left Weber psychologically unstable for a period of time and had lingering effects throughout his life.

YOUR JOB

Examine and clearly set out one or two of Weber's arguments in the first chapter ("Denomination and Social Stratification") in the first part ("The Problem") of *The Protes-*

tant Ethic and the Spirit of Capitalism, where he rules out certain proposed explanations for the disparity of wealth and position of Protestants and Catholics. On what evidence does he undermine these proposed explanations? Select either hypothetico-deductivism or falsificationism and explain the approach in detail. How would an advocate of either of those positions make sense of Weber's work? Are there aspects that do not fit in well with those understandings of the scientific method?

RESEARCH HINT

- Once you have located your source, make a photocopy of it. As you read it, make notes on your copy, underlining passages to quote and picking out the author's premises and conclusions. These notes will make writing up your exposition much easier, as you may forget what you had been thinking while immersed in the details of the work.

Economics Track

PAPER 3: SMITH, NATURAL PRICES, AND THE LAWS OF THE MARKET

THE CASE

Adam Smith admired the work of the physiocrats but disagreed with the central role in which they cast farming. He believed labor, not nature, to be the foundation of the economy, when the marketplace is examined, people may set out the rules that it obeys.

For Smith, the Newtonian picture of physics was a template to be copied in economics. Just as there were well-defined forces that behaved according to well-defined natural laws governing the movements of, say, balls on a billiard table, so there were such well-defined laws of the marketplace. Specification of the mass, location, and velocities of the balls on the table was sufficient to allow one to predict the state of the balls on the table at future moments, according to Newton's laws of motion. So, too, because of the rationality of self-interested participants, specification of economic factors would generate determinable quantities.

One such quantity is the price of goods. There are three component parts of price: rent, labor, and profit. What Smith called rent we now refer to as overhead, the cost of the materials and circumstances needed to produce the product. Labor counts both wages paid and one's own input. Profit is what the seller ultimately can claim as the added value. Sellers and buyers are both self-interested; sellers want to maximize profit and thereby price, while buyers want to minimize price. Since sellers are in competition, meaning that one only profits on that which one sells and a lower price will attract more buyers, there are natural laws governing price.

Prices, Smith therefore contended, were not arbitrary, but would move a natural value based upon supply and demand. Artificially high or low prices will be corrected, just as the level of mercury in a thermometer is adjusted by the ambient temperature. The price of corn, he pointed out, is much more variable than that of woolen cloth, because the number of spinners and weavers and the amount of cloth each can produce are fairly consistent, whereas the corn harvest is much more variable.

THE SCIENTIST

Adam Smith (1723–1790) lived in Edinburgh, Scotland. His interests included not only economics, but also moral theory, which he saw as an inextricable part of the same questions. A shy, odd, and absentminded man, he was close friends with the famous Scottish philosopher David Hume, who was well known as much more sociable, gregarious, and welcoming of controversy. Many stories of Smith's absentmindedness abound, chronicling long walks in his dressing gown and his being so lost in contemplative conversation as to fall into tanning pits.

YOUR JOB

Explain Smith's argument for the existence of natural and market prices (hint: see Book 1, chapter 7 of *The Wealth of Nations*). Select either hypothetico-deductivism or falsificationism and clearly explain the approach to science. How would an advocate of either of these views interpret Smith's reasoning here? What sorts of examples of predictions and tests does Smith set out? What aspects of Smith's system do not fit in well with these approaches?

RESEARCH HINT

- Once you have located your source, make a photocopy of it. As you read it, make notes on your copy, underlining passages to quote and picking out the author's premises and conclusions. These notes will make writing up your exposition much easier, as you may forget what you had been thinking while immersed in the details of the work.

HOLISTIC VIEW OF THEORIES

All of the competing accounts of the scientific method we have examined so far share a particular view of the nature of scientific theories. They all assert that scientific theories are sets of propositions that are individually testable. Philosophers of science group these models together under the name "the syntactic view of theories." The aim of science is to figure out the complete set of laws of nature, statements that are universal and true, laws which explain groupings of natural phenomena.

Having this picture of scientific theories shapes the approach to the scientific method. Deductivism and inductivism both contend that the purported laws of nature are derivable from a logic of discovery, although they disagree on the logical machinery employed. Hypothetico-deductivism and falsificationism allow for a free and open context of discovery but posit rigorous logics of justification that take each hypothesis and one by one see if it matches up to our experiences or experimental results coming from the lab.

For all of these differences, the ideal end result is the same: a set of universal generalizations, each with a ticket in its pocket marked "Inspected by #28" and a "Goodlabkeeping Seal of Approval." Each part can be looked at independently of all the other parts and determined to be in working condition or not. If a piece is shown to be faulty for logical or empirical reasons, it and it alone is removed and replaced with a new, better-designed component. These views may quibble over what sorts of tests are needed to see if a piece is functioning properly and the means by which one goes about designing its replacement, but that theories can be examined piece by piece is something they all have in common.

This view, however, is not shared by all philosophers and scientists. An alternative picture of scientific theories is called the "holistic view," wherein scientific theories are webs of interrelated sentences, each intertwined with every other. Holists argue that you can never isolate the pieces of a theory and test it without testing the rest of the theory at the same time.

Pierre Duhem, a forefather of this view, argued that for the syntactic view to work, individual hypotheses must have empirical, testable results, but that the scientific testing of any observable consequences that you get from a part of the theory, in fact, has the rest of the theory packed into it. Consider Ohm's law in electronics that relates the current flowing through a circuit to the voltage and the resistance. It seems simple enough, the sort of law that can simply be tested by itself. It's the sort of experiment every student of elementary electronics performs: set up a circuit with a battery of a known voltage and insert a resistor of known resistance, then measure the resulting current. "There," you might think to yourself, "I've tested a piece of the theory and only that piece."

But hold on. How did you measure the current? "With an ammeter, of course," you reply. And how do you know that the ammeter gives you the amperage? At this question, you begin to explain how the ammeter works, the circuitry inside until suddenly you realize that you need other parts of the theory governing electronics to explain how you can test this part. But how do you know these other parts are true? Testing them requires yet other parts, and so on.

And what happens if the test fails? Which of these parts is to blame? Do we reject Ohm's law or the principles governing the ammeter? What exactly got falsified by your negative result? It turns out that you could place the blame anywhere in the web of beliefs, if you are willing to make revisions elsewhere. Therefore, there is never a conclusive test of an individual piece of a theory. There are never crucial experiments where you are forced by empirical data to give up any given hypothesis.

Thomas Kuhn expanded this notion of a holistic approach by moving beyond theories as sets of sentences to "paradigms," which include not only statements taken as likely laws of nature but also the practices, tools, and procedures used in deriving and testing results and determines the meanings of its central terms and concepts. Science, Kuhn argued, is about puzzle solving, and by specifying the meaning of the terms of the paradigm's language, science determines what counts as a legitimate puzzle, what are considered the legitimate means for trying to solve puzzles, and what counts as legitimate answers.

The legitimacy inherent in the paradigm, he further argues, comes not from a logical foundation but a sociological one. Science occurs within a community and if you want to understand why scientists believe what they believe, say what they say, and do in the lab what they do, you need to understand the dynamics of the scientific community that serves to keep the paradigm in place and transmit it to the next generation.

The history of science, Kuhn argued, is a series of periods dominated by stable paradigms, normal science, dotted with occasional revolutions in which a once-dominant paradigm is rejected for a new one. Since paradigms define their terms and concepts, they impose a complete worldview in which to understand all observations. But this means that all of the sense we make of the world comes through the paradigm, there is no extra-paradigmatic standpoint from which to look at the world. As such, one cannot meaningfully talk across paradigms, because the words and concepts only have meaning within the coherent structure of the paradigm.

Occasionally, however, problems that are well formed according to the paradigm will resist solutions according to the tools deemed acceptable by the paradigm. These are termed anomalies. When a number of anomalies begin to pile up or significant anomalies surface, the community goes into a state of crisis, the only time scientists begin to look *at* the paradigm instead of looking through it. At some point the crisis will deepen enough that a critical mass begins to abandon the paradigm for a competitor. When the new paradigm becomes the accepted stance in the community, you have scientific revolution.

Since rationality is thus paradigm-dependent, you cannot comparison shop for a paradigm and there can never be good reason to switch paradigms (good reason only exists within a paradigm). As such, one cannot say that science progresses or advances by revolutions, merely that it changes direction.

But surely we do want to say that we have good reason to prefer, say, Einstein to Newton and that science does progress. At the same time, it also seems as if we need to save Kuhn's insights in expanding upon Duhem's work. Imre Lakatos, a student of Karl Popper, was deeply impressed by Kuhn's system and weaved together an intricate tapestry comprised of insights from both Kuhn and Popper.

Kuhn and Duhem, Lakatos contended, were correct that one could always hold onto any given belief if there was willingness to alter the others. In this way, he argued that a holistic approach like that of Kuhn and Duhem demonstrated why Popper's brand of falsification was naive. There are some central propositions so crucial to the worldview that they are protected while we have a more provisional sense of others. Thus, Lakatos changed Kuhn's paradigms into "research programs" by adding internal structure to them, delineating a hard core of principles held to be essential to the project and considered unfalsifiable from other less crucial statements that one would willingly abandon or alter that form a "protective belt" around the hard core.

But from Popper he took the idea that a good theory is one that is bold and simple. We ought to, Popper contended, prefer those theories that make risky predictions, especially those in areas for which the theory was not originally designed. Such increases in falsifiability should be taken as an indicator of rational superiority. But then Kuhn would be right that research programs cannot be directly compared. But this does not negate any meaningful sense of scientific progress when undergoing a paradigm shift. Rather, the paradigms' development can be viewed over time. If challenges arise that require modifications to the protective belt that are artificial patches specifically designed to save the program, it becomes less falsifiable and there degenerates. On the other hand, if the program is able to expand its reach and account for new additional phenomena without significant modification it is seen as progressive. Although one could never set up a crucial experiment that reaches across programs, one still can have the sense that one program is degenerate while the other is progressive and therefore have a historical sense that provides rationality for preferring one theory over another restoring a picture of scientific progress to a system with the advantages and insights of Kuhn and Duhem.

Pierre Duhem

. . . .

Aim and Structure of Physical Theory

AN EXPERIMENT IN PHYSICS CAN NEVER CONDEMN
AN ISOLATED HYPOTHESIS BUT ONLY A WHOLE
THEORETICAL GROUP

The physicist who carries out an experiment, or gives a report of one, implicitly recognizes the accuracy of a whole group of theories. Let us accept this principle and see what consequences we may deduce from it when we seek to estimate the role and logical import of a physical experiment.

In order to avoid any confusion we shall distinguish two sorts of experiments: experiments of *application,* which we shall first just mention, and experiments of *testing,* which will be our chief concern.

You are confronted with a problem in physics to be solved practically; in order to produce a certain effect you wish to make use of knowledge acquired by physicists; you wish to light an incandescent bulb; accepted theories indicate to you the means for solving the problem; but to make use of these means you have to secure certain information; you ought, I suppose, to determine the electromotive force of the battery of generators at your disposal; you measure this electromotive force: that is what I call an experiment of application. This experiment does not aim at discovering whether accepted theories are accurate or not; it merely intends to draw on these theories. In order to carry it out, you make use of instruments that these same theories legitimize; there is nothing to shock logic in this procedure.

But experiments of application are not the only ones the physicist has to perform; only with their aid can science aid practice, but it is not through them that science creates and develops itself; besides experiments of application, we have experiments of testing.

A physicist disputes a certain law; he calls into doubt a certain theoretical point. How will he justify these doubts? How will he demonstrate the inaccuracy of the law? From the proposition under indictment he will derive the prediction of an experimental fact; he will bring into existence the conditions

From Pierre Duhem, *The Aim and Structure of Physical Theory* (Princeton, NJ: Princeton University Press, 1954), 183–90. © 1954 Princeton University Press, 1982 renewed PUP. Reprinted by permission of Princeton University Press.

under which this fact should be produced; if the predicted fact is not produced, the proposition which served as the basis of the prediction will be irremediably condemned.

F. E. Neumann assumed that in a ray of polarized light the vibration is parallel to the plane of polarization, and many physicists have doubted this proposition. How did O. Wiener undertake to transform this doubt into a certainty in order to condemn Neumann's proposition? He deduced from this proposition the following consequence: If we cause a light beam reflected at 45° from a plate of glass to interfere with the incident beam polarized perpendicularly to the plane of incidence, there ought to appear alternately dark and light interference bands parallel to the reflecting surface; he brought about the conditions under which these bands should have been produced and showed that the predicted phenomenon did not appear, from which he concluded that Neumann's proposition is false, viz., that in a polarized ray of light the vibration is not parallel to the plane of polarization.

Such a mode of demonstration seems as convincing and as irrefutable as the proof by reduction to absurdity customary among mathematicians; moreover, this demonstration is copied from the reduction to absurdity, experimental contradiction playing the same role in one as logical contradiction plays in the other.

Indeed, the demonstrative value of experimental method is far from being so rigorous or absolute: the conditions under which it functions are much more complicated than is supposed in what we have just said; the evaluation of results is much more delicate and subject to caution.

A physicist decides to demonstrate the inaccuracy of a proposition; in order to deduce from this proposition the prediction of a phenomenon and institute the experiment which is to show whether this phenomenon is or is not produced, in order to interpret the results of this experiment and establish that the predicted phenomenon is not produced, he does not confine himself to making use of the proposition in question; he makes use also of a whole group of theories accepted by him as beyond dispute. The prediction of the phenomenon, whose non-production is to cut off debate, does not derive from the proposition challenged if taken by itself, but from the proposition at issue joined to that whole group of theories; if the predicted phenomenon is not produced, not only is the proposition questioned at fault, but so is the whole theoretical scaffolding used by the physicist. The only thing the experiment teaches us is that among the propositions used to predict the phenomenon and to establish whether it would be produced, there is at least one error; but where this error lies is just what it does not tell us. The physicist may declare that this error is contained in exactly the proposition he wishes to refute, but is he sure it is not in another proposition? If he is, he accepts implicitly the accuracy of all the other propositions he has used, and the validity of his conclusion is as great as the validity of his confidence.

Let us take as an example the experiment imagined by Zenker and carried out by O. Wiener. In order to predict the formation of bands in certain circumstances and to show that these did not appear, Wiener did not make use merely of the famous proposition of F. E. Neumann, the proposition which he wished to refute; he did not merely admit that in a polarized ray vibrations are parallel to the plane of polarization; but he used, besides this, propositions, laws, and hypotheses constituting the optics commonly accepted: he admitted that light consists in simple periodic vibrations, that these vibrations are normal to the light ray, that at each point the mean kinetic energy of the vibratory motion is a measure of the intensity of the light, that the more or less complete attack of the gelatin coating on a photographic plate indicates the various degrees of this intensity. By joining these propositions, and many others that would take too long to enumerate, to Neumann's proposition, Wiener was able to formulate a forecast and establish that the experiment belied it. If he attributed this solely to Neumann's proposition, if it alone bears the responsibility for the error this negative result has put in evidence, then Wiener was taking all the other propositions he invoked as beyond doubt. But this assurance is not imposed as a matter of logical necessity; nothing stops us from taking Neumann's proposition as accurate and shifting the weight of the experimental contradiction to some other proposition of the commonly accepted optics; as H. Poincaré has shown, we can very easily rescue Neumann's hypothesis from the grip of Wiener's experiment on the condition that we abandon in exchange the hypothesis which takes the mean kinetic energy as the measure of the light intensity; we may, without being contradicted by the experiment, let the vibration be parallel to the plane of polarization, provided that we measure the light intensity by the mean potential energy of the medium deforming the vibratory motion.

These principles are so important that it will be useful to apply them to another example; again we choose an experiment regarded as one of the most decisive ones in optics.

We know that Newton conceived the emission theory for optical phenomena. The emission theory supposes light to be formed of extremely thin projectiles, thrown out with very great speed by the sun and other sources of light; these projectiles penetrate all transparent bodies; on account of the various parts of the media through which they move, they undergo attractions and repulsions; when the distance separating the acting particles is very small these actions are very powerful, and they vanish when the masses between which they act are appreciably far from each other. These essential hypotheses joined to several others, which we pass over without mention, lead to the formulation of a complete theory of reflection and refraction of light: in particular, they imply the following proposition: The index of refraction of light passing from one medium into another is equal to the velocity of the light projectile within the medium it penetrates, divided by the velocity of the same projectile in the medium it leaves behind.

This is the proposition that Arago chose in order to show that the theory of emissions is in contradiction with the facts. From this proposition, a second follows: Light travels faster in water than in air. Now Arago had indicated an appropriate procedure for comparing the velocity on light in air with the velocity of light in water; the procedure, it is true, was inapplicable, but Foucault modified the experiment in such a way that it could be carried out; he found that the light was propagated less rapidly in water than in air. We may conclude from this, with Foucault, that the system of emission is incompatible with the facts.

I say the *system* of emission and not the *hypothesis* of emission; in fact, what the experiment declares stained with error is the whole group of propositions accepted by Newton, and after him by Laplace and Biot, that is, the whole theory from which we deduce the relation between the index of refraction and the velocity of light in the various media. But in condemning this system as a whole by declaring it stained with error, the experiment does not tell us where the error lies. Is it in the fundamental hypothesis that light consists in projectiles thrown out with great speed by luminous bodies? Is it in some other assumption concerning the actions experienced by the light corpuscles due to the media through which move? We know nothing about that. It would be rash to believe, as Arago seems to have thought, that Foucault's experiment condemns once and for all the very hypothesis of emission, i.e., the assimilation of a ray of light to a swarm of projectiles. If physicists had attached some value to this task, they would undoubtedly have succeeded in founding on this assumption a system of optics that would agree with Foucault's experiment.

In sum, the physicist can never subject an isolated hypothesis to experimental test, but only a whole group of hypotheses; when the experiment is in disagreement with his predictions, what he learns is that at least one of the hypotheses constituting this group is unacceptable and ought to be modified; but the experiment does not designate which one should be changed.

We have gone a long way from the conception of the experimental method arbitrarily held by persons unfamiliar with its actual functioning. People generally think that each one of the hypotheses employed in physics can be taken in isolation, checked by experiment, and then, when many varied tests have established its validity, given a definitive place in the system of physics. In reality, this is not the case. Physics is not a machine that lets itself be taken apart; we cannot try each piece in isolation and, in order to adjust it, wait until its solidity has been carefully checked. Physical science is a system that must be taken as a whole; it is an organism in which one part cannot be made to function except when the parts that are most remote from it are called into play, some more so than others, but all to some degree. If something goes wrong, if some discomfort is felt in the functioning of the organism, the physicist will have to ferret out through its effect on the entire system which organ needs to be remedied or modified without the possibility of isolating this organ and examining it apart. The watchmaker to whom you give a watch that has stopped separates all the wheelworks and examines them one by one until he finds the part that is defective or broken. The doctor to whom a patient appears cannot dissect him in order to establish his diagnosis; he has to guess the seat and cause of the ailment solely by inspecting disorders affecting the whole body. Now the physicist concerned with remedying a limping theory resembles the doctor and not the watchmaker.

A "CRUCIAL EXPERIMENT" IS IMPOSSIBLE IN PHYSICS

Let us press this point further, for we are touching on one of the essential features of the experimental method, as it is employed in physics.

Reduction to absurdity seems to be merely a means of refutation, but it may become a method of demonstration: in order to demonstrate the truth of a proposition it suffices to corner anyone who would admit the contradictory of the given proposition into admitting an absurd consequence. We know to what extent the Greek geometers drew heavily on this mode of demonstration.

Those who assimilate experimental contradiction to reduction to absurdity imagine that in physics we may use a line of argument similar to the one Euclid employed so frequently in geometry. Do you wish to obtain from a group of phenomena a theoretically certain and indisputable explanation? Enumerate all the hypotheses that can be made to account for this group of phenomena; then, by experimental contradiction eliminate all except one; the latter will no longer be a hypothesis, but will become a certainty.

Suppose, for instance, we are confronted with only two hypotheses. Seek experimental conditions such that one of the hypotheses forecasts the production of one phenomenon and the other the production of quite a different effect; bring these conditions into existence and observe what happens; depending on whether you observe the first or the second of the predicted phenomena, you will condemn the second or the first hypothesis; the hypothesis not condemned will be henceforth indisputable; debate will be cut off, and a new truth will be acquired by science. Such is the experimental test that the author of the *Novum Organum* called the *"fact of the cross,"* borrowing this expression from the crosses which at an intersection indicate the various roads."

We are confronted with two hypotheses concerning the nature of light; for Newton, Laplace, or Biot light consisted of projectiles hurled with extreme speed, but for Huygens, Young, or Fresnel light consisted of vibrations whose waves are propagated within an ether. These are the only two possible hypotheses as far as one can see: either the motion is carried away by the body it excites and remains attached to it, or else it passes from one body to another. Let us pursue the first hypothesis; it declares that light travels more quickly in water than air; but if we follow the second, it declares that light travels more quickly in air than in water. Let us set up Foucault's apparatus; we set into motion the turning mirror; we see two luminous spots formed before us, one colorless, the other greenish. If the greenish band is to the left of the colorless one, it means that light travels faster in water than in air, and that the hypothesis of vibrating waves is false. If, on the contrary, the greenish band is to the right of the colorless one, that means that light travels faster in air than in water, and the hypothesis of emissions is condemned. We look through the magnifying glass used to examine the luminous spots, and we notice that the greenish spot is to the right of the colorless one; the debate is over; light is not a body, but a vibratory wave motion propagated by the ether; the emission hypothesis has had its day; the wave hypothesis is a new article of the scientific credo.

What we have said in the foregoing paragraph shows how mistaken we should be to attribute to Foucault's experiment so simple a meaning and so decisive an importance; for it is not between two hypotheses, the emission and the wave hypotheses, that Foucault's experiment judges trenchantly; it decides rather between two sets of theories each of which has to be taken as a whole, i.e., between two entire systems, Newton's optics and Huygen's optics.

But let us admit for a moment that in each of these systems everything is compelled to be necessary by strict logic, except a single hypothesis; consequently, let us admit that the facts, in condemning one of the two systems condemn once and for all the single doubtful assumption it contains. Does it follow that we can find in the "crucial experiment" an irrefutable procedure for transforming one of the two hypotheses before us into a demonstrated truth? Between two contradictory theorems of geometry there is no room for a third judgment; if one is false, the other is necessarily true. Do two hypotheses in physics ever constitute such a strict dilemma? Shall we ever dare to assert that no other hypothesis is imaginable? Light may be a swarm of projectiles, or it may be a vibratory motion whose waves are propagated in a medium; is it forbidden to be anything else at all? Arago undoubtedly thought so when he formulated this incisive alternative: does light move more quickly in water than in air? "Light is a body. If the contrary is the case, then light is a wave." But it would be difficult for us to take such a decisive stand; Maxwell, in fact, showed that we might just as well attribute light to a periodical electrical disturbance that is propagated within a dielectric medium.

Unlike the reduction to absurdity employed by geometers, experimental contradiction does not have the power to transform a physical hypothesis into an indisputable truth; in order to confer this power on it, it would be necessary to enumerate completely the various hypotheses which may cover a determinate group of phenomena; but the physicist is never sure he has exhausted all the imaginable assumptions. The truth of a physical theory is not decided by heads or tails.

Thomas Kuhn

The Structure of Scientific Revolutions

THE ROUTE TO NORMAL SCIENCE

In this essay, 'normal science' means research firmly based upon one or more past scientific achievements that some particular scientific community acknowledges for a time as supplying the foundation for its further practice. Today such achievements are recounted, though seldom in their original form, by science textbooks, elementary and advanced. These textbooks expound the body of accepted theories, illustrate many or all of its successful applications, and compare these applications with exemplary observations and experiments. Before such books became popular in the nineteenth century (and until even more recently in the newly matured sciences), many of the famous classics of science fulfilled a similar function. Aristotle's *Physica*, Ptolemy's *Almagest*, Newton's *Principia* and *Optiks*, Franklin's *Electricity*, Lavoisier's *Chemistry*, and Lyell's *Geology*—these and many other works served for a time implicitly to define the legitimate problems and methods of a research field for succeeding generations of practitioners. They were able to do so because they shared two essential characteristics. Their achievement was sufficiently unprecedented to attract an enduring group of adherents away from competing modes of scientific activity. Simultaneously, it was sufficiently open-ended to leave all sort of problems for the redefined group of practitioners to resolve.

Achievements that share these two characteristics I shall henceforth refer to as 'paradigms,' a term that relates closely to 'normal science.' By choosing it, I mean to suggest that some accepted examples of actual scientific practice—examples which include law, theory, application, and instrumentation together—provide models from which spring particular coherent traditions of scientific research. These are the traditions which the historian describes under such rubrics as 'Ptolemaic astronomy' (or 'Copernican'), 'Aristotelian dynamics' (or 'Newtonian'), 'corpuscular optics' (or 'wave optics'), and so on. The study of paradigms, including many that are far more specialized than those named illustratively above, is what mainly prepares the student for membership in the particular scientific community with which he will late

From Thomas Kuhn, *The Structure of Scientific Revolutions* (Chicago: University of Chicago Press, 1962), 10–13, 16–17, 23–24, 35–38, 52–53, 67–69, 77–78, 81–83, 92–95, 157–59.

practice. Because he there joins men who learned the bases of their field from the same concrete models, his subsequent practice will seldom will seldom evoke overt disagreement over fundamentals. Men whose research is based on shared paradigms are committed to the same rules and standards for scientific practice. That commitment and the apparent consensus it produces are prerequisites for normal science, i.e., for the genesis and condition of a particular research tradition.

If the historian traces the scientific knowledge of any selected group of related phenomena backward in time, he is likely to encounter some minor variant of a pattern here illustrated from the history of physical optics. Today's physics textbooks tell the student that light is photons, i.e., quantum-mechanical entities that exhibit some characteristics of waves and some of particles. Research proceeds accordingly, or rather according to the more elaborate and mathematical characterization from which this usual verbalization is derived. That characterization of light is, however, scarcely half a century old. Before it was developed by Planck, Einstein, and others early in this century, physics texts taught that light was a transverse wave motion, a conception rooted in a paradigm that derived ultimately from the optical writings of Young and Fresnel in the early nineteenth century. Nor was the wave theory the first to be embraced by almost all practitioners of optical science. During the eighteenth century the paradigm for this field was provided by Newton's *Optiks*, which taught that light was material corpuscles. At that time, physicists sought evidence, as the early wave theorists had not, of the pressure exerted by light particles impinging on solid bodies.[1]

These transformations of the paradigms of physical optics are scientific revolutions, the successive transition from one paradigm to another via revolution is the usual developmental pattern of mature science. It is not, however, the pattern characteristic of the period before Newton's work, and that is the contrast that concerns us here. No period between remote antiquity and the end of the seventeenth century exhibited a single generally accepted view about the nature of light. Instead there were a number of competing schools and subschools, most of them espousing one variant or another of Epicurean, Aristotelian, or Platonic theory. One group took light to be particles emanating from material bodies; for another it was a modification of the medium that intervened between the body and the eye; still another explained light in terms of an interaction of the medium with an emanation of the eye; and there were other combinations and modifications besides. Each of the corresponding schools derived strength from its relation to some particular metaphysic, and each emphasized, as paradigmatic observations, the particular cluster of optical phenomena that its own theory could do most to explain. Other observations were dealt with by *ad hoc* elaborations, or they remained as outstanding problems for further research.[2]

At various times all these schools made significant contributions to the body of concepts, phenomena, and techniques from which Newton drew the first nearly uniformly accepted paradigm for physical optics. Any definition of the scientist that excludes at least the more creative members of these various schools will exclude their modern successors as well. Those men were scientists. Yet anyone examining a survey of physical optics before Newton may well conclude that, though the field's practitioners were scientists, the net result of their activity was something less than science. Being able to take no common body of belief for granted, each writer on physical optics felt forced to build his field anew from its foundations. In doing so, his choice of supporting observation and experiment was relatively free, for there was no standard set of methods or of phenomena that every optical writer felt forced to employ and explain. Under these circumstances, the dialogue of the resulting books was often directed as much to the members of other schools as it was to nature. That pattern is not unfamiliar in a number of creative fields today, nor is it incompatible with significant discovery and invention. It is not, however, the pattern of development that physical optics acquired after Newton and that other natural sciences make familiar today.

Though fact-collecting has been essential to the origin of many significant sciences, anyone who examines Pliny's encyclopedic writings or the Baconian natural histories of the seventeenth century will discover that it produces a morass. One somehow hesitates to call the literature that results scientific. The Baconian "histories" of heat, color, wind, mining, and so on, are filled with information, some of it recondite. But they juxtapose facts that will later prove revealing (e.g., heating by mixture) with others (e.g., the warmth of dung heaps) that will for some time remain too complex to be integrated with theory at all.

This is the situation that creates the schools characteristic of the early stages of a science's development. No natural history can be interpreted in the absence of at least some implicit body of intertwined theoretical and methodological belief that permits selection, evaluation, and criticism. If that body of beliefs is not already implicit in the collection of facts—in which case more than "mere facts" are at hand—it must be externally supplied, perhaps by a current metaphysic, by another science, or by personal and historical accident. No wonder, then, that in, the early stages of the development of any science different men confronting the same range of phenomena, describe and interpret them in different ways. What is surprising, and perhaps also unique in the degree to the fields we call science, is that such initial divergences should ever largely disappear.

For they do disappear to a very considerable extent and then apparently once and for all. Furthermore, their disappearance is usually caused by the triumph of one of the pre-paradigm schools, which, because of its own characteristic beliefs and preconceptions, emphasized only some special part of the too sizable and inchoate pool of information.

THE NATURE OF NORMAL SCIENCE AS PUZZLE-SOLVING

What then is the nature of the more professional and esoteric research that a group's reception of a single paradigm permits? If the paradigm represents work that has been done once and for all, what further problems does it leave the united group to resolve?

Paradigms gain their status because they are more successful than their competitors in solving a few problems that the group of practitioners has come to recognize as acute. To be more successful is not, however, to be either completely successful with a single problem or notably successful with any large number. The success of a paradigm—whether Aristotle's analysis of motion, Ptolemy's computations of planetary motion, Lavoisier's application of the balance, of Maxwell's mathematization of the electromagnetic field—is at the start largely a promise of success discoverable in selected and still incomplete examples. Normal science consists in the actualization of that promise, an actualization achieved by extending the knowledge of those facts that the paradigm displays as particularly revealing, by increasing the extent of the match between those facts and the paradigm's predictions, and by further articulation of the paradigm itself.

Few people who are not actually practitioners of a mature science realize how much mop-up work of this sort a paradigm leaves to be done or quite how fascinating such work can prove in the execution. And these points need to be understood. Mopping-up operations are what engage most scientists throughout their careers. They constitute what I am here calling normal science. Closely examined, whether historically or in the contemporary laboratory, that enterprise seems an attempt to force nature into the preformed and relatively inflexible box that the paradigm supplies. No part of the aim of normal science is to call forth new sorts of phenomena; indeed those that will not fit the box are often not seen at all. Nor do scientists normally aim to invent new theories, and they are often intolerant of those invented by others.[3] Instead, normal-scientific research is directed to the articulation of those phenomena and theories that the paradigm already supplies.

Perhaps these are defects. The areas investigated by normal science are, of course, miniscule; the enterprise now under discussion has drastically restricted vision. But those restrictions, born from confidence in a paradigm, turn out to be essential to the development of science. By focusing attention upon a small range of relatively esoteric problems, the paradigm forces scientists to investigate some part of nature in a detail and depth that would otherwise be unimaginable.

Perhaps the most striking feature of the normal research problems is how little they aim to produce major novelties, conceptual or phenomenal. Sometimes, as in a wave-length measurement, everything but the most esoteric detail of the result is known in advance, and the typical latitude of expectation is only somewhat wider. Coulomb's measurements need not, perhaps, have fitted an inverse square law; the men who worked on heating by compression were prepared for any one of several results. Yet, even in cases like these the range of anticipated, and thus assimilable, results is always small compared with the range that imagination can conceive. And the project whose outcome does not fall in that narrower range is usually just a research failure, one which reflects not on nature but on the scientist.

But if the aim of normal science is not major substantive novelties—if failure to come near the anticipated result is usually failure as a scientist—then why are these problems undertaken at all? Part of the answer has already been developed. To scientists, at least, the results gained in normal research are significant because they add to the scope and precision with which the paradigm can be applied. That answer, however, cannot account for the enthusiasm and devotion that scientists display for the problems of normal research. No one devotes years to, say, the development of a better spectrometer or the production to an improved solution to the problem of vibrating strings simply because of the importance of the information that will be obtained. The data to be gained by computing ephemerides or by further measurements with an existing instrument are often just as significant, but those activities are regularly spurned by scientists because they are so largely repetitions of procedures that have been carried through before. That rejection provides a clue to the fascination of the normal research problem. Though its outcome can be anticipated, often in detail so great that what remains to be known is itself uninteresting, the way to achieve that outcome remains very much in doubt. Bringing a normal research problem to a conclusion is achieving the anticipated in a new way, and it requires the solution of all sorts of complex instrumental, conceptual, and mathematical puzzles. The man who succeeds proves himself an expert puzzle-solver, and the challenges of the puzzle is an important part of what usually drives him on.

The terms 'puzzle' and 'puzzle-solver' highlight several of the themes that have become increasingly prominent in the preceding pages. Puzzles are, in the entirely standard meaning here employed, that special category of problems that can serve to test ingenuity or skill in solution. Dictionary illustrations are 'jigsaw puzzle' and 'crossword puzzle,' and it is the characteristics that these share with the problems of normal science that we now need to isolate. One of them has just been mentioned. It is no criterion of goodness in a puzzle that its outcome be intrinsically interesting or important. On the contrary, the really pressing problems, e.g., a cure for cancer or design of a lasting peace, are often not puzzles at all, largely because they may not have any solution. Consider the jigsaw puzzle whose pieces are selected at random from each of two different puzzle boxes. Since that problem is likely to defy (though it might not) even the most ingenious of men, it cannot serve as a test of skill in solution. In any usual sense it is not a puzzle at all. Though intrinsic value is no criterion for a puzzle, the assured existence of a solution is.

We have already seen, however, that one of the things a scientific community acquires with a paradigm is a criterion for choosing problems that, while the paradigm is taken for granted, can be assumed to have solutions. To a great extent these are the only problems that the community will admit as scientific or encourage its members to undertake. Other problems, including many that had previously been standard, are rejected as metaphysical, as the concern of another discipline, or sometimes as just too problematic to be worth the time. A paradigm can, for that matter, even insulate the community from those socially important problems that are not reducible to the puzzle form, because they cannot be stated in terms of the conceptual and instrumental tools the paradigm supplies. Such problems can be a distraction, a lesson brilliantly illustrated by several facets of the seventeenth-century Baconianism and by some of the contemporary social sciences. One of the reasons why normal science seems to progress so rapidly is that its practitioners concentrate on problems that only their own lack of ingenuity should keep them from solving.

If, however, the problems of normal science are puzzles in this sense, we need no longer ask why scientists attack them with such passion and devotion. A man may be attracted to science for all sorts of reasons. Among them are the desire to be useful, the excitement of exploring new territory, the hope of finding order, and the drive to test established knowledge. These motives and others besides also help to determine the particular problems that will later engage him. Furthermore, though the result is occasional frustration, there is good reason why motives like these should first attract him and then lead him on.[4] The scientific enterprise as a whole does from time to time prove useful, open up new territory, display order, and test long-accepted beliefs. Nevertheless, *the individual* engaged on a normal research problem *is almost never doing any one of these things.* Once engaged, his motivation is of a rather different sort. What then challenges him is the conviction that, if only he is skillful enough, he will succeed in solving a puzzle that no one before has solved or solved so well. Many of the greatest scientific minds have devoted all of their professional attention to demanding puzzles of this sort. On most occasions any particular field of specialization offers nothing else to do, a fact that makes it no less fascinating to the proper sort of addict.

ANOMALY AND THE EMERGENCE OF CRISIS

Normal science, the puzzle-solving activity we have just examined, is a highly cumulative enterprise, eminently successful in its aim, the steady extension of the scope and precision of scientific knowledge. In all these respects it fits with great precision the most usual image of scientific work. Yet one standard product of the scientific enterprise is missing. Normal science does not aim at novelties of fact or theory and, when successful, finds none. New and unexpected phenomena are, however, repeatedly uncovered by scientific research, and radical new theories have again and again been invented by scientists. History even suggests that the scientific enterprise has developed a uniquely powerful technique for producing surprises of this sort. If this characteristic of science is to be reconciled with what has already been said, then research under a paradigm must be a particularly effective way of inducing paradigm change. That is what fundamental novelties of fact and theory do. Produced inadvertently by a game played under one set of rules, their assimilation requires the elaboration of another set. After they have become parts of science, the enterprise, at least of those specialists in whose particular field the novelties lie, is never quite the same again.

We must now ask how changes of this sort can come about, considering first discoveries, or novelties of fact, and then inventions, or novelty of theory. That distinction between discovery and invention or between fact and theory will, however, immediately prove to be exceedingly artificial. Its artificiality is an important clue to several of this essay's main theses. We shall find quickly that discoveries are not isolated events but extended episodes with a regularly recurrent structure. Discovery commences with the awareness of anomaly, i.e., with the recognition that nature has somehow violated the paradigm-induced expectations that govern normal science. It then continues with a more or less extended exploration of the area of the anomaly. And it closes only when the paradigm theory has been adjusted so that the anomalous has become the expected. Assimilating a new sort of fact demands a more than additive adjustment of theory, and until that adjustment is completed—until the scientist has learned to see nature in a different way—the new fact is not quite a scientific fact at all.

If awareness of anomaly plays a role in the emergence of new sorts of phenomena, it should surprise no one that a similar but more profound awareness is prerequisite to all acceptable changes of theory. On this point, historical evidence is, I think entirely unequivocal. The state of Ptolemaic astronomy was scandalous before Copernicus' announcement.[5] Galileo's contributions to the study of motion depended closely upon difficulties discovered in Aristotle's theory by scholastic critics.[6] Newton's new theory of light and color originated in the discovery that none of the existing pre-paradigm theories would account for the length of the spectrum, and the wave theory that replaced Newton's was announced in the midst of growing concern about anomalies in the relation of diffraction and polarization effects to Newton's theory.[7]

Thermodynamics was born from the collision of two existing nineteenth century physical theories, and quantum mechanics from a variety of difficulties surrounding black-body radiation, specific heats, and the photoelectric effect.[8] Furthermore, in all these cases except that of Newton, the awareness of anomaly had lasted so long and penetrated so deep that one can appropriately describe the fields affected by it as in a state of growing crisis. Because it demands large-scale paradigm destruction and major shifts in the problems and techniques of normal science, the emergence of new theories is generally preceded by a period of pronounced professional insecurity. As one might expect, that insecurity is generated by the persistent failure of the puzzles of normal science to come out as they should. Failure of existing rules is the prelude to a search for new ones.

Look at a particularly famous case of paradigm change, the emergence of Copernican astronomy. When its predecessor, the Ptolemaic system, was first developed during the last two centuries before Christ and the first two after, it was admirably successful in predicting the changing positions both of stars and planets. No other ancient system had performed so well; for the stars Ptolemaic astronomy is still widely used today as an engineering approximation; for the planets, Ptolemy's predictions were as good as Copernicus'. But to be admirably successful is never, for a scientific theory, to be completely successful. With respect to both planetary position and to precession of the equinoxes, predictions made with Ptolemy's system never quite conformed with the best possible observations. Further reduction of those minor discrepancies constituted many of the principle problems of normal astronomical research for many of Ptolemy's successors, just as a similar attempt to bring celestial observation and Newtonian theory together provided normal research problems for Newton's eighteenth century successors. For some time astronomers had every reason to suppose that these attempts would be as successful as those that had led to Ptolemy's system. Given a particular discrepancy, astronomers were invariably able to eliminate it by making some particular adjustments in Ptolemy's system of compounded circles. But as time went on, a man looking at the net result of the normal research effort of many astronomers could observe that astronomy's complexity was increasing far more rapidly than its accuracy and that a discrepancy corrected in one place was likely to show up in another.[9] Because the astronomical tradition was repeatedly interrupted from outside and because, in the absence of printing, communication between astronomers was restricted, these difficulties were only slowly recognized. But awareness did come. By the thirteenth century Alfonso X could proclaim that if God had consulted him when creating the universe, he would have received good advice. In the sixteenth century, Copernicus' co-worker, Domenico da Novara, held that no system so cumbersome and inaccurate as the Ptolemaic had become could possibly be true of nature. And Copernicus himself wrote in the Preface to the *De Revolutionibus* that the astronomical tradition he inherited had finally created only a monster. By the early sixteenth century an increasing number of Europe's best astronomers were recognizing that the astronomical paradigm was failing in application to its own traditional problems. That recognition was prerequisite to Copernicus' rejection of the Ptolemaic paradigm and his search for a new one. His famous preface still provides one of the classic descriptions of a crisis state.[10]

THE RESPONSE TO CRISIS

Let us then assume that crises are a necessary precondition for the emergence of novel theories and ask next how scientists respond to their existence. Part of the answer, as obvious as it is important, can be discovered by noting first what scientists never do when confronted by even severe and prolonged anomalies. Though they may begin to lose faith and then to consider alternatives, they do not renounce the paradigm that has led them into crisis. They do not, that is, treat anomalies as counterinstances, though in the vocabulary of philosophy of science that is what they are. In part this generalization is simply a statement from historical fact, based upon examples like those given above. These hint what our later examination of paradigm rejection will disclose more fully: once it has achieved the status of paradigm, a scientific theory is declared invalid only if an alternate candidate is available to take its place. No process yet disclosed by the historical study of scientific development at all resembles the methodological stereotype of falsification by direct comparison with nature. That remark does not mean that scientists do not reject scientific theories, or that experience and experiment are not essential to the process in which they do so. But it does mean—what will ultimately be a central point—that the act of judgment that leads scientists to reject a previously accepted theory is always based upon more than a comparison of that theory with the world. The decision to reject one paradigm is always simultaneously the decision to accept another, and the judgment leading to that decision involves the comparison of both paradigms with nature *and* with each other.

There is, in addition, a second reason for doubting that scientists reject paradigms when confronted with anomalies or counterinstances. In developing it my argument will itself foreshadow another of this essay's main theses. The reasons for doubt sketched above were purely factual; they were, that is, themselves counterinstances to a prevalent epistemological theory. As such, if my present point is correct, they can at best help to create a crisis, or more accurately, to reinforce one that is already very much in existence. By themselves they cannot and will not falsify that philosophical theory, for its defenders, when confronted with anomaly, will devise numerous articulations and *ad hoc* modifications of their theory in order to eliminate any apparent conflict. Many of the relevant modifications and qualifications are, in fact, already in the literature. If, therefore, these epistemological counterinstances are to constitute more than a minor irritant, that will be because they help to permit the emergence of a new and different analysis of science within which they are no longer a source of trouble. Furthermore, if a typical pattern, which we shall later observe in scientific revolutions, is applicable here, these anomalies will then no longer seem to be simply facts. From within a new theory of scientific knowledge, they may instead seem very much like tautologies, statements of situations that could not conceivably have been otherwise.

It has often been observed, for example, that Newton's second law of motion, though it took centuries of difficult factual and theoretical research to achieve, behaves for those committed to Newton's theory very much like a purely logical statement that no amount of observation could refute.[11]

Though history is unlikely to record their names, some men have undoubtedly been driven to desert science because of their inability to tolerate crisis. Like artists, creative scientists must occasionally be able to live in a world out of joint. But that rejection of science in favor of another occupation is, I think, the only sort of paradigm rejection to which counterinstances by themselves can lead. Once a first paradigm through which to view nature has been found, there is no such thing as research in the absence of any paradigm. To reject one paradigm without simultaneously substituting another is to reject science itself. That act reflects not on the paradigm but on the man. Inevitably he will be seen by his colleagues as "the carpenter who blames his tools."

How, then, do scientists respond to the awareness of an anomaly in the fit between theory and nature? What has just been said indicates that even a discrepancy unaccountably larger than that experienced in other applications of the theory need not draw any very profound response. There are always some discrepancies. Even the most stubborn ones usually respond at last to normal practice. Very often, scientists are willing to wait, particularly if there are many problems available in other parts of the field. We have already noted, for example, that during the sixty years after Newton's original computation, the predicted motion of the moon's perigee remained only half of that observed. As Europe's best mathematical physicists continued to wrestle unsuccessfully with the well-known discrepancy, there were occasional proposals for a modification of Newton's inverse square law. But no one took these proposals very seriously, and in practice this patience with a major anomaly proved justified. Clairaut in 1750 was able to show that only the mathematics of the application had been wrong and that Newtonian theory could stand as before.[12] Even in cases where no mere mistake seems quite possible (perhaps because the mathematics involved is simpler or of a familiar and elsewhere successful sort), persistent and recognized anomaly does not always induce crisis. No one seriously questioned Newtonian theory because of the long-recognized discrepancies between predictions from that theory and both the speed of sound and the motion of Mercury. The first discrepancy was ultimately and quite unexpectedly resolved by experiments on heat undertaken for a very different purpose; the second vanished with the general theory of relativity after a crisis that it had no role in creating.[13] Apparently neither had seemed sufficiently fundamental to evoke the malaise that goes with crisis. They could be recognized as counterinstances and still be set aside for later work.

It follows that if an anomaly is to evoke crisis, it must usually be more than just an anomaly. There are always difficulties somewhere in the paradigm-nature fit; most of them are set right sooner or later, often by processes that could not have been foreseen. The scientist who pauses to examine every anomaly he notes will seldom get significant work done. We therefore have to ask what it is that makes an anomaly seem worth concerted scrutiny, and to that question there is probably no fully general answer. Sometimes an anomaly will clearly call into question explicit and fundamental generalizations of the paradigm, as the problem of ether drag did for those who accepted Maxwell's theory. Or, as in the Copernican revolution, an anomaly without apparent fundamental import may evoke crisis if the applications that it inhibits have a particular practical importance, in this case for calendar design and astrology. Or, as in eighteenth century chemistry, the development of normal science may transform an anomaly that had previously been only a vexation into a source of crisis: the problem of weight relations had a very different status after the evolution of pneumatic-chemical techniques.

When for these reasons or others like them, an anomaly comes to seem more than just another puzzle of normal science, the transition to crisis and to extraordinary science has begun. The anomaly itself now comes to be more generally recognized as such by the profession. More and more attention is devoted to it by more and more of the field's most eminent men. If it still continues to resist, as it usually does not, many of them may come to view its resolution as *the* subject matter of their discipline. For them the field will no longer look quite the same as it had earlier. Part of its difference in appearance results simply from the new fixation point of scientific scrutiny. An even more important source of change is the divergent nature of the numerous partial solutions that concerted attention to the problem has made available. The early attacks upon the resistant problem will have followed the paradigm rules quite closely. But with continuing resistance, more and more of the attacks upon it will have involved some minor or not so minor articulation of the paradigm, no two of them being quite alike, each partially successful, but none sufficiently so to be accepted as paradigm by the group. Through this proliferation of divergent articulations (more and more frequently they will come to be described as *ad hoc* adjustments), the rules of normal science become increasingly blurred. Though there still is a paradigm, few practitioners prove to be entirely agreed about what it is. Even formerly standard solutions of solved problems are called into question.

It is, I think, particularly in periods of acknowledged crisis that scientists have turned to philosophical analysis as a device for unlocking the riddles of their field. Scientists have not generally needed or wanted to be philosophers. Indeed, normal science usually holds creative philosophy at arm's length, and probably for good reason. To the extent that normal research work can be conducted by using the paradigm as a model, rules and assumptions need not be made explicit. The full set of rules sought by philosophical analysis need not even exist.

THE NATURE AND NECESSITY OF SCIENTIFIC REVOLUTIONS

What are scientific revolutions, and what is their function in scientific development? Why should a change of paradigm be called a revolution? In the face of the vast and essential differences between political and scientific development, what parallelism can justify the metaphor that finds revolutions in both?

One aspect of the parallelism must already be apparent. Political revolutions are inaugurated by a growing sense, often restricted to a segment of the political community, that existing institutions have ceased adequately to meet the problems posed by an environment that they have in part created. In much the same way, scientific revolutions are inaugurated by a growing sense, again often restricted to a narrow subdivision of the scientific community, that an existing paradigm has ceased to function adequately in the exploration of an aspect of nature to which the paradigm itself had previously led the way. In both political and scientific development the sense of malfunction that can lead to crisis is prerequisite to revolution.

Political revolutions aim to change political institutions in ways that those institutions themselves prohibit. Their success therefore necessitates the partial relinquishment of one set of institutions in favor of another, and in the interim, society is not fully governed by institutions at all. Initially it is crisis alone that attenuates the role of political institutions as we have already seen it attenuate the role of paradigms. In increasing numbers individuals become increasingly estranged from political life and behave more and more eccentrically within it. Then, as the crisis deepens, many of these individuals commit themselves to some concrete proposal for the reconstruction of society in a new institutional framework. At that point the society is divided into competing camps or parties, one seeking to defend the old institutional constellation, the other seeking to institute some new one. And, once that polarization has occurred, *political recourse fails*. Because they differ about the institutional matrix within which political change is to be achieved and evaluated, because they acknowledge no supra-institutional framework for the adjudication of revolutionary difference, the parties to a revolutionary conflict must finally resort to the techniques of mass persuasion, often including force. Though revolutions have had a vital role in the evolution of political institutions, that role depends upon their being partially extrapolitical or extrainstitutional events.

The historical study of paradigm change reveals every similar characteristics in the evolution of the sciences. Like the choice between competing political institutions, that between competing paradigms proves to be a choice between incompatible modes of community life. Because it has that character, the choice is not and cannot be determined merely by the evaluative procedures characteristic of normal science, for these depend in part upon a particular paradigm, and that paradigm is at issue. When paradigms enter, as they must, into a debate about paradigm choice, their role is necessarily circular. Each group uses its own paradigm to argue in that paradigm's defense.

The resulting circularity does not, of course, make the arguments wrong or even ineffectual. The man who premises a paradigm when arguing in its defense can nonetheless provide a clear exhibit of what scientific practice will be like for those who adopt the new view of nature. That exhibit can be immensely persuasive, often compellingly so. Yet, whatever its force, the status of the circular argument is only that of persuasion. It cannot be made logically or even probabilistically compelling for those who refuse to step into the circle. The premises and values shared by the two parties to a debate over paradigms are not sufficiently extensive for that. As in political revolutions, so in paradigm choice—there is no standard higher than the assent of the relevant community. To discover how scientific revolutions are effected, we shall therefore have to examine not only the impact of nature and of logic, but also the techniques of persuasive argumentation effective within the quite special groups that constitute the community of scientists.

Granting that paradigm rejection has been a historic fact, does it illuminate more than human credulity and confusion? Are there intrinsic reasons why the assimilation of either a new sort of phenomenon or a new scientific theory must demand the rejection of an older paradigm?

THE RESOLUTION OF REVOLUTIONS

Paradigm debates are not really about relative problem solving ability, though for good reason they are usually couched in those terms. Instead, the issue is which paradigm should in the future guide research on problems many of which neither competitor can yet claim to resolve completely. A decision between alternative ways of practicing science is called for. And in the circumstances that decision must be based less on past achievement than on future promise. The man who embraces a new paradigm at an early stage must often do so in defiance of the evidence provided by problem-solving. He must, that is, have faith that the new paradigm will succeed with the many large problems that confront it, knowing only that the older paradigm has failed with a few. A decision of that kind can only be made on faith.

That is one of the reasons that prior crisis proves so important. Scientists who have not experienced it will seldom renounce the hard evidence of problem-solving to follow what may easily prove and will be widely regarded as a will-o'-the-wisp. But crisis alone is not enough. There must also be a basis, though it need be neither rational nor ultimately correct, for faith in the particular candidate chosen. Something must make at least a few scientists feel that the new proposal is on the right track, and sometimes it is only personal and inarticulate aesthetic consideration that can do that. Men have been converted by them at times when most of the articulable technical arguments pointed the other way. When first introduced, neither Copernicus' astronomical theory nor De Broglie's theory of matter had many other significant grounds of appeal. Even today, Einstein's general theory attracts men principally on aesthetic grounds, an appeal that few people outside of mathematics have been able to feel.

This is not to suggest that new paradigms triumph ultimately through some mystical aesthetic. On the contrary, very few men desert a tradition for these reasons alone. Often those who turn out to have been misled. But if a paradigm is ever to triumph it must gain some first supporters, men who will develop it to then point where hardheaded arguments can be produced and multiplied. And even those arguments, when they come, are not individually decisive. Because scientists are reasonable men, one or another argument will ultimately persuade many of them. But there is no single argument that can or should persuade them all. Rather than a single group conversion, what occurs is an increasing shift in the distribution of professional allegiances.

At the start a new candidate for paradigm may have few supporters, and on occasions the supporters' motives may be suspect. Nevertheless, if they are competent, they will improve it, explore its possibilities, and show what it would be like to belong to the community guided by it. And as that goes on, if the paradigm is one destined to win its fight, the number and strength of the persuasive arguments in its favor will increase. More scientists will be converted, and the exploration of the new paradigm will go on. Gradually, the number of experiments, instruments, articles, and books based upon the paradigm will multiply. Still more men, convinced of the new view's fruitfulness, will adopt the new mode of practicing normal science, until at last only a few elderly hold-outs remain. And even they, we cannot say, are wrong. Though the historians can always find men—Priestly, for instance—who were unreasonable to resist for as long as they did, he will not find a point at which resistance becomes illogical or unscientific. At most he may wish to say that the man who continues to resist after his whole profession has been converted has *ipso facto* ceased to be a scientist.

NOTES

1. Joseph Priestly, *The History and Present State of Discoveries Relating to Vision, Light, and Colours* (London, 1772), 385–90.

2. Vasco Ronchi, *Histoire de la lumière*, trans. by Jean Taton (Paris, 1956), chaps. i–iv.

3. Bernard Barber, "Resistance by Scientists to Scientific Discovery," *Science* CXXXIV (1961): 596–602.

4. The frustrations induced by the conflict between the individual's role and the over-all pattern of scientific development can, however, occasionally be quite serious. On this subject, see Lawrence S. Kubie, "Some Unsolved Problems of the Scientific Career," *American Scientist* XLII (1954): 104–12.

5. A. R. Hall, *The Scientific Revolution, 1500–1800* (London, 1954), 16.

6. Marschall Claggett, *The Science of Mechanics in the Middle Ages* (Madison, Wis., 1959), parts II–III. A. Koyré displays a number of medieval elements in Galileo's thought in his *Etudes Galiléennes* (Paris 1931), particularly vol. I.

7. For Newton, see T. S. Kuhn, "Newton's Optical Papers," in *Isaac Newton's Papers and Letters in Natural Philosophy*, ed. I. B. Cohen (Cambridge, Mass., 1958), 27–45. For the prelude to the wave theory, See E. T. Whittaker, *A History of the Theories of Aether and Electricity*, I (2d ed.; London, 1951), 94–109; and W. Whewell, *History of the Inductive Sciences* (rev. ed.; London, 1847), II, 396–466.

8. For thermodynamics, see Silvanus P. Thompson, *The Life of William Thomson: Baron Kelvin of Largs* (London, 1910), I: 266–81. For the quantum theory, see Fritz Reiche, *The Quantum Theory*, trans. by H. S. Hatfield and H. L. Brose (London, 1922), chaps. i–ii.

9. J. L. E. Dreyer, *A History of Astronomy from Thales to Kepler* (2nd ed.; New York, 1953), chaps. xi–xii.

10. T. S. Kuhn, *The Copernican Revolution* (Cambridge, Mass., 1957), 135–43.

11. See particularly the discussion in N. R. Hanson, *Patterns of Discovery* (Cambridge, 1958), 99–105.

12. W. Whewell, *History of the Inductive Sciences.* (rev. ed.; London, 1947), II, 220–21.

13. For the speed of sound, see T. S. Kuhn, "The Caloric Theory of Adiabatic Compression," *Isis* XLIV (1958). For the secular shift in Mercury's perihelion, see E. T. Whittaker, *A History of the Theories of Aether and Electricity*, II: 151, 179.

Imre Lakatos

. . . .

The Methodology of Scientific Research Programmes
Falsification and the Methodology of
Scientific Research Programmes

In this paper I shall first show that in Popper's logic of scientific discovery two different positions are conflated. Kuhn understands only one of these, 'naïve falsificationism' (I prefer the term 'naïve methodological falsification-ism'); I think that his criticism of it is correct, and I shall even strengthen it. But Kuhn does not understand a more sophisticated position the rationality of which is not based on 'naïve' falsificationism. I shall try to explain—and further strengthen—this stronger Popperian position which, I think, may escape Kuhn's strictures and present scientific revolutions not as constituting religious conversions but rather as rational progress.

FALLIBILISM VERSUS FALSIFICATIONISM

To see the conflicting theses more clearly, we have to reconstruct the situation as it was in philosophy of science after the breakdown of 'justificationism.'

According to the 'justificationists' scientific knowledge consisted of proven propositions. Having recognized that strictly logical deductions enable us only to infer (transmit truth) but not to prove (establish truth), they disagreed about the nature of those propositions (axioms) whose truth can be proved by extralogical means. *Classical intellectualists* (or 'rationalists' in the narrow sense of the term) admitted very varied—and powerful—sorts of extralogical 'proofs' by revelation, intellectual intuition, experience. These, with the help of logic, enabled them to prove every sort of scientific proposition. *Classical empiricists* accepted as axioms only a relatively small set of 'factual propositions' which expressed the 'hard facts.' Their truth value was established by experience and they constituted the *empirical basis* of science. In order to prove scientific *theories* from nothing else but the narrow empirical basis, they needed a logic much more powerful than the deductive logic of the classical intellectualists:

From Imre Lakatos, *The Methodology of Scientific Research Programmes* (Cambridge: Cambridge University Press, 1978), 10–14, 16–18, 31–38, 47–52, 86, 90–93. Reprinted with the permission of Cambridge University Press.

'*inductive logic.*' All justificationists, whether intellectualists or empiricists, agreed that a singular statement expressing a 'hard fact' may *disprove* a universal theory; but few of them thought that a finite conjunction of factual propositions might be sufficient to *prove* 'inductively' a universal theory.

Justificationism, that is, the identification of knowledge with proven knowledge, was the dominant tradition in rational thought throughout the ages. Scepticism did not deny justificationism: it only claimed that there was (and could be) no proven knowledge and *therefore* no knowledge whatsoever. For the sceptics 'knowledge' was nothing but animal belief. Thus justificationist skepticism ridiculed objective thought and opened the door to irrationalism, mysticism, superstition.

This situation explains the enormous effort invested by classical rationalists in trying to save the synthetic *a priori* principles of intellectualism and by classical empiricists in trying to save the certainty of an empirical basis and the validity of inductive inference. For all of them *scientific honesty demanded that one assert nothing that is unproven.* However, both were defeated: Kantians by non-Euclidean geometry and by non-Newtonian physics, and empiricists by the logical impossibility of establishing an empirical basis (as Kantians pointed out, facts cannot prove propositions) and of establishing an inductive logic (no logic can infallibly increase content). It turned out that *all theories are equally unprovable.*

Philosophers were slow to recognize this, for obvious reasons: classical justificationists feared that once they conceded that theoretical science is unprovable, they would have also to conclude that it is sophistry and illusion, a dishonest fraud. The philosophical importance of *probabilism* (or '*neojustificationism*') lies in the denial that such a conclusion is necessary.

Probabilism was elaborated by a group of Cambridge philosophers who thought that although scientific theories are equally unprovable, they have different degrees of probability (in the sense of the calculus of probability) relative to the available empirical evidence. *Scientific honesty then requires less than had been thought: it consists in uttering only highly probable theories; or in merely specifying, for each scientific theory, the evidence, and the probability of the theory in light of this evidence.*

Of course, replacing proof by probability was a major retreat for justificationist thought. But even this retreat turned out to be insufficient. It was soon shown, mainly by Popper's persistent efforts, that under very general conditions all theories have zero probability, whatever the evidence; *all theories are not only equally unprovable but also equally improbable.*

Many philosophers still argue that the failure to obtain at least a probabilistic solution of the problem of induction means that we 'throw over almost everything that is regarded as knowledge by science and common sense.'[1] It is against this background that one must appreciate the dramatic change brought about by falsificationism in evaluating theories, and in general, in the standards of intellectual honesty. Falsificationism was, in a sense, a new and considerable retreat for rational thought. But since it was a retreat from a utopian standard, it cleared away much hypocrisy and muddled thought, and thus, in fact, it represents an advance.

DOGMATIC (OR NATURALISTIC) FALSIFICATIONISM.
THE EMPIRICAL BASIS.

First I shall discuss a most important brand of falsificationism: dogmatic (or 'naturalistic') falsificationism. Dogmatic falsificationism admits the fallibility of *all* scientific theories, but it retains a sort of infallible empirical basis. It is strictly empiricist without being inductivist: it denies that the certainty of the empirical basis can be transmitted to theories. *Thus dogmatic falsificationism is the weakest brand of falsificationism* . . .

The hallmark of dogmatic falsificationism is the recognition that all theories are equally conjectural. Science cannot *prove*, it can only *disprove*: it 'can perform with complete logical certainty [the act of] repudiation of what is false,'[2] that is, there is an absolutely firm empirical basis of facts which can be used to disprove theories. Falsificationists provide new—very modest—standards of scientific honesty: they are willing to regard a proposition as 'scientific' not only if it is a proven factual proposition, but even if it is nothing more than a falsifiable one, that is, if there are experimental and mathematical techniques available at the time which designate certain statements as potential falsifiers.[3]

Scientific honesty then consists of specifying, in advance, an experiment such that if the result contradicts the theory, the theory has to be given up.[4] The falsificationist demands that once a proposition is disproved, there must be no prevarication: the proposition must be unconditionally rejected. To (non-tautologous) unfalsifiable propositions the dogmatic falsificationist gives short shrift; he brands them 'metaphysical' and denies them scientific standing.

Dogmatic falsificationists draw a sharp demarcation between the theoretician and the experimenter: the theoretician proposes, the experimenter—in the name of Nature—disposes. As Weyl put it: 'I wish to record my unbounded admiration for the work of the experimenter in his struggle to wrest interpretable facts from an unyielding Nature who knows so well how to meet our theories with a decisive *No*—or with an inaudible *Yes*.'[5]

According to the logic of dogmatic falsificationism, science grows by repeated overthrow of theories with the help of hard facts. For instance, according to this view, Descartes's vortex theory of gravity was refuted—and eliminated—by the *fact* that planets move in ellipses rather than in Cartesian circles; Newton's theory, however, explained successfully the then available facts, both those which had been explained by Descartes's theory and those which refuted it. Therefore Newton's theory replaced Descartes's theory. Analogously, as seen by falsificationists, Newton's theory was, in turn, refuted—proved false—by the anomalous perihelion of Mercury, while Einstein's explained that too. Thus science proceeds by bold speculations, which are never proved or even made probable, but some of which are later eliminated by hard, conclusive refutations and then replaced by still bolder, new and, at least at the start, un-refuted speculations.

Dogmatic falsificationism, however is untenable ... [because it is] useless for eliminating the most important class of what are commonly regarded as scientific theories. For even if experiments *could* prove experimental reports, their disproving power would still be miserably restricted: *exactly the most admired scientific theories simply fail to forbid any observable state of affairs.*

To support this contention, I shall first tell a characteristic story and then propose a general argument.

The story is about an imaginary case of planetary misbehavior. A physicist of the pre-Einsteinian era takes Newton's mechanics and his law of gravitation, (N), the accepted initial conditions, I, and calculates, with their help, the path of a newly discovered small planet, p. But the planet deviates from the calculated path. Does our Newtonian physicist consider that the derivation was forbidden by Newton's theory and therefore that, once established, it refutes the theory N? No. He suggests that there must be a hitherto unknown planet p' which perturbs the path of p. He calculates the mass orbit, etc., of this hypothetical planet and then asks an experimental astronomer to test his hypothesis. The planet p' is so small even the biggest available telescopes cannot possibly observe it: the experimental astronomer applies for a research grant to build yet a bigger one.[6] In three years' time, the new telescope is ready. Were the unknown planet p' to be discovered, it would be hailed as a new victory of Newtonian science. But it is not. Does our scientists abandon Newton's theory and his idea of the perturbing planet? No. He suggests a cloud of cosmic dust hides the planet from us. He calculates the location and properties of this cloud and asks for a research grant to send up a satellite to test his calculations. Were the satellite's instruments (possibly new ones, based on a little-tested theory) to record the existence of the con-

jectural cloud, the result would be hailed as an outstanding victory for Newtonian science. But the cloud is not found. Does our scientist abandon Newton's theory, together with the idea of the perturbing planet and the idea of a
cloud that hides it? No. He suggests that there is some magnetic field in that
region of the universe which disturbed the instruments of the satellite. A new
satellite is sent up. Were the magnetic field to be found, Newtonians would
celebrate a sensational victory. But it is not. Is this regarded as a refutation of
Newtonian science? No. Either yet another ingenious auxiliary hypothesis is
proposed or . . . the whole story is buried in the dusty volumes of periodicals
and the story never mentioned again.[7]

This story strongly suggests that even a most respected scientific theory,
like Newton's dynamics and theory of gravitation, may fail to forbid any observable state of affairs.[8] Indeed, *some specified finite spatio-temporal region (or
briefly, a 'singular event') only on the condition that no other factor* (possibly hidden in some distant and unspecified spatio-temporal corner of the universe)
has any influence on it. But then *such theories never alone contradict a 'basic'
statement:* they contradict at most a conjunction of a basic statement describing a spatio-temporally singular event and of a universal non-existence statement saying that no other relevant cause is at work anywhere in the universe.
And the dogmatic falsificationists cannot possibly claim that such universal
non-existence statements belong to the empirical basis: that they can be observed and proved by experience.

SOPHISTICATED VERSUS NAÏVE METHODOLOGICAL FALSIFICATIONISM. PROGRESSIVE AND DEGENERATING PROBLEMSHIFTS.

Sophisticated falsificationism differs from naïve falsificationism both in its
rules of *acceptance* (or 'demarcation criterion') and its rules of *falsification* or
elimination.

For the naïve falsificationist any theory which can be interpreted as experimentally falsifiable, is 'acceptable' or 'scientific.' For the sophisticated falsificationist a theory is 'acceptable' or 'scientific' only if it has corroborated excess
empirical content over its predecessor (or rival), that is, only if it leads to the
discovery of novel facts. This condition can be analysed into two clauses: that
the new theory has excess empirical content ('*acceptability*$_1$') and that some of
this excess content is verified ('*acceptability*$_2$'). The first clause can be checked
instantly by *a priori* logical analysis; the second can be checked only empirically and this may take an indefinite time.

For the naïve falsificationist a theory is *falsified* by an observational state-ment which conflicts with it (or which he decides to interpret as conflicting with it). For the sophisticated falsificationist a scientific theory T is *falsified* if and only if another theory T' has been proposed with the following charac-teristics: (1) T' has excess empirical content over T: that is: it predicts *novel* facts, that is facts improbable in the light of, or even forbidden by, T[9]; (2) T' explains the previous success of T, that is, all the unrefuted content of T is in-cluded (within the limits of observational error) in the content of T'; and (3) some of the excess content of T' is corroborated.

In order to be able to appraise these definitions we need to understand their problem background and their consequences. First, we have to remember the conventionalists' methodological discovery that no experimental result can ever kill a theory: nay, theory can be saved from counterinstances either by some auxiliary hypothesis or by a suitable reinterpretation of its terms. Naïve falsificationists solved this problem by relegating—in crucial contexts—the auxiliary hypotheses to the realm of unproblematic background knowledge, eliminating them from the deductive model of the test-situation and thereby *forcing* the chosen theory into logical isolation, in which it becomes a sitting target for the attack of test-experiments. But since this procedure did not offer a suitable guide for a rational reconstruction for the history of science, we may just as well completely rethink our approach. Why aim at falsification at any price? Why not rather impose certain standards on the theoretical adjust-ments by which one is allowed to save a theory? Indeed, some such standards have been well-known for centuries, and we find them expressed in age-old wisecracks against *ad hoc* explanations, empty prevarications, face-saving, lin-guistic tricks.[10] We have already seen that Duhem adumbrated such standards in terms of 'simplicity' and 'good sense.' But *when* does lack of 'simplicity' in the protective belt of theoretical adjustments reach the point at which the theory *must* be abandoned?[11] In what sense was the Copernican theory, for instance, 'simpler' than the Ptolemaic?[12] The vague notion of Duhemian 'sim-plicity' leaves, as the naïve falsificationist correctly argued, the decision very much to taste and fashion.

Can one improve on Duhem's approach? Popper did. His solution—a so-phisticated version of methodological falsificationism—is more objective and more rigorous. Popper agrees with the conventionalists that theories and factual propositions can always be harmonized with the help of auxiliary hy-potheses which satisfy certain well-defined conditions represents scientific progress; but saving a theory with the help of auxiliary hypotheses which do not, represents degeneration. Popper calls such inadmissible auxiliary hypotheses *ad hoc* hypotheses, mere linguistic devices, 'conventionalist stratagems.' But then any theory has to be appraised together with its auxiliary hypotheses, initial conditions, etc., and, especially, together with its predecessors so that we may see by what sort of *change* it was brought about. Then, of course, what we appraise is a *series of theories* rather than isolated *theories*.

Now we can easily understand why we formulated the criteria of acceptance and rejection of sophisticated methodological falsificationism as we did. But it may be worthwhile to reformulate them slightly, couching them explicitly in terms of *series of theories*.

Let us take a series of theories, $T_1, T_2, T_3, \ldots$ where each subsequent theory results from adding auxiliary clauses to (or from semantical reinterpretations of) the previous theory in order to accommodate some anomaly, each theory have at least as much content as the unrefuted content of its predecessor. Let us say that such a series of theories is *theoretically progressive* (or '*constitutes a theoretically progressive problemshift*') if each new theory has some excess empirical content over its predecessor, that is if it predicts some novel, hitherto unexpected fact. Let us say that a theoretically progressive series of theories is also *empirically progressive* (or '*constitutes an empirically progressive problemshift*') if some of this excess empirical content is also corroborated, that is, if each new theory leads us to the actual discovery of some *new* fact.[13] Finally, let us call a problemshift *progressive* if it is both theoretically and empirically progressive, and *degenerating* if it is not. We '*accept*' problemshifts as 'scientific' only if they are at least theoretically progressive; if they are not, we '*reject*' them as 'pseudoscientific.' Progress is measured by the degree to which a problem shift is progressive, by the degree to which the series of theories leads us to the discovery of novel facts. We regard a theory in a series 'falsified' when it is superceded by a theory with a higher corroborated content.

This demarcation between progressive and degenerating problemshifts sheds new light on the appraisal of *scientific—or, rather, progressive—explanations*. If we put forward a theory to resolve a contradiction between a previous theory and a counterexample in such a way that the new theory, instead of offering a content-increasing (scientific) *explanation*, only offers a content-decreasing (linguistic) *reinterpretation*, the contradiction is resolved in a merely semantical, unscientific way. *A given fact is explained scientifically only if a new fact is also explained with it.*

Sophisticated falsificationism thus shifts the problem of how to appraise *theories* to the problem of how to appraise *series of theories*. Not an isolated the-

ory, but only a series of theories can be said to be scientific or unsc
apply the term 'scientific' to one *single* theory is a category mistak

The time honored empirical criterion for a satisfactory theor
ment with the observed facts. Our empirical criterion for a serie
is that it should produce new facts. *The idea of growth and the concept of empir.*
cal character are soldered into one.

This revised form of methodological falsificationism has many new features. First, it denies that 'in the case of a scientific theory, our decision depends upon the results of experiments. If these confirm the theory, we may
accept it until we find a better one. If they contradict the theory, we reject it.'[14]
It denies that 'what ultimately decides the fate of a theory is the result of a test,
i.e., an agreement about basic sentences.'[15] Contrary to naïve falsificationism,
no experiment, experimental report, observation statement or well-corroborated
low-level falsifying hypothesis alone can lead to falsification. There is no falsifica-
tion before the emergence of a better theory. But then the distinctly negative character of naïve falsificationism vanishes; criticism becomes more difficult, and
also positive, constructive. But, of course, if falsificationism depends on the
emergence of better theories, on the invention of theories which anticipate
new facts, then falsificationism is *not* simply a relation between a theory and
the empirical basis, but a multiple relation between competing theories, the
original 'empirical basis,' and the empirical growth resulting from the competition. Falsification can thus be said to have a *'historical character.'* Moreover,
some of the theories which bring about falsification are frequently proposed
after the 'counterevidence.' This may sound paradoxical for some people in-
doctrinated with naïve falsificationism. Indeed, this epistemological theory
of the relation between theory and experiment differs sharply from the epistemological theory of naïve falsificationism. The very term 'counterevidence'
has to be abandoned in the same sense that no experimental result must be
interpreted directly as 'counterevidence.' If we still want to retain the time-
honoured term, we have to redefine it like this: 'counterevidence to T_1' is a
corroborating instance to T_2 which is either inconsistent with or independent
of T_1 (with the *proviso* that T_2 is a theory which satisfactorily explains the empirical success of T_1). This shows that *'crucial counterevidence'*—or *'crucial ex-*
periments'—can be recognized as such among the scores of anomalies only
with hindsight, in the light of some superceding theory.'[16]

Thus the crucial element in falsification is whether the *new theory* offers any
novel, excess information compared with its predecessor and whether some
of this excess information is corroborated. Justificationists valued 'confirm-

.ng' instances of a theory; naïve falsificationists stressed 'refuting' instances; for the methodological falsificationist it is the—rather rare—corroborating instances of the *excess* information which are the crucial ones; these receive all the attention. We are no longer interested in the thousands of trivial verifying instances nor in the hundreds of readily available anomalies: the few crucial *excess-verifying instances* are decisive. This consideration rehabilitates—and reinterprets—the old proverb: *Exemplum docet, exempla obscurant.*

'Falsification' in the sense of naïve falsificationism (corroborated counter-evidence) is not a *sufficient* condition for eliminating a specific theory: in spite of hundreds of known anomalies we do not regard it as falsified (that is, eliminated) until we have a better one.[17] Nor is 'falsification' in the naïve sense *necessary* for falsification in the sophisticated sense: a progressive problemshift does not have to be interspersed with 'refutations.' Science can grow without any 'refutations' leading the way. Naïve falsificationists suggest a linear growth of science, in the sense that theories are followed by powerful refutations which eliminate them; these refutations in turn are followed by new theories.[18] It is perfectly *possible* that theories be put forward 'progressively' in such a rapid succession that the 'refutation' of the *n*th appears only as the corroboration of the $(n\text{th} + 1)$th. The problem fever of science is raised by proliferation of rival theories rather than counterexamples or anomalies.

This shows that the slogan of *proliferation of theories* is much more important for sophisticated than for naïve falsificationism. For the naïve falsificationist science grows through repeated experimental overthrow of theories; new rival theories proposed before such 'overthrows' may speed up growth but are not absolutely necessary; constant proliferation of theories is optional but not mandatory. For the sophisticated falsificationist proliferation of theories cannot wait until the accepted theories are 'refuted' (or until their proponents get into a Kuhnian crisis of confidence). While naïve falsificationism stresses 'the urgency of replacing a *falsified* hypothesis by a better one,' sophisticated falsificationism stresses the urgency of replacing *any* hypothesis by a better. Falsification cannot, 'compel the theorist to search for a better theory,'[19] simply because falsification cannot precede the better theory.

A METHODOLOGY OF SCIENTIFIC RESEARCH PROGRAMMES

I have discussed the problem of objective appraisal of scientific growth in terms of progressive and degenerating problemshifts in series of scientific theories. The most important such series in the growth of science are characterized by a certain *continuity* which connects their members. This continuity evolves from a genuine research programme adumbrated at the start. The programme consists of methodological rules: some tell us what paths to avoid (*negative heuristic*), and others what paths to pursue (*positive heuristic*).

Even science as a whole can be regarded as a huge research programme with Popper's supreme heuristic rule: 'devise conjectures which have more empirical content than their predecessors.' Such methodological rules may be formulated, as Popper pointed out, as metaphysical principles.[20]

But what I have in mind is not science as a whole, but rather *particular* research programmes, such as the one known as 'Cartesian metaphysics.' Cartesian metaphysics, that is, the mechanistic theory of the universe—according to which the universe is a huge clockwork (and system of vortices) with push as the only cause of motion—functioned as a powerful heuristic principle. It discouraged work on scientific theories—like (the 'essentialist' version of) Newton's theory of action at a distance—which were inconsistent with it (*negative heuristic*). On the other hand, it encouraged work on auxiliary hypotheses which might have saved it from apparent counterevidence—like Keplerian ellipses (*positive heuristic*).

(A) NEGATIVE HEURISTIC:
THE 'HARD CORE' OF THE PROGRAMME

All scientific research programmes may be characterized by their 'hard core.' The negative heuristic of the programme forbids us to direct the *modus tollens* at this 'hard core.' Instead, we must use our ingenuity to articulate or even invent 'auxiliary hypotheses,' which form a *protective belt* around this core, and we must redirect the *modus tollens* to *these*. It is this protective belt of auxiliary hypotheses which has to bear the brunt of tests and get adjusted and readjusted, or even completely replaced, to defend the thus-hardened core. A research programme is successful if all this leads to a progressive problemshift; unsuccessful if it leads to a degenerating problemshift.

The classical example of a successful research programme is Newton's gravitational theory: possibly the most successful research programme ever. When it was first produced, it was submerged in an ocean of 'anomalies' (or, if you wish, 'counterexamples'), and opposed by the observational theories supporting these anomalies. But Newtonians turned, with brilliant tenacity and ingenuity, one counterinstance after another into corroborating instances, primarily by overthrowing the original observational theories in the light of which this 'contrary evidence' was established. In the process they themselves produced new counter-examples which they again resolved. They 'turned each new difficulty into a new victory of their programme.'[21]

In Newton's programme the negative heuristic bids us to divert the *modus tollens* from Newton's three laws of dynamics and his law of gravitation. This 'core' is 'irrefutable' by the methodological decision of its proponents: anomalies must lead to changes only in the 'protective belt' of auxiliary, 'observational' hypotheses and initial conditions.[22]

(B) POSITIVE HEURISTIC: THE CONSTRUCTION OF THE 'PROTECTIVE BELT' AND THE RELATIVE AUTONOMY OF THEORETICAL SCIENCE

Research programmes, beside their negative heuristic, are also characterized by their positive heuristic.

Even the most rapidly and consistently progressive research programmes can digest their 'counter-evidence' only piecemeal: anomalies are never completely exhausted. But it should not be thought that yet unexplained anomalies—'puzzles' as Kuhn might call them—are taken in random order, and the protective belt built up in an eclectic fashion, without any preconceived order. The order is usually decided in the theoretician's cabinet, independently of the *known* anomalies. Few theoretical scientists engaged in a research programme pay undue attention to 'refutations.' They have a long-term research policy which anticipates these refutations. This research policy, or order of research, is set out—in more or less detail—in the *positive heuristic* of the research programme. The negative heuristic specifies the 'hard core' of the programme which is 'irrefutable' by the methodological decision of its proponents; the positive heuristic consists of a partially articulated set of suggestions or hints on how to change, develop the 'refutable variants' of the research programme, how to modify, sophisticate, the 'refutable' protective belt.

The positive heuristic of the programme saves the scientist from becoming confused by the ocean of anomalies. The positive heuristic sets out a programme which lists a chain of ever more complicated *models* simulating reality: the scientist's attention is riveted on building his models following instructions which are laid down in the positive part of his programme. He ignores the *actual* counterexamples, the available '*data*.' Newton first worked out his program for a planetary system with a fixed point-like sun and one single point-like planet. It was in this model that he derived his inverse square law for Kepler's ellipse. But this model was forbidden by Newton's own third law of dynamics, therefore the model had to be replaced by one in which both sun and planet revolved around their common center of gravity. This change was not motivated by any observation (the data did not suggest an 'anomaly') here) but by a theoretical difficulty in developing the programme. Then he worked out the programme for more planets as if there were only heliocentric but no interplanetary forces. Then he worked out the case where the sun and planets were not mass-points but mass-*balls*. Again, for this change, he did not *need* the observation of an anomaly; infinite density was forbidden by an (inarticulated) touchstone theory, therefore planets *had* to be extended. This change involved considerable mathematical difficulties, held up Newton's work—and delayed the publication of the *Principia* by more than a decade. Having solved this 'puzzle,' he started work on *spinning balls* and their wobbles. He admitted interplanetary forces and started work on *perturbations*. At this point he started to look more anxiously at the facts. Many of them were beautifully explained (qualitatively) by this model, many were not. It was then that he started to work on *bulging* planets, rather than round planets, etc.

Most, if not all Newtonian 'puzzles,' leading to a series of new variants superceding each other, were foreseeable at the time of Newton's first naïve model and no doubt Newton and his colleagues *did* foresee them: Newton must have been fully aware of the blatant falsity of his first variants. Nothing shows the existence of a positive heuristic of a research programme clearer than this fact: this is why one speaks of 'models' in research programmes. A *'model'* is a set of initial conditions (possibly together with some of the observational theories) which one knows is *bound* to be replaced during the further development of the programme, and one even knows, more or less, how. This shows once more how irrelevant 'refutations' of any specific variant are in a research programme: their existence is fully expected, the positive heuristic is there both as a strategy for predicting (producing) and digesting them. Indeed, if the positive heuristic is clearly spelt out, the difficulties of the programme are mathematical rather than empirical.

One may formulate the 'positive heuristic' of a research programme as a 'metaphysical' principle. For instance one may formulate Newton's programme like this: 'the planets are essentially gravitating spinning-tops of roughly spherical shape.' This idea was never *rigidly* maintained: the planets are not *just* gravitational, they have also, for example, electro-magnetic characteristics which may influence their motion. Positive heuristic is thus in general more flexible than negative heuristic. Moreover, it occasionally happens that when a research programme gets into a degenerating phase, a little revolution or a *creative shift* in its positive heuristic may push it forward again.[23] It is better therefore to separate the 'hard core' from the more flexible metaphysical principles expressing the positive heuristic.

Our considerations show that the positive heuristic forges ahead with almost complete disregard of 'refutations': it may seem that it is the 'verifications'[24] rather than the refutations which provide the contact points with reality.

We may appraise research programmes, even after their 'elimination,' for their *heuristic power:* how many new facts did they produce, how great was 'their capacity to explain their refutations in the course of their growth'?

Thus the methodology of scientific research programmes accounts for the *relative autonomy of theoretical science:* a historical fact whose rationality cannot be explained by the earlier falsificationists. Which problems scientists working in powerful research programmes rationally choose, is determined by the positive heuristic of then programme rather than by psychologically worrying (or technologically urgent) anomalies. The anomalies are listed but shoved aside in the hope that they will turn, in due course, into corroborations of the programme. Only those scientists have to rivet their attention on anomalies who are either engaged in trial and error exercises or who work in a degenerating phase of a research programme when the positive heuristic ran out of steam.

THE POPPERIAN VERSUS THE KUHNIAN
RESEARCH PROGRAMME

Let us now sum up the Kuhn-Popper controversy.

We have shown that Kuhn is right in objecting to naïve falsificationism, and also in stressing the *continuity* of scientific growth, the *tenacity* of some scientific theories. But Kuhn is wrong in thinking that by discarding naïve falsificationism, he has discarded thereby all brands of falsificationism. Kuhn objects to the entire Popperian research programme, and he excludes *any* possibility of a rational reconstruction of the growth of science. *In Kuhn's view there can be no logic, but only psychology of discovery.* For instance, in Kuhn's conception, anomalies, inconsistencies *always* abound in science, but in 'normal' periods the dominant paradigm secures a pattern of growth which is eventually overthrown by a 'crisis.' There is no particular rational cause for the appearance of a Kuhnian 'crisis.' 'Crisis' is a psychological concept; it is a contagious panic. Then a new 'paradigm' emerges, incommensurable with its predecessor. There are no rational standards for their comparison. Each paradigm contains its own standards. The crisis sweeps away not only the old theories and rules but also the standards which made us respect them. The new paradigm brings a totally new rationality. There are no super-paradigmatic standards.

The change is a bandwagon effect. Thus *in Kuhn's view scientific revolution is irrational, a matter for mob psychology.*

But Kuhn overlooked Popper's sophisticated falsificationism and the research programme he initiated. Popper replaced the central problem of classical rationality, *the old problem of foundations,* with the *new problem of fallible-critical growth* and started to elaborate objective standards of this growth. In this paper, I have tried to develop his programme a step further. I think this small development is sufficient to escape Kuhn's strictures.

The reconstruction of scientific progress as proliferation of rival research programmes and progressive and degenerating problemshifts gives a picture of the scientific enterprise which is in many ways different from the picture provided by its reconstruction as a succession of bold theories and their dramatic overthrows. Its main aspects were developed from Popper's ideas and, in particular, from his ban on 'conventionalist,' that is, content-decreasing stratagems. The main difference from Popper's original version is, I think, that in my conception criticism does not—and must not—kill as fast as Popper imagined. *Purely negative, destructive criticism, like 'refutation' or demonstration of an inconsistency does not eliminate a programme. Criticism of a programme is a long and often frustrating process and one must treat budding programmes leniently.* One may, of course, show up the degeneration of a research programme, but it is only *constructive criticism* which, with the help of rival research programmes, can achieve real success; and dramatic spectacular results become visible only with hindsight and rational reconstruction.

NOTES

1. Bertrand Russell, "Reply to Critics" in P.A. Schilpp (ed.) *The Philosophy of Betrand Russell* (Chicago: Open Court, 1943). p. 683.

2. P.B. Medawar, *The Art of the Soluble.* (London: Metheun, 1967). p. 144.

3. This discussion already indicates the vital importance of a demarcation between provable factual and unprovable theoretical propositions for the dogmatic falsificationist.

4. '*Criteria of refutation* have to be laid down beforehand: it must be agreed which observable situations, if actually observed, mean that the theory is refuted' (*Conjectures and Refutations,* Karl Popper. (London: Routledge and Kegan Paul, 1963), p. 38, n. 3).

5. Quoted in *Logic of Scientific Discovery.* Karl Popper (Vienna: Springer, 1934), section 85, with Popper's comment: 'I fully agree.'

6. If the tiny conjectural planet were out of the reach of even the biggest *possible* optical telescopes, he might try some quite novel; instrument (like a radio telescope) in order to enable him to 'observe' it, that is, to ask Nature about it, even if only indirectly. (The new 'observational' the-

ory may itself not be properly articulated, let alone severely tested, but he would care no more than Galileo did.

7. At least not until a new research programme supercedes Newton's programme which happens to explain this previously recalcitrant phenomenon. In this case, the phenomenon will be unearthed and enthroned as a 'crucial experiment.'

8. Popper asks: 'What kind of clinical responses would refute to the satisfaction of the analyst not merely a particular diagnosis, but psychoanalysis itself?' (*Conjectures*, p. 38, n. 3.) But what kind of observation would refute to the satisfaction of the Newtonian not merely a particular version but Newtonian theory itself?

9. I use 'prediction' in a wide sense that includes 'postdiction.'

10. Molière, for instance, ridiculed the doctors of his *Malade Imaginaire*, who offered the *virtus dormitiva* of opium as the answer to the question as to why opium produced sleep. One might even argue that Newton's famous dictum *hypotheses non fingo* was really directed against *ad hoc* explanations—like his own explanation of gravitational forces by an aether-model in order to meet Cartesian objections.

11. Incidentally, Duhem agreed with Bernard that experiments alone—without simplicity considerations—can decide the fate of theories in physiology. But in physics, he argued, they cannot (*The Aim and Structure of Physical Theory*, chapter VI, section 1).

12. Koestler correctly points out that only Galileo created the myth that the Copernican theory was simple (*The Sleepwalkers*. A. Koestler. (London: Hutchinson, 1959), p.476); in fact; 'the motion of the earth [had not] done much to simplify the old theories, for although the old equants had disappeared, the system was still bristling with auxiliary circles' (*History of the Planetary Systems from Thales to Kepler*. J. Dreyer. (New York: Dover, 1906), chapter XIII.

13. If I already know P_1: 'Swan A is white,' P_ω: 'All swans are white' represents no progress, because it may only lead to the discovery of such further similar facts as P_2: ' Swan B is white.' So-called 'empirical generalizations' constitute no progress. A *new* fact must be improbable or even impossible in the light of previous knowledge.

14. *The Open Society and Its Enemies*. Karl Popper. (London: Routledge and Kegan Paul, 1945), Vol. II, p. 233. Popper's more sophisticated attitude surfaces in the remark that 'concrete and practical consequences can be *more* directly tested by experiment' (*ibid.*, my italics).

15. *Logic of Scientific Discovery*, section 30.

16. In the distorting mirror of naïve falsificationism, new theories which replace old refuted ones, are themselves born unrefuted. Therefore they do not believe that there is a relevant difference between anomalies and crucial counterevidence. For them, anomaly is a dishonest euphemism for counterevidence. But in actual history new theories are born refuted: they inherit many anomalies of the old theory. Moreover, frequently it is *only* the new theory which dramatically predicts that fact which will function as crucial counterevidence against its predecessor, while the 'old' anomalies may well stay on as 'new' anomalies.

17. It is clear that the theory T' may have corroborated empirical content over another theory T even if both T and T' are refuted. Empirical content has nothing to do with truth or falsity. Corroborated contents can also be compared irrespective of the refuted content. Thus we may see the rationality of the elimination of Newton's theory in favour of Einstein's, even though Ein-

stein's theory may be said to have been born—like Newton's—'refuted.' We have only to remember that 'qualitative confirmation' is a euphemism for 'quantitative disconfirmation.'

18. Cf. *Logic of Scientific Discovery*, section 85, p. 279 of the English translation.

19. *Logic of Scientific Discovery*, section 30.

20. *Logic of Scientific Discovery*, sections 11 and 70. I use 'metaphysical' as a technical term of naïve falsificationism: a contingent proposition is 'metaphysical' if it has no 'potential falsifiers.'

21. Exposition du Systeme du Monde. (Paris: Bachelier, 1824), livre IV, chapter 11.

22. The actual hard core of a programme does not actually emerge fully armoured like Athene from the head of Zeus. It develops slowly, by a long, preliminary process of trial and error. In this paper, this process is not discussed.

23. Soddy's contribution to Prout's programme or Pauli's to Bohr's (old quantum theory) programme are typical examples of such creative shifts.

24. A 'verification' is a corroboration of excess content in the expanding programme. But, of course, a 'verification' does not *verify* a programme: it shows only its heuristic power.

Case Studies

. . . .

Astronomy Track

PAPER 4: KEPLER, NEWTON, AND HALLEY
ON PLANETARY ORBITS

THE CASE

Galileo's contemporary Johannes Kepler was a very gifted mathematician who had been hired by the Dutch astronomer Tycho Brahe as his assistant. Tycho, one of the most meticulous astronomical observers in history, had acquired a wealth of the best astronomical data to be collected up to that point in scientific history. Kepler was assigned to focus on the planet Mars, something that had been of particular interest to Tycho because its path that was unexpected in certain ways. Because his data was so good and so hard to acquire, Tycho was very guarded with it, and did not allow Kepler to copy or see all of it. But after his death, Kepler was finally given full access to Tycho's results.

With these observations in hand, Kepler posed the question, what shape would planetary orbits have to be if one wanted to do away with epicycles completely? It was clear that perfectly circular orbits were insufficient to account for actual observations, but what more complex trajectory would have to substituted in order to account for the data that Tycho spent his life collecting with the utmost care?

After a considerable number of failed attempts with many possible shapes, Kepler hit upon the one that worked. If one assumed that planetary orbits were elliptical with the sun at one focus of the ellipse, then the data would fit. From that data, Kepler also found two other general rules. The first is that while planets move faster when closer to the sun and slower when farther away, the area of the ellipse swept out by the line connecting the sun and planet would always be the same for any given period of time. The other is that the time it takes a planet to go around the sun equals the square root of length of the longer radius of the ellipse cubed.

While Kepler himself put forward no explanation for these regularities, they were later shown by Isaac Newton to follow from the three laws of motion and the law of universal gravitation Newton put forth in the *Principia*. By setting out gravitation as an attractive force between any two bodies proportional to the product of the masses and inversely proportional to the square of the distance between them and showing that such a system of bodies would rotate around the combined center of mass—something that would be closer to the heavier body (indeed, within the heavier body if it was heavy enough)—the view from one of the bodies would make the path elliptical in exactly the ways Kepler had discovered.

Indeed, it was this success that led to its initial acceptance. However, it was Newton's friend Edmond Halley's use of the rules to predict the appearance of a comet in late December 1758 that the Keplerian account of planetary motion, supported by Newtonian mechanics, became the almost universally accepted scientific position.

THE SCIENTISTS

Johannes Kepler (1571–1630) was a mathematician who became fascinated early with the large mysteries of the universe. Originally intending to join the clergy, Kepler began to look for mathematical regularities in nature as a way of exposing the divine structure of the universe. He spent much effort not only deriving his famous results, but also attempting to relate the ratios of the radii of the planetary orbits to regular geometric solids. He worked for a brief time with Tycho, during and after which he became the court astrologer for Holy Roman Emperor Rudolph II, something he found distasteful but which allowed him sufficient income to conduct his scientific works.

Isaac Newton (1643–1727) was extremely shy in his youth. His father died before his birth, and Newton was raised by his mother then sent away to school where he had few friends. Finding solace in science and alchemy, Newton studied intensely and attended university at Cambridge, where he remained alienated from those around him. In 1665–1666, the university closed due to plague and Newton spent his time at home working on questions of mechanics and gravitation. When Edmond Halley discovered that Newton had accounted for Kepler's laws, Halley asked that Newton write up his work. The printing of the resulting book, *The Mathematical Principles of Natural Philosophy*, was financed by Halley. The book shaped all of science for the next several centuries. After its publication, Newton became one of the most famous and powerful men in Europe, knighted for his work and made the head of the Royal Society, the most prestigious scientific organization in England.

YOUR JOB

There is no doubt that Newton reshaped science, both in terms of what was understood and how it was done. Explain how Newton's three laws of motion and his law of universal gravitation account for Kepler's laws. Select either Kuhn's or Lakatos's version of the holistic approach to theories and explain this account of the scientific method in detail. Consider Newton's work. Surely it was revolutionary in Kuhn's sense, but where does the revolution begin? Was Copernicus working in the same paradigm/research programme as Newton? Was Kepler? How could Kuhn or Lakatos account for the historical importance of the sighting of Halley's comet?

- For this paper, focus on using secondary sources, that is, the works of scholars who are writing about the case rather than the scientists themselves. A good secondary source will not only explain your case in a fashion that is often clearer than the original primary material (which you should still try to use and quote from) but will also cite other resources that will be helpful. Scholars read other scholars and their work will give you a sense of the larger historical background and debates in the community of smart people who are thinking hard about what happened in this case.

Physics Track

PAPER 4: THE BOHR MODEL

THE CASE

The classical atom was a great success. It accounted for much of the observable phenomena in atomic physics and chemistry. There were a few remaining difficulties that seemed to require some fancy footwork, but even the when the best minds bumped up against them they remained obstinately unresolved.

Two such problems involved blackbody radiation and the photoelectric effect. When a hollow body was heated, the amount of radiation of given wavelengths in the cavity failed to match the predictions from classical theory. Indeed, the classical approach, which gave a result that posited infinite energy at certain wavelengths, could not be right. Similarly, there was a bizarre result when ultraviolet light was shined upon a piece of metal. Electrons were kicked out, something that made sense at first, but when a brighter light was focused on the same metal, more electrons came off—but not electrons at higher speeds, as classical theory demanded.

Max Planck and Albert Einstein solved the problems of blackbody radiation and the photoelectric effect, respectively, by employing an interesting mathematical trick. If energy was treated not as a continuously varying property that could take any possible value but rather as a property that could take only certain numerical values and not others, the problems could be solved. You can buy any amount of butter you like, but eggs you have to buy in ones—you can't buy a third of an egg in the way you can buy a third of a stick of butter. Energy, Planck and Einstein proposed, was like eggs and not butter, as we had always thought. This quantizing of energy was the first step towards a new picture of the workings of the world that came to be called quantum mechanics.

Niels Bohr took this quantum approach and gave us a new picture of the hydrogen atom. The so-called Bohr atom had a nucleus with one proton and an orbiting electron, much as Ernest Rutherford had proposed. But in the Bohr model, there were only certain orbits that the electron was allowed to occupy. By quantizing the electron paths, Bohr was able to solve two problems.

First, each element has an optical fingerprint, that is, each element gives off particular wavelengths of light when it is electrically excited. These and only these wavelengths come from this particular type of atom. By having a unique set of orbits that the electron can jump between, it would give off precise quantities of energy whenever transitioning between two orbits and therefore emit a particular frequency of light. If the quantized orbits were particular to types of atoms, so too would be the light they gave off.

Secondly, there was a problem with rotating electrons. A rotating charge gives off energy. As an electron circled the nucleus, it should be giving off energy and should be constantly emitting light and thereby losing energy. Under the classical model, all electrons should be spiraling in towards the nucleus. Yet atoms are stable. Bohr's model produced stable atoms that would not implode, something that allowed us to sleep easier at night.

THE SCIENTIST

Niels Bohr (1885–1962) was a Danish physicist. His early investigations on the properties of metals brought him into contact with Planck's work, allowing him to be amongst the first to seriously consider the ramifications of the quantum hypothesis. As a young scientist, he spent time at Cambridge, working in Rutherford's lab on questions surrounding radiation where he applied the quantum machinery to the problems of the atom. This work was fruitful enough to win Bohr the Nobel Prize and to convince Denmark to create an institute in Copenhagen to study theoretical physics, which Bohr oversaw for the rest of his life and whose members included, among others, the young Werner Heisenberg, Wolfgang Pauli, and George Gamow. During World War II, however, when the working conditions for the half-Jewish Bohr necessitated his relocation to the United States, he and Albert Einstein engaged in their famous friendly, but adversarial, debates about the truth and completeness of quantum mechanics.

YOUR JOB

Explain how the Bohr model works and how it differs from the classical model of the hydrogen atom. Explain in detail how it solves the problem of the stability of the atom and the light spectrum problem. Select either Kuhn's or Lakatos's version of the holistic view of theories and explain this approach to the scientific method in detail. If you choose Kuhn, explain whether Bohr's quantized hydrogen atom is part of a different paradigm than Rutherford's. If it is a different paradigm, what is different? Couldn't Rutherford, with his classical paradigm, still understand the solution to the problems proposed by Bohr? Doesn't this raise problems for Kuhn's claims of incommensurability? If you select Lakatos, is Bohr's model in a different research programme from Rutherford's? If so, what is the difference between the theories' hard cores? If not, is it just an adjustment to the protective belt and then what exactly is being adjusted? What made the quantum programme progressive and the classical programme degenerate in this case?

RESEARCH HINT

- For this paper, focus on using secondary sources, that is, the works of scholars who are writing about the case rather than the scientists themselves. A good secondary source will not only explain your case in a fashion that is often clearer than the original primary material (which you should still try to use and quote from) but will also cite other resources that will be helpful. Scholars read other scholars and their work will give you a sense of the larger historical background and debates in the community of smart people who are thinking hard about what happened in this case.

Chemistry Track

PAPER 4: DALTON, MULTIPLE PROPORTIONS, AND ATOMIC THEORY

THE CASE

The notion of heat as an element was challenged in the community of physicists, where Newtonian theory naturally led to a mechanical picture of a world containing interacting particles. Chemists, on the other hand, were loath to adopt an atomistic viewpoint where their practical science would be set out in terms of invisible particles with magical properties. Such talk seemed a step backward to the mystical metaphysics of alchemy. Major chemists like Lavoisier refused to take atoms seriously, arguing that such notions were unscientific and not the sort of thing a serious empirical researcher discussed.

A major step forward for the atomic approach in chemistry came with the work of John Dalton, who discovered a property of atoms that was measurable in the laboratory: relative weight. Once measurable properties could be attributed, Dalton held that it could be reasonably argued that the atoms were real. As such, he contended that the atomistic picture now became one that empirically minded scientists could buy into.

From Lavoisier, it was known that compounds were made up of a few more basic types of substance that did not seem to be made of something even more basic. From Joseph Proust, it was known that substances combined in simple proportions to form these compounds.

Dalton concluded from the fact that substances combined in simple ratios that there must be an underlying mechanism of chemical combination, and the one that seemed most obvious was that there are individual atoms of different types combining. This opened up questions about how they combine. Dalton assumed that simplicity would be the guiding principle, so that where only one combination of the substances existed, it would be thought to be a combination of one atom of each. In cases in which there was more than one combination, there would be at least three different possible substances: one with a single atom of each, one with two of one substance combined with one of another, and a third with one of the first and two of the second.

These rules, coupled with the presupposition that atoms could never be destroyed or changed into atoms of a different type, allowed Dalton to infer from careful measurements of the compounds and constituents the relative weights of the elements. Taking hydrogen, the lightest element, to have a mass of 1, he was then able to determine the relative weight of the others, thereby giving a measurable property to the unobservable atoms. Qualities of the atoms could be empirically determined, even if they could not directly be seen.

Additionally, it allowed Dalton to predict the existence of substances that were not commonly found in nature and therefore not known at the time. When some of these substances were created in the lab and had some of the properties Dalton's theory predicted, the novel substances were seen as confirmation of the atomic hypothesis.

While his rules ultimately turned out to be wrong, his ability to link the previously unobservable and seemingly metaphysical atom of the ancients with the most modern means of chemical measurement of the time contributed to the evolution in the way chemists looked at reactions and the sort of explanations and questions they asked. Chemistry as a result of scientists like Dalton became the study of the joining of atoms.

THE SCIENTIST

John Dalton (1766–1844) began his scientific career studying meteorology, formulating what became known as Dalton's law of partial pressures, which argues that, in terms of pressure, mixed gasses are invisible to each other, that each exerts the pressure it would if the other was not there, and the measured pressure is then the combination of these unrelated partial pressures. He also became interested in the composition of the atmosphere and these studies led him to chemistry. His scientific interests were broad and he wrote on topics including color blindness. As a teacher, his most famous student was the physicist James Prescott Joule, for whom the unit of energy is named.

Explain Dalton's argument for atoms. What led Dalton to believe that atoms were real, and how did his atomic presupposition allow him to calculate the relative atomic weights of the elements? Select either Kuhn's or Lakatos's holistic account and explain this approach to the scientific method in detail. If you select Kuhn, was Dalton's work revolutionary science or was it a continuation of the atomic paradigm of the ancient Greeks? How did normal chemistry after Dalton look different from the chemistry of Priestley's era? If you select Lakatos, is Dalton's work in a different research programme from ancient atomists? What is in the hard core and what is in the protective belt? What made the atomists' research programme progressive? Is it still progressive? If so, why? If not, why not?

RESEARCH HINT

- For this paper, focus on using secondary sources, that is, the works of scholars who are writing about the case rather than the scientists themselves. A good secondary source will not only explain your case in a fashion that is often clearer than the original primary material (which you should still try to use and quote from) but will also cite other resources that will be helpful. Scholars read other scholars and their work will give you a sense of the larger historical background and debates in the community of smart people who are thinking hard about what happened in this case.

Genetics Track

PAPER 4: CRICK, WATSON, AND THE DISCOVERY OF DNA

THE CASE

While there was little doubt that genes contained information that guided the growth and development of the organism, how they did so remained unclear. What was the mechanism by which the organism's cells knew what to do and how to do it? Where was the information encoded that would form the genetic blueprint?

James Watson and Francis Crick were young scientists at the Cavendish Laboratory in Cambridge who had a hunch that the answer resided in understanding the role of the part of the cells called deoxyribonucleic acid, or DNA. At the same time, a group of scientists at Oxford working under physicist-turned-biologist Maurice Wilkins were also trying to puzzle out the structure of DNA to figure out how it was built and what its role in reproduction might be. Chief among these scientists was Rosalind Franklin, an expert in x-ray diffraction who was undertaking a very careful study of DNA and coming up with the best data in the world.

Working from Franklin's data, Crick and Watson tried to construct a model of the molecule, taking what they knew of its biochemical composition and trying to determine how those pieces came together in a coherent, operative architecture. From this structure, they hoped it would become clear how information could be coded in.

While their initial attempt failed miserably—the result of Watson incorrectly remembering what Franklin reported from her painstaking work—they continued to try to put together a scale model of the molecule that would conform to all known chemical and biological theories. Realizing from the work of Erwin Chargaff that there is a relation between pairs of nucleotides, they tried to figure how these base pairs would bond. When a colleague pointed out that they were using the wrong form of the molecules, it became clear how they fit together. That insight, combined with Franklin's x-ray diffraction images, made clear that DNA had a double helical structure which could be separated and reconstructed in a way that passes information about the parent cell to its divided offspring. The simplicity and coherence of Crick and Watson's model altered what questions could be asked in genetics and how to go about answering them changed forever.

THE SCIENTISTS

James Watson (1928–) is an American biologist whose groundbreaking work on the structure of DNA earned him a Nobel Prize. His book about the ordeal, *The Double Helix,* generated plenty of controversy for its portrayal of the central characters of the episode. Well known for being outspoken on many issues, Watson became the first head of the Human Genome Project.

Francis Crick (1916–2004) was an English scientist who made the turn from physics to biology in search of big questions to answer. As a graduate student working on a separate question, he was lured into questions about the structure of DNA when James Watson was assigned to share an office during his postdoctoral work at Cambridge. His work on DNA only spurred his interest in questions at the boundary of science and philosophy.

Rosalind Franklin (1920–1958) was an English biologist who worked in the *Laboratoire central des services chimiques de l'État* in Paris on x-ray diffraction before moving to Oxford to work in Maurice Wilkins's lab. Applying the techniques of x-ray diffraction to samples of DNA and painstakingly analyzing the results, she produced the data that would lead to the determination of the molecule's structure. She died at age 38 of cancer.

YOUR JOB

Explain how Crick and Watson discovered the structure of DNA. What role did Franklin's data play? Select either Kuhn's or Lakatos's approach and explain his view of scientific methodology. If you select Kuhn, discuss whether Crick and Watson's discovery was an example of revolutionary science. If so, what role did Franklin's data play?

Franklin had no clue about the structure of DNA during her work, so how could data from a method that was part of a previous paradigm be useful in Crick and Watson's work if it is deemed revolutionary? How would a Kuhnian answer this question? Is the answer convincing? If you select Lakatos, did Crick and Watson's discovery create a new research programme? If so, what is in the hard core and what is in the protective belt? How would an advocate of Lakatos's view argue that there was evidence that showed the model to be part of a progressive research programme?

RESEARCH HINT

- For this paper, focus on using secondary sources, that is, the works of scholars who are writing about the case rather than the scientists themselves. A good secondary source will not only explain your case in a fashion that is often clearer than the original primary material (which you should still try to use and quote from) but will also cite other resources that will be helpful. Scholars read other scholars and their work will give you a sense of the larger historical background and debates in the community of smart people who are thinking hard about what happened in this case.

Evolutionary Biology Track

PAPER 4: DARWIN'S NATURAL SELECTION

THE CASE

Charles Darwin's five-year voyage on the HMS *Beagle* is a veritable science legend. As the captain was charting the waters and surveying the coastline of South America, Darwin observed what he recognized to be distinct but clearly related species of birds in the Galapagos Islands. He noted that they were very similar to a species found in North America, but the birds on different islands had quite distinct features.

Using the domestic breeding of plants and livestock as an example, he realized that heredity could play a significant role in changing the characteristics of offspring. By selective breeding, certain properties could be artificially selected for, allowing even very old varieties to yield new and improved ones.

On the farm, these decisions could be made intentionally by a rational breeder. Where could the natural variations have come from, especially given that the differences happened to make the individuals better suited to the particulars of their environment?

The clue came from Thomas Malthus's principle of population, which asserts that the innate drive to reproduce at much as possible will cause the population to increase faster than our ability to produce food. As such, hunger and starvation could not be eliminated. Indeed, good-hearted attempts to help the hungry avoid starvation would only cause them to have more children, further increasing the pressure on the now even more limited food supply. Struggle for the means of survival and starvation and poverty were not only natural but unavoidable features of society.

Using Malthus as a tip, Darwin posited that in nature, survival was necessarily a struggle. This meant that not every individual lived to reproduce, and those that survived the longest would reproduce the most, passing on those characteristics that allowed it to survive in the particulars of its environmental surroundings to their offspring, just as breeders' new plant varieties were better designed for the breeders' needs. Isolated populations, subject to different natural situations, therefore, would diverge as a result of the different pressures leading different properties to be selected for. This mechanism would come to be known as natural selection and would explain the origin and distribution of species around the world. Different species had common ancestors but developed in isolation under different pressures, and thereby developed significant anatomical differences.

THE SCIENTIST

Charles Darwin (1809–1882) was the grandson of Josiah Wedgewood, the famous potter, and Erasmus Darwin, who himself had formulated a theory of evolution. He was asked to join the voyage of the *Beagle* because the captain, Robert Fitzroy, thought it the trip would be more interesting with an educated companion along for conversation. Having taken a copy of Charles Lyell's *Principles of Geology* along with him to read, much of Darwin's actual scientific work during the voyage was geological and not biological. Working back in England after his return, he devised his theory of evolution through natural selection at the same time as another naturalist, Alfred Russel Wallace, developed the same theory. Darwin had both of their papers published together in a volume in 1859. His theory challenged traditional views of divine Creation of species, but it was when he applied the theory to the descent of humans that his work truly ruffled feathers.

YOUR JOB

Explain Darwin's theory of natural selection. How do different species emerge? Select either Kuhn's or Lakatos's view and explain this version of the scientific method in detail. If you select Kuhn, is Darwin's theory to be considered revolutionary science? If so, is Malthus's work in the same paradigm, even though it precedes the revolution? What of the practical knowledge of plant and animal breeders? Is that part of Darwin's paradigm or the one that preceded it? If you select Lakatos's view, is Malthus part of the same research programme? Is the practical knowledge of the plant and animal breeders also part of the same research prrogramme or prescience? If so, are they part of the hard core or protective belt?

- For this paper, focus on using secondary sources, that is, the works of scholars who are writing about the case rather than the scientists themselves. A good secondary source will not only explain your case in a fashion that is often clearer than the original primary material (which you should still try to use and quote from) but will also cite other resources that will be helpful. Scholars read other scholars and their work will give you a sense of the larger historical background and debates in the community of smart people who are thinking hard about what happened in this case.

Geology Track

PAPER 4: WEGENER, PLATE TECTONICS, AND MODELS

THE CASE

In 1910, Alfred Wegener noticed that the western coast of Africa and eastern coast of South America seemed to fit together as if they were jigsaw puzzle pieces. But he didn't think much of it until a year later when he read of archaeological evidence connecting the two regions. Fossils of a plant called *Glossopteris* and a reptile called *Mesosaurus* were found in both regions. The similarities led some to suppose there had been a land bridge at one time connecting the two regions, but Wegener's geometric insight provided a different theory.

Wegener argued that instead of thinking of the continents as stable and an additional bit of land having disappeared between them, it should be considered that continents have been displaced over time. Once a single supercontinent termed Pangaea, it broke into plates that shifted across the globe. There current arrangement is the result of what came to be called "continental drift."

It had been discovered that the continents were lighter than Earth's crust, beneath which floated a dense, viscous layer. As such, the sinking of a land bridge was unlikely, since the land, being lighter, would rise rather than sink. But if landmasses could move up or down, why not horizontally?

Mountains like the Alps, Andes, or Himalayas, on this view, could be seen as a buckling up of land as the result of plates pushing against one another. There were stark similarities between rock strata along the African and South American coastlines that indicated a single origin. Regions in Argentina matched up with one segment of the African geological profile, while part of Brazil connected with that another corresponding region.

Based on geological, paleontological, and geometrical arguments, Wegener argued that the earth ought to be thought of as a set of plates floating on a thick liquid, bouncing off of each other according to Newton's laws of physics.

Alfred Wegener (1880–1930) was not a geologist, rather, he was trained as an astronomer and worked as a meteorologist. He was fascinated with the island of Greenland
and made three successful expeditions there, one of which completed the longest ever
crossing of the ice cap on foot. Wegener, an outsider to the geological community, proposed his theory of continental drift in 1912, although it was not until the 1960s that
it became widely accepted. He died of a heart attack during his final expedition to the
glaciers of Greenland.

YOUR JOB

Explain the origin and evidence of Wegener's theory of continental drift. Pick either
Kuhn's or Lakatos's version of the holistic view of theories and explain how an advocate of that view would account for Wegener's work. If you pick Kuhn, is the theory
of plate tectonics revolutionary? If so, what is in the paradigm? If you select Lakatos,
what would go in the hard core of Wegener's research programme and what resides in
the protective belt? What evidence would Wegener cite to show that his research programme is progressive?

RESEARCH HINT

- For this paper, focus on using secondary sources, that is, the works of scholars
  who are writing about the case rather than the scientists themselves. A good
  secondary source will not only explain your case in a fashion that is often clearer
  than the original primary material (which you should still try to use and quote
  from) but will also cite other resources that will be helpful. Scholars read other
  scholars and their work will give you a sense of the larger historical background
  and debates in the community of smart people who are thinking hard about what
  happened in this case.

Psychology Track

PAPER 4: FREUD, THE RAT MAN, AND REVOLUTION

THE CASE

While many psychologists were examining the functioning of the mind, others were
interested in trying to set out its structure. The great figure in this movement was Sigmund Freud, who argued that the mind is like an iceberg, with much of its workings
hidden below the surface. The conscious mind and its intentional, rational decisions
are only a tiny piece of the explanation for our actions.

Freud argued that the mind has a tripartite structure, with the ego being the conscious aspect we are aware of. Above this is the superego, which functions as our higher selves, and the id, which is the repository for desires and instincts. When our instincts and wishes go unfulfilled, the mind sets up an internal conflict that plays itself out in our lives in ways that we are not aware of, appearing in tics and strange behaviors, such as the so-called Freudian slip. Indeed, insults to our being that leave our natural desires unfulfilled will often remain as wounds in the subconscious, continually taking a toll on our lives. This is, Freud argues, especially true of obstacles in childhood that interfered with normal psychological development.

Psychoanalysis is the process of peeling back the conscious mind to reveal more of the psychic iceberg beneath the surface. Dreams, for example, are one entryway into the subconscious, as it is one place in which we can become aware of the workings of the mind that are generally hidden, eclipsed by the workings of consciousness. Understanding personal history is also essential to locating potential places where natural urges may have been foiled, leaving scars that continue to be operative in choices made.

Freud had many famous cases wherein he used psychoanalytic methods to treat mentally ill patients. One famous case is the so-called Rat Man. Among his abnormal behaviors, the Rat Man one day suddenly became obsessed with losing weight, skipping meals and compulsively exercising in extreme heat to the point of exhaustion. Freud traced the source of this behavior to his sexual desire for a woman who had become involved with his American cousin named Richard, who went by the nickname Dick. In German, *dick* means fat, and Freud contended that the Rat Man's obsession with losing weight was really his way of trying to murder his cousin; by losing *dick* he would be removing Richard from the situation. The reasons for the Rat Man's actions were not ones he may have thought they were. His true motivations were buried within his subconscious.

THE SCIENTIST

Sigmund Freud (1856–1939) lived in Vienna, where he was a trained physician. He was disappointed by the methods of the day for treating hysteria and began to use the technique of free association, wherein the patient would respond to random words quickly, without thinking. When patients hesitated, displaying difficulties making associations, Freud realized that there was an issue beneath the surface that needed digging out. Many of these issues, he came to realize, were sexual. He reported that hysteria was the result of premature sexual experiences in childhood reported by his patients; only later did he find out that many of these reports were false, but at the time, the patients truly believed they had happened. In considering why such fantasies would become false memories, Freud developed his picture of the mind.

Explain Freud's picture of the human mind and why psychoanalysis was thought by him to be an effective therapy. Show how this view of the mind was applied to the case of the Rat Man. Select either Kuhn or Lakatos and explain the associated view of the scientific method. If you select Kuhn, would Freud be considered revolutionary science? If so, what is in the new paradigm? If you select Lakatos, what in Freud's theory would be in the hard core and what in the protective belt? Why did Freudians consider their research programme to be progressive?

RESEARCH HINT

- For this paper, focus on using secondary sources, that is, the works of scholars who are writing about the case rather than the scientists themselves. A good secondary source will not only explain your case in a fashion that is often clearer than the original primary material (which you should still try to use and quote from) but will also cite other resources that will be helpful. Scholars read other scholars and their work will give you a sense of the larger historical background and debates in the community of smart people who are thinking hard about what happened in this case.

Sociology Track

PAPER 4: SOROKIN, THE IDEATIONAL, AND THE SENSATE

THE CASE

Émile Durkheim and Max Weber disagreed over the nature of sociological research. Durkheim contended that the process began with social facts from which one derived general conclusions. Social scientific methodology should be modeled upon that of the natural sciences. Weber argued that the goals of social science were not that of natural science; sociology, in particular, was not out to determine with equation-like certainty how independent and dependent social variables interact. Rather, the questions put before the sociologist were much like those put before the historian—how do we understand what happened here? And like the historian, the answer could be narrative; indeed multiple historians will tell multiple narratives, each stressing a different set of operative factors, all containing insightful aspects of the truth.

This tension between the empirical and the humanistic are incorporated into the sociological approach of Pitirim Sorokin. In his multivolume masterwork, *Social and Cultural Dynamics*, Sorokin argued that there are two ways to construct reality, the sensate and the ideational. The sensate approach is to take the world to consist of that which can be experienced through the senses. What is real is held to be what is experienced. All social institutions are geared towards this view. The ideational, on the other hand, contends that what is real is that which goes beyond the senses, a transcendent, spiritual truth. Art, science, and social codes take on a completely different flavor as a result.

Societies could be structured around one or the other of these approaches and it would shape their understanding of truth and every aspect of the products of the culture, from its ethics to its economics, from its religion to its science, from it art to its laws. The notion of truth in an ideational culture is radically different from the concept in a sensate culture. The worlds they inhabit are different, and so too is the meaning they make of it and the structures they create in it.

But societies are not stable in their orientation, Sorokin argued. There will always be tensions, and over time, cultures cycle between being sensate and being ideational. Such transformations take significant time, but they are revolutionary. In contrast to others at the time who advocated an evolutionary picture of the development of society, in particular, his Harvard colleague Talcott Parsons, Sorokin contended that there was what he termed a creative recurrence. Societies will be led by a sensate or ideational mentality, but will eventually reach a saturation point and elements within will steer it back in a novel, creative fashion giving rise to "new variations of old themes."

THE SCIENTIST

Pitirim Sorokin (1889–1968) was born in Russia. His work there was not popular with the authorities; he was arrested by the forces of the czar, and later, after the revolution, by the Communists. When released, he emigrated to the United States, where he took a position at the University of Minnesota but eventually moved to Harvard University, where he was the institution's first professor of sociology.

YOUR JOB

Explain Sorokin's notion of the sensate and ideational mentalities and his doctrine of creative recurrence. Select either Kuhn's or Lakatos's version of the holistic view of theories and explain this account of the scientific method in detail. If you select Kuhn, would Sorokin's work be considered revolutionary science or normal science within the paradigm of Durkheim and Weber? What would be contained within this paradigm? If you select Lakatos, what would be contained in the hard core of the research programme and what would be contained in the protective belt? What evidence would his supporters produce to contend that his research programme is progressive? Are there any aspects of Sorokin's work that are not well accounted for by this approach?

RESEARCH HINT

- For this paper, focus on using secondary sources, that is, the works of scholars who are writing about the case rather than the scientists themselves. A good secondary source will not only explain your case in a fashion that is often clearer than the original primary material (which you should still try to use and quote from) but will also cite other resources that will be helpful. Scholars read other scholars and their work will give you a sense of the larger historical background and debates in the community of smart people who are thinking hard about what happened in this case.

Economics Track

PAPER 4: MARX, DIALECTICAL MATERIALISM, AND ECONOMIC REVOLUTION

THE CASE

Adam Smith's picture of the economy had human beings as rational billiard balls, as atomic agents, each acting in his or her own self-interest. These interests then align to give an order, a structure guided by an invisible hand, but it is the small-scale interaction that gives us the laws of the marketplace. Karl Marx saw things differently.

Marx's dialectical materialism takes its cue from the work of G. W. F. Hegel, in which everything must be seen in the long view through the lens of history. We do not exist as autonomous individuals but as instantiations of the spirit of the times. We do not define ourselves but are defined by our place in the historical structure that Marx reinterpreted to mean an integrated structure of economics and politics, one that was to be understood completely in terms of class and class struggles. There would be one class that controlled the means of production and thereby the reigns of power, and another class who was ruled by them.

This would inevitably give rise to class struggle, conflict, and ultimately revolution that would overthrow the structure and replace it with another. The new system would advance a step beyond the last as new class lines would be drawn and a new class comes to power. But this, in turn, would again divide the society, giving rise to class tensions and on and on. The process would repeat itself, each time people on all sides gaining a bit more dignity, a bit more autonomy, a bit more humanity, until it reached a utopian end phase.

Marx's approach forced political concerns and economic concerns to be seen as inseparable and required interpreting all economic reasoning through the lens of class. Where feudal systems divided the population into land-owners and servants, capitalism divided them into the bourgeoisie, who profited from industrial production and the proletariat who created those profits through their labor which they had to sell in order to acquire the necessities of life. The capitalists increased their interests, that is, their profits, through technological advancement, which decreased labor costs; but that advancement also held the seeds of their own demise as they set the stage for an image of life that did not require them and which would allow for increased stability through central planning and more equitable distributions of goods. Such a conversion to socialism and then ultimately to communism was a Hegelian inevitability, it was a truth of the large scale, unavoidable process of history.

THE SCIENTIST

Karl Marx (1818–1883) came from a Prussian Jewish family of some social standing. His was a prominent lawyer despite his progressive leanings. Marx opted for philosophy over law, becoming a member of the Young Hegelians, a move that sunk his academic career; he became a journalist and activist, In the aftermath of the workers' uprisings across Europe in 1848–1849, he fled to London, where his family lived off of the wages of his collaborator Friedrich Engels, who labored in various positions in the Manchester textile factory partly owned by Engel's father.

YOUR JOB

Explain Marx's dialectical materialism. How does it differ from the economic thought before it? Select either Kuhn or Lakatos and explain their view of science in detail. If you select Kuhn, was Marx a revolutionary thinker in economics? If so, what is to be included in the paradigm? If you select Lakatos, what is in the hard core and what is in the protective belt of his research programme? What evidence would a Marxist cite that the research programme is progressive? Are there any aspects of the Marxist project that do not fit in well with the view you selected?

RESEARCH HINT

- For this paper, focus on using secondary sources, that is, the works of scholars who are writing about the case rather than the scientists themselves. A good secondary source will not only explain your case in a fashion that is often clearer than the original primary material (which you should still try to use and quote from) but will also cite other resources that will be helpful. Scholars read other scholars and their work will give you a sense of the larger historical background and debates in the community of smart people who are thinking hard about what happened in this case.

SEMANTIC VIEW OF THEORIES

Both the syntactic and holistic views share the belief that scientific theories are collections of sentences we have good reason to believe to be true. It may be that they are individually testable as the syntactic theorists hold, or testable as a group as Pierre Duhem argued, or only testable (and held to be meaningfully true) within the confines of a paradigm or research program, but science is about finding probably true general statements, laws of nature.

This assertion is precisely where advocates of the semantic view of theories take a different path. The goal of science is to describe the way the world works, to explain puzzling natural phenomena. The problem with the syntactic and holistic approaches is that these views make the assumption that the first requires finding propositions that are likely to be true. What they really need, semantic theory advocates contend, are good models of the phenomena we want to explain.

Think about how you would go about explaining to someone how to drive to your house. If you just told them the directions, they might be able to get there. But the best thing to do is to draw them a map. On the map, you can show major intersections and landmarks to watch out for, represent relative distances to show where they'll be on one road for a long while but another only briefly, distinguish right from left turns, and provide other important information. Your house, of course, is not the little square with a star in it that you've drawn, it is just represented by it. The map is not true or false, but a better or worse representation of the area your friend will have to drive through to get there. But while neither true nor false, it is helpful in explaining how to get to your place and explaining why a wrong turn is wrong.

If, instead of drawing it, you printed a map from an online mapping service, the map would be a better map. It would have more detail, the distances shown would be more proportional, and the landmarks more accurately shown. It would be a better representation. Not that the new map is true either, but it is better and perhaps more useful. We can talk about better and worse here in terms of the accuracy of the representation and the ability for it to explain why certain things turn out in certain ways and other things turn out differently, even if we no longer speak of true and false.

Scientific theories, according to the semantic view, are much like maps in that way. Theories are not sets of true or false propositions but rather sets of models. We want models that give us good explanations, explanations that work and allow us to make predictions, and for this we can jettison much of the talk about true and false propositions and use language like better or worse representations.

The connection between models and semantics in scientific theories was made by Marshall Spector. The term "semantic," as used in "the semantic view of theories," refers to the meaning of words. When you hear someone say that something "is a matter of semantics," usually they are saying that a conversation that is being held out as concerning an important issue is actually trivial, nothing but a silly disagreement over definitions. But the truth is that there is nothing trivial about semantics, especially when we talk about the meanings of terms in scientific theories.

Theorists who hold on to the syntactic view of theories see theories as composed of testable sentences. Those sentences—usually expressed as equations (mathematical sentences)—are comprised of words or symbols. Some of those symbols represent things like length, mass, or color that can be directly observed with your eyes or read off a meter. The meanings of these terms are therefore clear; they have what philosophers call "operational definitions," that is, their meanings come from the operations or processes used by the scientist to measure them. Other terms, like equal signs, parentheses, or minus signs, are part of the grammar of the mathematical language we use in science, and these symbols are defined by their mathematical usage. But then there is a third group of terms that also appear in scientific theories. "Theoretical terms" are things like electrons, gravitational fields, and entropy that cannot be directly observed or measured yet are not simply part of the mathematical formalism. What do these terms mean?

Some syntactic advocates like Rudolf Carnap and R. B. Braithwaite argued that they mean nothing in and of themselves and only acquire indirect meanings through their usages in connecting terms that could be observed. These theoretical terms are not world referring on their own, but are shorthand for combinations of real properties that are directly observed. In this way, theories are only partially interpreted, that is, only the observable and mathematical parts are truly meaningful, while the rest of the theory—the theoretical part—is comprised of terms we use for convenience, but they do not really describe the actual furniture of the universe.

We may come up with models that give us ways of picturing these terms to ourselves in certain ways—space-filling elastic tubes as fields, electric currents as flowing rivers of charge, atoms as little solar systems—but these models are just intellectual crutches, toys we use to feel comfortable with the uninterpreted abstractions of science. They are not really used in determining the acceptability of scientific theories which comes from the testability of the sentences making up the theories.

Spector argued that this view is wrong. Models are not just silly illustrations used to help the uninitiated come to grips with the theory. In some cases, he contends, they are a legitimate part of the full meaning of the theory and have been used in classic examples of legitimate, even legendary, scientific reasoning to advance theories and make great discoveries about how the world actually is. The models have given us a way of finding out about parts of the world we cannot directly observe.

Consider, for example, James Clerk Maxwell's kinetic theory of gases in which a gas is seen as being made up of tiny particles that interact with each other and their container through elastic collisions. Such a model could account for observable behaviors captured by, say, the ideal gas law. But, of course, the ideal gas law is not actually correct—it is, as the name suggests, an idealization. When we add other elements to our model—allowing the particles to interact through electric charges and not being perfectly spherical, for example, the corrections bring us closer and closer to observation. This seems to give us reason to believe that the world is like our model. The use of the model in advancing our understanding was good science and the theoretical terms that have meaning within the model keep those meanings when we fully interpret the theory for the real world. Models give us a new scientific methodology and give us new meanings for our theoretical terms, a new route to scientific semantics.

In "Models and Archetypes," Max Black continued to discuss the uses of models in science, explaining that there are several types of models. Scale models are familiar to anyone who every built a snap-together fire engine or fighter jet as a child, but also to anyone who has taken organic chemistry. These are representations of systems whose scale or size has been altered, but whose internal relations are still to be represented in relative order. Analog models are used when we take one system that we understand better to stand in for another that we seek to know better. A good analog model will have mathematical relations that are analogous, allowing us to think of electricity in a wire in terms of the flow of water through a pipe. Mathematical models are the mapping of a material system into sets of equations. We often speak as if our equations are the real thing and not themselves models. For example, we will often say things like "c is the speed of light," when we really mean "c represents the speed of light." But this is a confusion of the model for the thing.

Ronald Giere filled in the views by examining what we mean by models in science and the way truth and falsity enter the picture. When we say that theories are sets of models, what do we mean? Logicians have a technical sense of models as that which satisfies a set of axioms. In this way, they speak of "models of non-euclidean geometry." But we also use the word "model" in other ways. Think of model cars, or computer models, or fashion models. A model car is a representation designed to resemble a car but at a different scale. We do this sort of thing in science as well; think of Crick and Watson's double helix model, which literally was a model of the molecule built out of rods and clips. Computer models take certain mathematical representations of actual factors in a natural system and shows that if we think of interactions in terms of mathematical functions and limit the world to the inputted factors; we can see how the system would develop over time. Of course, it is only a model, but in population genetics and the earth sciences, for example, these sorts of models are standard operating procedure. Similarly, fashion models give us idealizations of what we would look like in certain outfits. Of course, in real life we never actually look like the models, but like the ideal gas law, it gives us a more perfect approximation.

But once we have a model, we can make hypotheses about the degree to which the model represents the natural system well. We can say that the model is effective at predicting observations $o_1, o_2, o_3, \ldots, o_n$, and this claim can be true or false. Models themselves are better or worse fits, but the hypotheses we make about the way they represent the actual system is the point where truth and falsity reenter the picture.

The semantic view of theories contends that theories are not sets of sentences, but sets of models. Models are not just pretty illustrations that can be thrown away if we want to just deal with the "real theory" in all its abstraction. Models are an essential part of scientific reasoning. But even if we do not want to follow the semantic theorists all the way down this road, they still allow us to ask the deeply interesting question about how we are to understand the meanings of the terms we use in science. Does "entropy," "gravitational field," or "potential energy" actually refer to something in the world? Do scientific theories tell us about what is real, but invisible?

Marshall Spector

· ·

"Models and Theories"

1. INTRODUCTION

In this paper I will attempt to show that an important currently held view as to the nature of a model for physical theory is in error. Indirectly, I shall be concerned to show that a certain related thesis about the nature of physical theories themselves is infected with serious difficulties. This will be done in the context of a modification of the former view which I shall offer.

It will be essential to have before us a careful statement of the latter thesis before we begin, for it seems to me that much of its apparent plausibility rests on the fact that presentations of it are not always very clear. (In fact, I believe that several influential attempts to *refute* this analysis of physical theories also rest on unclear and misleading presentations of it.)

According to this thesis, a physical theory is to be analysed as an empirically interpreted hypothetico-deductive system or formal calculus—in Rudolf Carnap's terms, a 'semantical system.'[1] Its most basic assumption is that a general distinction can be drawn between two types of terms occurring in physical theories—*observation* terms and *theoretical* terms.[2] The former, terms like 'green,' 'desk,' 'longer than,' refer to observable objects, properties, relations, and events, and can be understood independently of any physical theory. The latter, expressions like 'electron,' 'magnetic field,' 'spin angular momentum,' refer to unobservable (=theoretical) objects, properties, etc., and can be understood only in the context of the theories in which they occur. The attractiveness of the view I shall be examining lies in its claim to present a general, well-articulated schema for showing just how we are to understand theories which seem to talk about objects which we have never observed, and perhaps never will directly observe. In outline, the analysis is as follows.

It is maintained that we can distinguish two components in any theory. The first is its *calculus,* which is the logical skeleton of the theory, considered as devoid of any empirical meaning. This will consist of a set of primitive formulas, i.e. sentences which are taken as postulates of the calculus; and other formulas which are obtained by derivation from the postulates in accordance

Marshall Spector, "Models and Theories," *British Journal for the History of Science* 62 (1965): 121–42. Reprinted with the permission of Cambridge University Press.

with specified rules of transformation. Two types of terms appearing in the calculus may be distinguished: the primitive terms, i.e. terms which are not defined on the basis of other terms within the calculus; and non-primitive terms, i.e. those which are introduced on the basis of the primitives. (This distinction is not identical with that between observation terms and theoretical terms.) When this calculus, or syntactical system, is given an empirical interpretation, or meaning, it becomes a system of empirical statements having the structure of a hypothetico-deductive system. The primitive formulas become empirical statements which will be true if the hypotheses are true.

To illustrate some of these ideas, consider the kinetic theory of gases. The postulates of this theory will contain such expressions as 'molecule,' 'mass of a molecule,' and 'position of a molecule.' These might be considered as primitives. Other expressions, such as 'momentum of a molecule,' and 'mean kinetic energy of a group of molecules,' will be introduced on the basis of the primitives. One of the postulates might be 'All gases are composed of molecules.' A typical derived formula might be 'If the pressure of a gas is increased while its temperature remains constant, its volume will decrease.'

The second component of a theory, the empirical interpretation, is given to the calculus by *semantical rules* for terms of the calculus, i.e. rules which are formulated in a suitable metalanguage (usually 'ordinary' English, or German, etc.) and provides the meaning of the terms by stating what properties, relations, or individuals the terms designate.[3] An example is 'The term "P" of the calculus designates the pressure of a sample of gas.'

Now the authors we shall be considering maintain that

(*a*) not all terms of the calculus of a theory *need* be given semantical rules, and

(*b*) not all terms of the calculus *can* be given semantical rules.

(It has not been clearly recognised that these are *distinct* claims. The importance of noting this distinction will become clear as we proceed.) Only the observation terms of the unanalysed theory are thus 'directly interpreted.' That is if we look upon the calculus of a semantical system as the uninterpreted logical skeleton of a theory, for which the semantical construction provides a *reconstruction*, only those terms in the calculus which represent the observation terms of the unreconstructed theory will be given semantical rules in the completed reconstruction. Theoretical terms of the unanalysed theory will not be given semantical rules in the semantical system reconstruction. It is claimed that such terms *cannot* be 'understood in themselves,' but must be understood—given their meaning—in an 'indirect' manner, through the role they play in the theory. Such terms obtain an empirical meaning if and only if they appear in sentences in the calculus which also contain terms which *are given semantical rules*—the observation terms. Such sentences are known as *correspondence rules*.

An example of a correspondence rule is the postulate stated earlier 'All gases are composed of molecules.' Symbolically, this would read '$(x)(Gx \rightarrow Qx)$.' The term 'G' is given the semantical rule '"G" designates the property of being a sample of gas,' and the term 'Q' (which is a symbolic translation of the theoretical expression 'is composed of molecules') is not given a semantical rule, but is said to obtain a partial meaning indirectly by virtue of its occurrence in a sentence which contains a term ('G') whose meaning is given directly and completely by a semantical rule. (It is important to notice that *all* the terms of a theory are found in the calculus, including the observation terms. If this is not kept in mind, there arises a tendency to confuse semantical rules with correspondence rules.)

This, in outline, is the analysis of physical theories which underlies the analysis of the concept of a model for a theory which I shall be examining. For reasons which should be apparent, I shall refer to it as the *partial interpretation thesis*.[4]

The clearest and most precise explication of the concept of a model for a physical theory offered by a proponent of the partial interpretation thesis is that given by Braithwaite,[5] so the burden of my analysis will be directed towards his position. A model for a theory, according to Braithwaite, is to be understood as *another interpretation* of the theory's calculus, in which the *theoretical* terms are directly interpreted (by semantical rules).[6] It is sometimes also stated that these direct interpretations must be in terms of familiar concepts. But this need not be stated as a separate condition if we remember Braithwaite's (and Carnap's) injunction to the effect that this is the only way direct interpretations can usefully be given.[7] Now, if we use the term 'model' to designate not a domain of non-linguistic entities, but rather the statements about such entities, we can say that a theory and a model for the theory are two sets of statements which share the same calculus, but with the epistemological order of the two reversed. As Braithwaite puts it:

> A theory and a model for it . . . have the same formal structure, since theory and model are both represented by the same calculus . . . But the theory and the model have different epistemological structures; in the model the logically prior premises determine the meaning of the terms occurring in the representation in the calculus of the conclusion; in the

theory the logically posterior consequences determine the meaning of the theoretical terms in the representation in the calculus of the premises.[8]

Braithwaite offers this explication of the concept of a model not as an arbitrary definition, but rather as

an attempt to make more precise the notion of a model for a scientific theory widely current in discussions of the philosophy of science [ibid.].

Braithwaite also points out, quite emphatically, what models allegedly are *not*. Braithwaite claims that one of the chief 'dangers' in the use of models is the tendency that

the theory will be identified with a model for it, so that the objects with which the model is concerned . . . will be supposed actually to be the same as the theoretical concepts of the theory [op. cit. p. 93].

Nagel also warns against confusing a model with the theory itself. After commenting on the possibility that a model may be 'an obstacle to the fruitful development of a theory', he writes:

The only point that can be afforded with confidence is that a model for a theory is not the theory itself.[9]

Braithwaite goes on to say that

Thinking of scientific theories by means of model sis always *as-if* thinking; hydrogen atoms behave (in certain respects) as if they were solar systems each with an electronic planet revolving round a protonic sun. But hydrogen atoms are not solar systems; it is only useful to think of them as if they were such systems if one remembers all the time that they are not.[10]

According to the explication offered by Braithwaite, then, the objects of the model cannot be identified with the theoretical objects of the theory. Such an identification would be a logical error; the possibility of this identification (i.e. the question of the 'reality' of the objects of the model) cannot even arise.[11]

However, as I shall argue, these questions *do* legitimately arise for some systems which physicists would recognise as models; and since Braithwaite's explication of the concept of model cannot allow for this, it is inadequate. To show this, I shall begin by distinguishing four types of domains. They will be quite different in important respects; yet Braithwaite's explication will not be able to distinguish them. Then I will offer a modification of Braithwaite's explication which will be able to distinguish these types of domains. I shall conclude with an analysis of this modification on the partial interpretation thesis itself.

2. FOUR 'BRAITHWAITEAN' MODELS

(1) Suppose we have a geometrical interpretation of the calculus of some physical theory. (Braithwaite frequently uses geometrical interpretations of his 'factor-theories' as examples of models.) Such an interpretation would be a model for the theory in Braithwaite's sense, and yet the possible identification of the 'objects' of this system with the theoretical objects of the theory would not even arise for the physicist. There would be no question of the lines, triangles, circles, etc., being identified with, or being similar to, the theoretical objects of the theory. Only the formal structure would be relevant. If there are, in fact, triangles, etc., which satisfy the postulates of a given physical theory (its calculus), then the objects of the system are 'real'; but this does not mean that the theoretical objects of the theory are triangles, etc. This sort of identification indeed would be a mistake, but a strange sort of mistake that only a modern-day Pythagorean might make. (It is certainly quite different from the mistake of identifying the theoretical objects of the kinetic theory of gases with, say, billiard balls.) In this sense, a model has nothing whatsoever to do with the *domain* of the theory. Here, the system is another interpretation of the theory's calculus in a very strong sense of 'another.' The objects of the 'model' are of a different logical type from the objects of the domain of the theory. A physicist would probably not call this a model at all.

(2) It has been recognized that there is a rather thorough-going analogy between some of the laws of acoustical theory and those of electrical circuit theory. Put another way, there is an important correspondence between acoustical systems and electric circuits. Let us consider a specific instance of this. If we have a series of circuits consisting of a resistance R, a capacitance C, and an inductance L, with a periodic electromotance ε equal to $E \cos{(wt)}$, the charge on the capacitor, q, will satisfy the equation

$$L\, d^2q/dt^2 + R\, dq/dt + 1/C\, q = E \cos(wt).$$

Now consider a Helmholtz resonator, which is a 'flask' of volume V with a neck of length d, radius of a, and cross-sectional area $S = \pi a^2$. A sound wave of amplitude P impinges on the resonator opening, so that the driving force at the neck is given by $SP \cos(wt)$. The air displacement z at the neck will then satisfy the equation

$$(pd'S)\, d^2z/dt^2 + (pck^2S^2/2\pi)\, dz/dt + (pc^2S^2/V)q = SP\cos(wt),$$

where $d' = d + 16a/2\pi$, $p =$ mean air density, $c =$ sound wave velocity, and $k = w/c$. these two equations are of the same form:

$$\alpha\, d^2\xi/dt^2 + \beta\, d\xi/dt + \gamma\,\xi = \delta \cos(wt).$$

For other acoustical systems, it is also possible to find electric circuit 'analogues.' In fact, this can be done in a general way, and certain combinations of acoustical parameters are given names taken from electric circuit theory. In the system just described, for example, the quantity $pck^2/2\pi$ is called the 'acoustical resistance,' R; V/pc^2 is called the 'compliance,' C; and pd'/S is called the 'inertance,' M. Moreover, these names are not given merely on the basis of the expressions appearing in the same place in the acoustical equation as the corresponding expressions in the equation for the series circuit; the analogy is more than formal. Thus, the inductance of an electric circuit is a measure of the tendency of the circuit to resist changes in current ($= d\xi/dt$), while the inertance of the acoustical system is a measure of the tendency of the system to resist changes in air velocity ($= dz/dt$). Similarly, the capacitance of a circuit is a measure of the ability of charge to 'pile up,' so to speak, in one place in the circuit, while the compliance in an acoustical system, e.g. the resonator, is a measure of just how far back the air in the neck will 'allow itself' to be pushed. (If it is claimed that the analogy is nevertheless a formal one only, this would strengthen the point I shall presently make.)

Finally, the analogy is actually used to solve practical problems. It is not merely a heuristic aid—allowing one theory to be taught in terms of the other. To quote from a textbook in acoustics:

Consideration of the equivalent electric circuit offers many advantages in solving practical engineering problems of applied acoustics. For example, many acoustical systems are so complicated that their mathematical analysis is very difficult, if not impossible, and their design by a cut-and-try [sic] experimental method is extremely tedious, as each change involves constructing a new part. On the other hand, if it is possible to set up an equivalent electric circuit, the electrical constants of this network may readily be varied [actually—not just in mathematical analysis] to obtain the desired experimental characteristics, and the constants of the mechanical system may then be calculated from their electronic equivalents. This technique has been used in the design of loudspeaker systems and other acoustical devices.[12]

Now, given all of this, is electric circuit theory a model for acoustical theory? (Or, are electric circuits models for acoustic theory?). Usually, the term 'analogue' is used here, though one would be understood if the above relation were referred to as a modeling relation. The important point is, again, that there is no question of the identification of the subject matter of the two theories. Acoustics deals with small vibrations in air (generally, elastic fluids), whereas electric circuit theory deals with the movement of electric charge in certain types of systems. The analogy is perhaps not completely formal but the substantive similarities are rather weak.

(3) Here I have in mind systems which physicists would certainly recognize as models—systems for which the question of 'reality' *does* arise. But here, the physicist knows that the objects of the model cannot be identified with the theoretical objects of the theory for *definite physical reasons*, although there is some *substantive similarity*. As an example of this, consider the following passage from the arch-modellist William Thompson (Lord Kelvin). (I have italicized the parts to which I wish attention drawn.)

> To think of ponderable matter, imagine for a moment that we make a *rude* mechanical *model*. Let this be . . . [Here Kelvin describes a rather elaborate contraption, as Duhem might have called it] . . . you will have a *crude model*, as it were, of what Helmholtz makes the subject of his paper on anomalous dispersion . . . If we had only dispersion to deal with there would be no difficulty in getting a full explanation by putting this *not* in a *rude* mechanical form, but in a form which would commend itself to our judgment as presenting the *actual* mode of action of the particles, of gross matter, whatever they may be upon the luminiferous ether. . . . It seems that there must be something in this molecular hypothesis, but a *reality*. But alas for the difficulties of the undulatory theory of light . . .[13]

Kelvin has offered a model, which is only a 'rude' model, *because* it will not work in certain important, fully specified situations. If it were not for these other circumstances, which contradict the phenomena that could be expected on the basis of the model, this would not be a rude model, but a 'reality.' Consider also the following passage:

> The luminiferous ether we must imagine to be a substance which *so far as* luminiferous vibrations are concerned move *as if* it were an elastic solid.

That it moves *as if* it were an elastic solid *in respect to* the luminerous vibrations, is the fundamental assumption of the wave theory of light [*loc. cit.* p.9. (italics mine).]

Kelvin does not identify his model with the domain of the relevant theory for *empirical* reasons — not because of a general philosophical point about the relation between models and theories. For Kelvin goes on to show that there are *other specific* phenomena with respect to which the ether does *not* behave like an elastic solid, (for example, the fact that material bodies move through it, showing that it sometimes behaves as if it were a fluid).[14] I think it is quite clear from the quoted passages that if it were not for these other, specific, intransient phenomena, Kelvin would have dropped the as-if, rude-model, terminology.[15]

Notice that there is no question of identifying the ether with the actually gadgetry in a constructed model, just as for a 19th century physicist there is no question of identifying the molecules of the kinetic theory of gases with actual billiard balls (which are made of ivory, have numbers on them, etc.). The proposed identity would be between molecules and elastic spheres, which are exemplified by billiard balls; or between the ether and an array of springs, etc., which is exemplified by the gadget on the laboratory bench.

Failure to make this distinction is what lends plausibility to Braithwaite's statement about hydrogen atoms and solar systems quoted near the end of the introduction to this paper. Of course hydrogen atoms are not solar systems; the nucleus of the hydrogen atom is not a star and the electron is not a planet — just as gas molecules are not billiard balls. But is it as obvious that the hydrogen atom is not a system in which something is going around something else? (Unfortunately there is no standard term 'elastic sphere' to describe billiard balls and other such objects.) Bohr believed this to be the case; we no longer do for definite physical reason, although the similarity is still there — hence the 'as-if.'

(4) Finally, there are systems which are no longer *merely* models. Here, the question of the reality of the model (i.e. the identification of the objects of the model with the theoretical objects of the theory) is not only significant, but is answered in the affirmative. A system may be originally introduced as a model in sense (3), but may eventually be modified to such an extent that we will finally speak of the identity of the objects of this system with the theoretical objects of the theory. When this happens, we may or may not continue to speak of the system as being a *model*. Actual usage may depend on factors such as the historical development of the theory. For example, we may still speak of

the discrete particle model for the kinetic theory of gases, even though we have identified the concept of a molecule from the kinetic theory with the concept of a discrete particle (or 'object'). Here, the model is not *another* interpretation of the theory's calculus, but a *filling out of the original interpretation*. (This point will be made clearer in the following sections.)

Now, the first serious problem with Braithwaite's explication of the concept of a model as simply another interpretation of the theory's calculus is that it cannot distinguish between the four types of systems just discussed—and these *are* distinguished by physicists. For Braithwaite, all four would be models. However, we found that the first would not be recognized as such by physicists. The second would usually be described as an analogue. The fourth would probably not be described as a model either, but for quite a different reason—it is too good, as it were. 'Questions of reality' arise for some of these systems, but not for others; whereas for Braithwaite, it is claimed that they should not arise for any of them. In other words, Braithwaite is correct in saying that identity of logical structure alone is not enough to be able to claim identity between the objects of the model and the theoretical objects of the theory. But I take this as showing that Braithwaite's explication of the concept of model is inadequate, and not as an insight into the nature of models for physical theories.

It could be objected at this point that reliance on what physicists say in such situations is not sufficient for establishing a point about the logic of the situation (or for refuting such a point, as I am attempting to do). Perhaps not; I shall have more to say about this below. But suppose that we could amend Braithwaite's explication of the concept of a model in such a way as to take account of what physicists say—which *would* leave a place for the distinctions drawn above. And suppose that this emendation would be such that it could be accepted by the partial interpretation theorists as being within the letter and spirit of their analysis of the structure of physical theories. Clearly this would be a gain in understanding. I shall, in the next section, offer such a modification of Braithwaite's explication. I shall also attempt to show that this modification will be able to overcome certain further difficulties inherent in Braithwaite's explication.

3. AN EMENDATION OF BRAITHWAITE'S EXPLICATION WITH AN EXAMPLE

As long as only identity of formal structure is required between a theory and a model for it, there can be no relevant connection between the domain or subject matter of the theory and the domain of the model. A geometrical model for the calculus of a physical theory provides a striking example of this, and the electric circuit analogue for acoustical theory provides an example where the domains are both of physical objects, yet quite different. In each case, we do not have a model in the physicists' sense, and there is no question of identification of the objects of the model with the theoretical objects of the theory.

If, however, the observable properties of the domain of the theory—the designata of the observation terms—are similar to the properties of the model presented by these same terms when the calculus is interpreted in the domain of the model, then the possibility arises of comparing the properties of the model represented by the theoretical terms of the calculus with the theoretical objects of the theory. That is, *we can argue by analogy to the nature of the theoretical properties.* A good example of this is provided by the elementary kinetic theory of gases and its usual model. It will be useful to compare in some detail the interpretations of the calculus of the kinetic theory of gases in the theory itself and in the elastic sphere model for the theory. (Notice that this distinction itself sounds somewhat strange, for this theory is usually presented by means of the model. The reasons for this will become clear as we proceed.)

In the theory, certain *defined* terms of the calculus are interpreted as designating the observable properties of a sample of gas in a container at equilibrium—volume, pressure, and temperature.[16] In the model, on the other hand, certain *primitive* terms of the calculus are interpreted as designating a group of elastic spheres in a container and some of their properties. Some of the primitive formulas express propositions describing these elastic spheres (their masses, for example), and others are interpreted as expressing the laws of classical dynamics (the laws governing their motion).

Now we notice that the designata of some of the defined terms of the calculus, when used to represent the theory, are similar (in this case, identical) to the corresponding designata of these terms, when the calculus is used to represent statements about the model. Thus, for example, there is an expression in the calculus, which by the interpretation given the calculus in the model, represents the total rate of momentum transfer per unit area to the walls of the container in which the elastic spheres are moving. This expression, when the calculus is used to represent the theory, appears in a correspondence rule of biconditional form (a definition, in Carnap's sense) with the observation term '*P*,' which designates the pressure of the gas. But according to classical dynamics, rate of momentum transfer per unit area is equal to force per unit area, which is (by definition) the pressure on the wall of the container. Thus we have not only a formal identity—a shared calculus, but also a *substantive* identity of two properties, one from the domain of the theory and one from the domain of the model.[17]

The same is trivially true in the case of volume, thus giving us two sets of *identical properties*—out of a possible three.[18] We are left with temperature in the theory and mean kinetic energy of the elastic spheres in the model as the designata of a defined term ('T') in the calculus when used in the theory and in the model, respectively. Are these last two properties also similar? Reasoning by analogy on the basis of the first two similarities (identities, in this case), we would expect so, which would amount to suspecting that gases are, in fact, *composed of* elastic spheres (suspecting that molecules are elastic spheres— that the model 'corresponds with reality'). At first sight, however, it would seem that the temperature of a gas and the mean kinetic energy of a swarm of elastic spheres are quite dissimilar. But we are not concerned here with temperature as a felt quality of bodies or as a sensation (hot-cold). Rather, we are concerned with the level of a column of mercury (for example—assuming that we are using a mercury thermometer) in a capillary tube. Now we notice that if liquids were also composed of elastic spheres,[19] they would expand if placed in contact with a gas, so composed, if (and only if) the mean kinetic energy of the particles of the gas were higher than in the liquid immediately before contact. In fact, an equilibrium state would soon be reached in which the mean kinetic energies of the two would be the same, at which time the liquid would cease expanding. All of this follows from classical dynamics. But this is just what is observed. When a mercury thermometer is inserted in a gas, the column will rise (or fall) for a time, and reach a stationary level.

On the basis of this, and the two identities mentioned earlier, we can conclude at least that gases behave just as we would expect them to behave if they were in fact composed of small elastic spheres in incessant motion. That is, we can tentatively identify the objects of the model with the theoretical objects of the theory. Gases behave as if they were composed of elastic spheres. The 'as-if' does not here indicate that we have any contrary information (cf. Kelvin) but rather that we may still fell that there are other tests to which we would like to put this hypothesis before committing ourselves. If we had not the slightest idea of what other tests would be relevant, or if we could satisfy ourselves that other tests were impossible, there would be no point in the 'as-if.' In this case, there are such further tests, and they follow directly from a consideration of the elastic sphere model. (The theory itself—considered as the partially interpreted calculus with semantic rules for the observation terms only—affords no reason whatsoever to expect these other phenomena.) I have in mind the so-called 'molecular beam' experiments.

We open a slit in one wall of the container of gas, and by a suitable experimental arrangement[20] we can see whether the results are those which would be expected (quantitatively as well as qualitatively) if a stream of small (unobserved) particles having a certain mean kinetic energy were to issue from the slit in accordance with the dynamics of such particles. As is well known, such experiments do in fact confirm the hypothesis that gases are composed as described—that we can identify the theoretical objects of the theory with the objects of the model. The measured velocities of the particles are what they would have to be in order to identify the temperature of the gas with the mean kinetic energies of these particles, thus completing the analogy.

Imagine a physicist who now says: 'Well, it has seemed as if gases might be composed of these small elastic spheres, considering the similarities you mentioned; now I'm convinced that they really are. After all, the results of this experiment too were exactly what could be expected if they were.' Has he committed an error? Has he succumbed to one of the 'dangers' involved in model thinking? If he had relied only on the *formal* characteristics of the model, he would have indeed made a mistake, although in this case a fortunate one. But the argument was based on the *substantive* characteristics of the model, and as such is perfectly good analogical reasoning. And it would still have been good reasoning even if the molecular beam experiment had given wholly different results, thus refuting the identification of the theoretical objects with the objects of the model. (Although we might then doubt the reliability of the apparatus rather than the model!) His mistake would then merely be that he had chosen a false hypothesis—but not a meaningless one, or an improbable one, or one which indicated a basic misunderstanding of the very use of models. The identification on the basis of the first two sets of identical properties alone, is at least probable, and the results of the molecular beam experiment increase the probability.

Notice that it is not necessary at this point to produce a satisfactory theory of confirmation. Any explication of the notion of the probability of a scientific hypothesis which did not reflect the above would have to be rejected, as the reasoning sketched is as good a paradigm as can be found of one type of sound scientific reasoning.[21]

Thus, for a model to qualify as a 'candidate for reality' (or, for a system to qualify as a model in the physicist's sense) it must be more than just some other interpretation of the theory's calculus. There must be a substantive similarity between the designata of the defined terms of the calculus when used to represent the observable properties in the domain of the theory and when used to represent the (formally) corresponding properties in the domain of the model. If there is, moreover, an identity of these properties, the model is no longer 'merely' a model; the theoretical objects and properties simply are. . . .

This then is the emendation of Braithwaite's explication which will allow one to distinguish between the four types of domains sketched earlier, and which will accommodate an important type of reasoning based on models which physicists do in fact use. This emended form of Braithwaite's explication can also account for some of the uses to which models are put by physicists in modifying and extending a theory. I shall attempt to show this in the next section. But at the same time, it will become apparent that this emendation cannot be accepted by one who holds the partial interpretation thesis. We shall see that the modification I have suggested is in contradiction with one of the most basic assumptions of this type of analysis of the structure of physical theories.

4. MODELS AND MODIFICATIONS OF A THEORY

The laws deduced from the simplified kinetic theory of gases sketched above are not as accurate as we would like them to be, and we would like to modify the theory to remedy this situation. In this case, the elastic sphere model 'points to its own extension,'[22] providing leads as to how it can be reasonably be modified. For example, we might argue as follows:

If gases are composed of elastic spheres, perhaps we should take their radii into account (i.e. if they are not just point masses). Also, they may exert forces on each other when not in actual contact. If we make these assumptions, and modify the calculus of the theory accordingly, we will be able to derive a formula which, when interpreted term by term in the domain of the theory, is the van der Waals equation of state,

$$(P + a/V^2)(V - b) = cT,$$

where a is a constant for a particular type of gas depending on the force law, and b is a constant for each type of gas depending on the radii of the spheres. This is a more accurate description of the behavior of gases than the original ideal gas law, $PV = kT$.

The new primitive formulas of the calculus (or, the primitive formulas of the new calculus) which allowed for the derivation of this law were essentially 'read off' from the model. By this I mean that a domain of objects was described which obeys the laws of another familiar theory—the dynamics of rigid spheres attracting each other in accordance with a stated force law (involving a). The statements of the description, together with the statements of the laws of this other theory, upon disinterpretation, become the primitive formulas of the modified calculus. (In the original model, the other theory was the dynamics of mass-points interacting only by contact. This was an idealisation, in an obvious sense, hence the name 'ideal gas law' for the equation derived from the associated theory. Note also how this is a reduction of part of thermodynamics to classical dynamics.)

In this modification of the model, the identity among derived properties in the model and in the theory, spoken of earlier, still obtains. Thus if we had started with the van der Waals equation of state, we could have used the same sort of reasoning by analogy described earlier to establish, or make probable, the identification of the primitive properties of this model with the theoretical properties of the theory. And once again, this sort of reasoning would not be valid if we had considered only the formal similarity (as Braithwaite correctly recognizes when he argues against identifying the model with the theory). Moreover, if we had started with the van der Waals equation, it would have been impossible to *construct* the calculus of the kinetic theory without thinking of the model. (The very form in which the equation is stated betrays its origin in the model.) And if perchance someone hit upon the right modification of the original calculus without considering the substantive aspects of the model, he would have no justification for believing it probable.

This difference between what can be done with the 'theory itself' and with the model is even more striking when we consider the partial interpretation thesis view that theoretical terms obtain their meaning by being connected with observation terms via correspondence rules. For now we have a much more complicated calculus just in order that the directly interpreted formula relating 'P,' 'V,' and 'T' can be more accurate for a certain range of conditions. Now suppose (contrary to fact) that the ideal gas law were more accurate than the van der Waals laws. Before dropping the modified theory in favour of the simpler one, physicists would, I think, look for experimental errors, because the modified model, associated with the modified theory, appears to be a *more plausible* representation of the theoretical objects than is the original model. But on the basis of the partial interpretation thesis, and its associated explication of the notion of a model, this would be an irrational procedure, stemming from a 'misunderstanding' of what a model is.

Also, the derivation of the results of the molecular beam experiment could not be carried out with either the simplified or the full theory without a change in the observation language itself. We would need new terms in the calculus designating the distance between the container of gas and a sheet of film located inside a rotating cylinder, the angular velocity of the cylinder, the degree of blackening of the film, etc.[23] But if we introduce these new terms and the new correspondence rules needed for them, and at the same time accept the partial interpretation thesis as to how theoretical terms obtain their empirical meaning, we will have the very strange result that the meanings of the terms 'molecule,' 'mass,' 'velocity,' etc., have changed. After all, the theoretical terms are supposed to obtain their meaning from the observation terms through correspondence rules, and we have added new observation terms and correspondence rules. In the model, on the other hand, there is no such change in meaning. 'Elastic sphere' means the same when the model is first conceived as it does after we realize that this model points a new *test* of the theory in the molecular beam experiment. But we found that we could identify the objects of the model with the theoretical objects of the theory, through analogical reasoning based on substantive similarities in the two domains. Therefore, on the basis of this identification we are forced to conclude (what is more plausible on its own merits) that the addition of the new test with its corresponding set of new observation terms and correspondence rules does not change the meaning of the theoretical terms of the theory, but rather gives new ways to test the truth of the theoretical statements (whose meanings are given in some other way—directly, as will soon become apparent.)

The view that the meanings of the theoretical terms change with the addition of new observation terms and correspondence rules has a similar strange consequence when we consider the sort of unification of physical theories that is accomplished by reasoning based on the use of a common model. For example, Niels Bohr, in his 1913 paper, 'On the Constitution of Atoms and Molecules,'[24] using reasoning based on a model of the atom (notice it is not called a model for atomic theory) which can be considered as a detailing of the elastic sphere model of kinetic theory, explains why certain lines in the spectrum of hydrogen gas had been missing, and predicts under what conditions we can expect to observe them. According to the partial interpretation thesis, even though the Bohr theory contains the terms 'atom,' 'mass,' etc., these terms have a different meaning from the that in the kinetic theory of gases. The observation language of the Bohr theory is different from that of the kinetic theory. As are (of course) the correspondence rules; and his reasoning, which depends on the identities of the meanings of these terms in the two theories, would be invalid.

In cases like this, it is the model which is the heart of the physicist's investigations. If we disinterpret the calculus that is read off from the model, and re-interpret it 'from the bottom up,' and call the result the theory itself, looking now upon the model as merely another possible interpretation of the theory's calculus, we must accept the conclusion that arguments such as Bohr's are co-lossal logical blunders—fortunate blunders—showing a lack of understand-ing of how theoretical terms get their meaning, and a lack of understanding of what a physical theory is.

Now it may well be the case that much of what physicists *say* about the methodology of their science is of dubious value—I do not wish to argue this point here. But the partial interpretation thesis, coupled with Braithwaite's explication of the concept of a model (which leaves out substantive similari-ties), is not only in contradiction with this; it is also in contradiction with ac-tual theory construction and theory modification—what physicists *do*. Now Braithwaite, Carnap, and Nagel say that they are not interested in construct-ing a logic of discovery. Rather, they are interested in analyzing the final prod-uct. But their position is not neutral with respect to the former; it instead de-clares certain types of reasoning which are paradigms of physical genius to be (fortunate) logical mistakes of a very basic sort. (The parenthetical adjective alone should make one suspicious.)

We would not even describe the kinetic theory of gases as a reduction of thermodynamics to (statistical) dynamics. We could only notice that there is a formal similarity between some of the primitive formulas of the kinetic theory and the laws of classical dynamics. But these laws would not have the same meaning in each case. In the kinetic theory, the meanings of the terms 'force,' 'mass,' 'momentum,' etc., would have to be analysed in terms of the ob-servation terms 'pressure,' 'temperature,' 'and 'volume.'

What has happened here? I think the basic point is that the suggested emendation of Braithwaite's explication of the notion of a model—i.e. taking into account substantive features of the model—as plausible as it may seem, cannot be accepted by one who holds the partial interpretation thesis. That is, even though my suggested emendation eliminates the difficulties I have been pointing out, and apparently does so while remaining within the letter and spirit of the programme of the partial interpretation theorists, the emen-dation contradicts a basic assumption of the partial interpretation thesis. For analogical reasoning from *substantive* similarities in the designata of the terms in the derived formulas in the model and in the theory to substantive simi-larities in the theoretical (primitive) properties amounts to a direct interpre-tation of the theoretical terms in the theory. Thus, in the case of an *identity* of derived properties, the completion of the analogy is tantamount to *giv-*

ing semantical rules for theoretical terms. (If there is only a similarity, in which case we have one type of 'as-if thinking,' we are giving qualified semantical rules, so to speak—cf. the earlier quotations from Kelvin.) But according to the partial interpretation thesis, we *cannot* give a consistent direct interpretation (semantical rules) for the theoretical terms of the calculus when it is used to express the theory, for 'we could not understand them' (Carnap); the theoretical terms are 'not understood in themselves,' but only as a part of the whole system (Braithwaite). Theoretical terms gain what empirical meaning they have *only* 'from below'; the 'zipper' moves from the bottom up in giving meaning to the theoretical terms. Thus, if the thesis is held as to how theoretical terms become meaningful, one is forced to consider a model as some *other* interpretation of the calculus. To accept our results is to accept the possibility of giving meaning directly for theoretical terms.

5. CONCLUSIONS

Our suggested emendation of Braithwaite's explication of the concept of a model has implied that theoretical terms *can* be given direct interpretations—semantical rules, where the metalanguage used to state them *is* 'understood.' It is in fact the same metalanguage used to interpret the observation terms in the calculus. That is, if we wish to analyse physical theories in terms of interpreted formal calculi, we *can* give semantical rules for theoretical terms (thus, for *all* terms of the calculus). Notice that this still gives the meanings of the theoretical terms 'on the basis of observations,' but by statements outside the calculus rather than by correspondence rules.

I have also shown that theoretical terms *must be given semantical* rules, to allow for certain types of physical reasoning. Certain extension and modifications of a theory were based on the use of a model *in the emended sense*, which we saw is tantamount to interpreting the theoretical terms of the theory. But this refutes what I referred to as assumptions (a) and (b) of the partial interpretation thesis in the introduction to this paper.[25]

All of this also implies that theoretical or unobservable *objects* may (in some cases) be described by observational *predicates*. Let us see why this is so. If a semantical rule is to be successful in giving meaning to a theoretical term of the calculus, it must be stated in a metalanguage which is 'already understood.' That is, it must supply a designatum which is understood independently of the theory being reconstructed. Otherwise, it would be, as Carnap has put it, a useless transcription 'from a symbol in a symbolic calculus to a corresponding word expression in a calculus of words.'[26] This requirement has usually been put in the form of demanding that the properties, things, and relations which are to be the designata of the terms of the calculus must be observable, i.e. the metalanguage must be a theory-uninfected observation language. (This observation language is not to be confused with the observation language of the theory's calculus, which is part of the object language being analysed.) It would seem that familiarity should be sufficient; but let us grant the stronger requirement of observability for the present. Now, if we give semantical rules for the theoretical predicates (say) of the calculus, and do so in accordance with the just mentioned condition on the metalanguage, we will have the result that *unobservable objects may be characterized by observable predicates.* More accurately, we will have a sentence of the form '$P(a)$,' with 'a' designating an unobservable (=theoretical) object and 'P' designating a property of which it is possible to observe instances applying to observable objects (e.g. 'having weight' as applied to an electron or atom).

This is the same conclusion reached earlier from a consideration of the concept of a model for a theory. We saw that the identification of the objects (and properties) of the model with the designata of the theoretical terms 'of the theory itself' was tantamount to giving semantical rules for theoretical terms and thus applying observational predicates to unobservable objects. In the example of the kinetic theory of gases, this amounted to maintaining that the unobservable atoms could have observational properties—or at least familiar properties (mass, velocity) from another theory (classical dynamics).

Moreover, we saw that there is nothing unintelligible about unobservable objects being characterized by observational predicates. The reasoning that leads a physicist to impute observational properties to unobservable objects was seen to be perfectly acceptable analogical reasoning.

These conclusions, however, conflict with the most basic presupposition of the proponents of the semantical system approach. This is the assumption that we can distinguish in a general manner two types of terms or concepts (and statements) in physical theories: the observational terms, which designate observable properties, relations, events, and objects *only;* and the theoretical terms, which purport to designate unobservable properties, relations, events, and objects *only.* This 'dual language model' of the vocabulary of science has been expressed by Carnap as follows:

[We accept] the customary and useful [division of] language of
science into two parts, the observation language uses terms designating
observable properties for the description of observable things and
events [, and] the theoretical language [uses] terms which may refer to
unobservable aspects or features of events, e.g. to micro-particles . . .[27]

We have seen, therefore, that what are perhaps the three most basic assumptions of the partial interpretation thesis are in error.

Now these assumptions involve two crucial notions—that of an observation term, and that of a theoretical term. In this paper, I have treated these notions as if they were clear and unproblematic. They are not, however, and a full evaluation of the partial interpretation thesis—and the semantical system approach itself—requires a careful analysis of them. This I hope to do in future papers.

NOTES

1. The oldest and most precise formulation of this position is found in Rudolf Carnap, 'Foundations of Logic and Mathematics,' I, no. 3, of the *International Encyclopedia of Unified Science*, Chicago, 1939. Later statements of this view are found in other writings of Carnap, in various papers by Carl Hempel, in Ernest Nagel's *The Structure of Science*, New York, 1961, in R.B. Braithwaite's *Scientific Explanation*, Cambridge, 1953, Arthur Pap's *Introduction of the Philosophy of Science*, Glencoe, Ill., 1962, Ernest Hutten's *The Language of Modern Physics*, London, 1956, Peter Caws's *The Philosophy of Science*, Princeton, 1965. Further references may be found in the works just cited.

2. See Carnap, loc. cit., p. 203, Carnap, 'The Methodological Character of Theoretical Concepts,' in *Minnesota Studies in the Philosophy of Science*, I, p. 38, Nagel, op. cit. pp. 81 ff, Braithwaite, op. cit. p. 51. The same assumption is made by the other authors mentioned in note 2. Most of them hasten to say that the distinction may not be a sharp one.

3. See Carnap, 'Foundation of Logic and Mathematics,' loc. Cit. p. 153, Carnap, *Meaning and Necessity*, Chicago, 1946, pp.4 f. See also the introductory sections of Alonzo Church, *Introduction to Mathematical Logic*, Princeton, 1956.

4. I have borrowed this expression from Professor Peter Achinstein of the Johns Hopkins University.

5. See his book, *Scientific Explanation*, as well as his more recent paper, 'Models in Empirical Sciences.' In Nagel, Suppes, and Tarski (eds.), *Logic, Methodology, and Philosophy of Science*, Stanford, 1960.

6. It will not usually matter whether we use the term 'model' to refer to the domain of objects which the interpreted calculus makes statements about, or to the interpretative sentences themselves.

7. Cf. 'Foundations of Logic and Mathematics,' p. 204. The metalanguage must be already understood. The familiarity of the concepts need not entail their observational character. They may be from another, 'better understood' theory, but eventually, the chain must end with observational concepts. (See Braithwaite, 'Models in the Empirical Sciences,' loc. Cit. p. 227.)

8. *Scientific Explanation*, p. 90.

9. *The Structure of Science*, p. 116.

10. *Scientific Explanation*, p. 93.

11. Compare Nagel, op. cit. p. 116.

12. L.E. Kinsler and A.R. Frey, *Fundamentals of Acoustics*, New York, 1950, p. 33.

13. *Baltimore Lectures on Molecular Dynamics and the Wave Theory of Light,* London, 1904, pp. 12–14.

14. Thus it was much like the product once on the market in the United States known as 'Silly Putty.' This was a substance which could be molded like clay, but which would also bounce if dropped several feet. It behaved like a fluid under some circumstances (slowly applied force), but like an elastic solid under others (rapidly applied force). Ice is another example, glaciers flow, although ice is 'usually' brittle.

15. See also Edmund Whittaker, *A History of Theories of Aether and Electricity,* New York, 1960, I, ch. IX, 'Models of the Aether.' Whittaker describes a series of (mechanical) models of the ether, and it is apparent that the 'question of reality' *did* arise for these models, and was taken quite seriously. In each of these cases, there was some *substantive* similarity between the domain of the model and the domain of the theory (electro-dynamics). And in each case, the model was considered only as an 'as-if' because it didn't work—i.e. certain phenomena (though not all) expected on the basis of the model were not observed.

16. If temperature and pressure are not considered as sufficiently elementary or observable, the analysis could be carried out in terms of the observed heights of mercury columns; but this would be at the expense of clarity without changing the results.

17. Here, the identity is established on the basis of classical dynamics, rather than being an observed identity. But, as Braithwaite has noted, the familiarity need not necessarily involve observability.

18. We are at present interested in only three 'observable' properties of the theory—those which enter into the ideal gas law, $PV = kT$.

19. But where the spheres also exert an attractive force on one another. This is admittedly crude, but putting in all of the details would only complicate matters without helping (or hindering) the point I wish to make.

20. See F.W. Sears, *Thermodynamics,* New York, 1955, p. 241—or almost any other elementary book on thermodynamics.

21. This type of reasoning is fruitfully compared with C.S. Pierce's 'abductive reasoning'; 'The surprising fact, C, is observed; but if A were true, C would be a matter of course. Hence, there is reason to suspect A is true.' (*Collected Papers of C.S. Peirce,* Cambridge, Mass, 1935, 5, paragraph 189.)

Einstein's explanation of the Brownian motion on the basis of the kinetic theory of gases is a more striking example of this. It was this which convinced many doubters that the atomic theory was not just a 'convenient fiction.'

22. This is Braithwaite's phrase in 'Models in the Empirical Sciences,' loc. cit. p. 229.

23. See Sears, op. cit.

24. *Philosophical Magazine,* 26, 1913, pp. 9–10.

25. I should point out here that in one sense the conclusions I have so far drawn, and those I shall presently draw, are not entirely new. (see, for example, R. Harré, *An Introduction to the Logic of the Sciences,* London, 1960, ch. 4; M.B. Hesse, *Models and Analogies in Science,* London, 1963; N.R. Campbell, *Physics, The Elements,* Cambridge, 1920, ch. 6 (recently reprinted in a paperback edition by Dover entitled *Foundation of Science*); H. Putnam, 'What Theories Are Not,'

in *Proceedings of the Congress of Logic, Methodology, and Philosophy of Science,* Stanford, 1962, pp. 240–51.)

However, I believe that I have argued for these conclusions in an importantly novel way. I have been concerned to show how the conclusions can be *generated out of* the 'partial interpretation' analysis of the structure of physical theories and comes the closest to this sort of undertaking.) My *general* point of view, which may be called roughly, 'realistic,' is of course nothing new at all.

26. 'Foundations of Logic and Mathematics,' p. 210.

27. 'The Methodological Character of Theoretical Concepts' in *Minnesota Studies in the Philosophy of Science,* I, Minneapolis, 1956, p. 38.

Max Black

"Models and Archetypes"

Scientists often speak of using models but seldom pause to consider the presuppositions and the implications of their practice. I shall find it convenient to distinguish between a number of operations, ranging from the familiar and trivial, to the farfetched but important, all of which are sometimes called "the use of models." I hope that even this rapid survey of a vast territory may permit a well-grounded verdict on the value of recourse to cognitive models.

To speak of "models" in connection with a scientific theory already smacks of the metaphorical. Were we are called upon to provide a perfectly clear and uncontroversial example of model, in the literal sense of that word, none of us, I imagine, would think of offering Bohr's model of the atom, or a Keynesian model of an economic system.

Typical examples of models in the literal sense of the word might include: the ship displayed in the showcase of a travel agency ("a model of the *Queen Mary*"), the airplane that emerges from a small boy's construction kit, the Stone Age village in the museum of natural history. That is to say, the standard cases are three-dimensional miniatures, more or less "true to scale," of some existing or imaginary thing depicted by a model of the *original* of that model.

We also use the word "model" to stand for a type of design (the dress designer's "spring models," the 1959 model Ford)—or to mean some exemplar (a model husband, a model solution of an equation). The senses in which a model is a type of design—or, on the other hand, something worthy of imitation—can usually be ignored in what follows.

It seems arbitrary to restrict the idea of a model to something *smaller* than its original. A natural extension is to admit magnification, as in a larger-than-life-size likeness of a mosquito. A further natural extension is to admit proportional change of scale in any relevant dimension, such as time.

In all such cases, I shall speak of *scale models*. This label will cover all likenesses of material objects, systems, or processes, whether real or imaginary,

Max Black, "Models and Archetypes," from *Both Human and Humane: The Humanities and Social Sciences in Graduate Education*, ed. Charles E. Boewe and Roy F. Nichols (Philadelphia: University of Pennsylvania Press, 1960), 39–51. Reprinted with the permission of the University of Pennsylvania Press.

discuss scale models

that preserve relative proportions. They include experiments in which chemicals or biological processes are artificially decelerated ("slow motion experiments") and those in which an attempt is made to imitate social processes in miniature.

The following points about scale models seem uncontroversial:

1. A scale model is always a model of something. The notion of a scale model is relational, an, indeed, asymmetrically so: If *A* is a scale model of *B, B* is not a scale model of *A*.

2. A scale model is designed to serve a purpose, to be a means to some end. It is to show how the ship looks, or how the machine will work, or what laws govern the interplay of parts in the original; the model is intended to be enjoyed for its own sake only in the limiting case where the hobbyist indulges a harmless fetishism.

3. A scale model is a representation of the real or imaginary thing for which it stands; its use is for "reading off" properties of the original from the directly presented properties of the model.

4. It follows that some features of the model are irrelevant or unimportant, while others are pertinent and essential, to the representation in question. There is no such thing as a perfectly faithful model; only by being unfaithful in *some* respect can a model represent its original.

5. As with all representations, there are underlying conventions of interpretation—correct ways of "reading" the model.

6. The conventions of interpretation rest upon partial identity of properties coupled with the invariance of proportionality. In making a scale model, we try on the one hand to make it resemble the original by reproduction of some features (the color of the ship's hull, the shape and rigidity of the airfoil) and on the other hand to preserve the *relative* proportions between relevant magnitudes. In Peirce's terminology, the model is an *icon*, literally embodying the features of interest in the original. It says, as it were: "This is how the original is."

In making scale models, our purpose is to reproduce, in a relatively manipulable or accessible embodiment, selected features of the "original": we want to see how the new house will look, or to find out how the airplane will fly, or to learn how the chromosome changes occur. We try to bring the remote and unknown to our own level of middle-sized existence.[1]

There is, however, something self-defeating in this aim, since change of scale must introduce irrelevance and distortion. We are forced to replace living tissue by some inadequate substitute, and sheer change of size may upset the balance of factors in the original. Too small a model of a uranium bomb will fail to explode, too large a reproduction of a housefly will never get off the ground, and the solar system cannot be expected to look like its planetarium model. Inferences from scale model to original are intrinsically precarious and in need of supplementary validation and correction.

Let us now consider models involving *change of medium.* I am thinking of such examples as hydraulic models of economic systems, or the use of electrical circuits in computers. In such cases, I propose to speak of *analogue models.*

An analogue model is some material object, system, or process designed to reproduce as faithfully as possible in some new medium the *structure* or web of relationships in an original. Many of our previous comments about scale models also apply to the new case. The analogue model, like the scale model, is a symbolic representation for making accurate inferences from the relevant features of the model.

The crucial difference between the two types of models is in the corresponding methods of interpretation. Scale models, as we have seen, rely markedly upon identity: their aim is to imitate the original, except where the need for manipulability enforces a departure from the sheer reproduction. And when this happens the deviation is held to a minimum, as it were: geometrical magnitudes in the original are still *reproduced,* though with a constant change of ratio. On the other hand, the making of analogue models is guided by the more abstract aim of reproducing the *structure* of the original.

An adequate analogue model will manifest a point-by-point correspondence between the relations *it* embodies and those embodied in the original: every incidence of a relation in the original must be echoed by a corresponding incidence of a correlated relation in the analogue model. To put the matter another way: there must be rules for translating the terminology applicable to the model in such a way as to conserve truth value.

Thus, the dominating principle of the analogue model is what mathematicians call "isomorphism." We can, if we please, regard the analogue model as iconic of its original, as we did in the case of the scale model, but if we do so we must remember that the former is "iconic" in a more abstract way than the latter. The analogue model shares with its original not a set of features or an identical proportionality of magnitudes but, more abstractly, the same structure or pattern of relationships. Now identity of structure is compatible with the widest variety of contents—hence the possibilities for construction of analogue models are endless.

The remarkable fact that the same pattern of relationships, the same structure, can be embodied in an endless variety of different media makes a powerful and a dangerous thing of the analogue model. The risks of fallacious inference from inevitable irrelevancies and distortions in the model are now present in aggravated measure. Any would-be scientific use of an analogue model demands independent confirmation. Analogue models furnish plausible hypotheses, not proofs.

I now make something of a digression to consider "mathematical models." This expression has become very popular among social scientists, who will characteristically speak of "mapping" an "object system" upon one or another of a number of "mathematical systems or models."

When used unemphatically, "model" in such contexts is often no more than a pretentious substitute for "theory" or "mathematical treatment." Usually, however, there are at least the following three additional suggestions: The original field is thought of as "projected" upon the abstract domain of sets, functions, and the like that is the subject matter of the correlated mathematical theory; thus social forces are said to be "modeled" by relations between mathematical entities. The model is conceived to be *simpler* and *more abstract* than the original. Often there is suggestion of the model's being a kind of ethereal analogue model, as if mathematical equations referred to an invisible mechanism whose operation illustrates or even partially explains the operation of the original social system under investigation. This last suggestion must be rejected as an illusion.

The procedures involved in using a "mathematical model" seem to be the following:

1. In some original field of investigation, a number of relevant variables are identified, either on the basis of common sense or by reason of more sophisticated theoretical considerations. (For example, in the study of population growth we may decide that variation of population with time depends upon the number of individuals born in that time, the number dying, the number entering the area, and the number leaving. I suppose these choices of variables are made at the level of common sense.)

2. Empirical hypotheses are framed concerning the imputed relations between the selected variables. (In population theory, common sense, supported by statistics, suggests that the numbers of births and deaths during any brief period of time are proportional both to that time and to the initial size of the population.)

3. Simplifications, often drastic, are introduced for the sake of facilitating mathematical formulation and manipulation of the variables. (Changes in a population are treated as if they were continuous; the simplest differential equation consonant with the original empirical data are adopted.)

4. An effort is now made to solve the resulting mathematical equations—or, failing that, to study the *global* features of the mathematical systems constructed. (The mathematical equations of population theory yield the so-called "logistic function," whose properties can be specified completely. More commonly, the mathematical treatment of social data leads at best to "plausible topology," to use Kenneth Boulding's happy phrase;[2] i.e., qualitative conclusions concerning distributions of maxima, minima, and so forth. This result is connected with the fact that the original data are in most case at best *ordinal* in character.)

5. An effort is made to extrapolate to testable consequences in the original field. (Thus the prediction can be made that an isolated population tends toward a limiting size independent of the initial size of that population.)

6. Removing some of the initial restrictions imposed upon the component functions in the interest of simplicity (e.g., their linearity) may lead to some increase in generality of the theory.

The advantages of the foregoing procedures are those usually arising from the introduction of mathematical analysis into any domain of empirical investigation, among them precision in formulating relations, ease of inference via mathematical calculation, and intuitive grasp of the structures revealed (e.g., the emergence of the "logistic function" as an organizing and mnemonic device.)

The attendant dangers are equally obvious. The drastic simplifications demanded for success of the mathematical analysis entail a serious risk of confusing accuracy of the mathematics with strength of empirical verification in the original field. Especially important is it to remember that the mathematical treatment furnishes no *explanations*. Mathematics can be expected to do no more than draw consequences from the original empirical assumptions. If the functions and equations have a familiar form, there may be a background of pure mathematical research readily applicable to the illustration at hand. We may say, if we like, that the pure mathematics provides the *form* of an explanation, by showing what *kinds* of functions would approximately fit the known data. But *causal* explanations must be sought elsewhere. In their inability to suggest explanations, "mathematical models" differ markedly from the theoretical models now to be discussed.[3]

In order now to form a clear conception of the scientific use of "theoretical models," I shall take as my paradigm Clerk Maxwell's celebrated representation of an electrical field in terms of the properties of an imaginary incompressible fluid. In this instance we can draw upon the articulate reflections of the scientist himself. Here is Maxwell's own account of his procedure:

> "The first process therefore in the effectual study of the science must be one of simplification and reduction of the results of previous investigation to a form in which the mind can grasp them. The results of this simplification may take the form of a purely mathematical formula or of a physical hypothesis. In the first case, we entirely lose sight of the phenomena to be explained; and though we may trace out the consequences of given laws, we can never obtain more extended views of the connexions of the subject. If, on the other hand, we adopt a physical hypothesis, we see the phenomena only through a medium, and are liable to that blindness to facts and rashness in assumption which a partial explanation encourages. We must therefore discover some method of investigation which allows the mind at every step to lay hold of a clear physical conception, without being committed to any theory founded on the physical science from which that conception is borrowed, so that it is neither drawn aside from the subject in pursuit of analytical subtleties, nor carried beyond the truth by a favourite hypothesis."[4]

Later comments of Maxwell's explain what he has in mind:

> "By referring everything to the purely geometrical idea of the motion of an imaginary fluid, I hope to attain generality and precision, and to avoid the dangers arising from a premature theory professing to explain the cause of the phenomena . . . The substance here treated of . . . is not even a hypothetical fluid which is introduced to explain actual phenomena. It is merely a collection of imaginary properties which may be employed for establishing certain theorems in pure mathematics in a way more intelligible to many minds and more applicable to physical problems than that in which algebraic symbols alone are used."[5]

Points that deserve special notice are Maxwell's emphasis upon obtaining a "clear physical conception" that is both "intelligible" and "applicable to physical problems," his desire to abstain from "premature theory," and above all, his insistence upon the "imaginary" character of the fluid invoked in his investigations. In his later elaboration of the procedure sketched above, the fluid seems at first to play the part merely of a mnemonic device for grasping mathematical relations more precisely expressed by algebraic equations held in reserve. The "exact mental image"[6] he professes to be seeking seems little more than a surrogate for facility with algebraic symbols.

Before long, however, Maxwell advances much farther toward ontological commitment. In his paper on action at a distance, he speaks of the "wonderful medium" filling all space and no longer regards Faraday's lines of force as "purely geometrical conceptions."[7] Now he says forthrightly that they "must not be regarded as mere mathematical abstractions. They are the directions in which the medium is exerting a tension like that of a rope, or rather, like that of our own muscles."[8] Certainly this is no way to talk about a collocation of imaginary properties. The purely geometrical medium has become very substantial.

A great contemporary of Maxwell is still more firmly committed to the realistic idiom. We find Lord Kelvin saying:

> "real matter between us and the remotest stars I believe there is, and that light consists of real motions of that matter . . . We know the luminiferous ether better than we know any other kind of matter in some particulars. We know it for its elasticity; we know it in respect to the consistency of the velocity of propagation of light for different periods . . . Luminiferous ether must be a substance of the most extreme simplicity. We might imagine it to be a material whose ultimate property is to be incompressible; to have a definite rigidity for vibrations in times less than a certain limit, and yet to have the absolute yielding character that we recognize in wax-like bodies when the force is continued for a sufficient time."[9]

There is certainly a vast difference between treating the ether as a mere heuristic convenience, as Maxwell's first remarks require, and treating it in Kelvin's fashion as "real matter" having definite—though, to be sure, paradoxical—properties independent of our imagination. The difference is between thinking of the electrical field *as if* it were filled with a material medium, and thinking of it *as being* such a medium. One approach uses a detached comparison reminiscent of simile and argument from analogy; the other requires an identification typical of metaphor.

In *as if* thinking there is a willing suspension of ontological unbelief, and the price paid, as Maxwell insists, is absence of explanatory power. Here we might speak of the use of models as *heuristic fictions*. In risking existential statements, however, we reap the advantages of an explanation but are exposed to the dangers of self-deception by myths (as the subsequent history of the ether sufficiently illustrates).

The *existential use of models* seems to me characteristic of the practice of the great theorists in physics. Whether we consider Kelvin's "rude mechanical models,"[10] Rutherford's solar system, or Bohr's model of the atom, we can hardly avoid concluding that these physicists conceived themselves to be describing the atom *as it is,* and not merely offering mathematical formulas in fancy dress. In using theoretical models, they were not comparing two domains from a position neutral to both. They used language appropriate to the model in thinking about the domain of application: they worked not *by* analogy, but *through* and by means of an underlying analogy. Their models were conceived to be more than expository or heuristic devices.

Whether the fictitious or the existential interpretation be adopted, there is one crucial respect in which the sense of "model" here in question sharply diverges from those previously discussed in this paper. Scale models and analogue models must be actually put together: a merely "hypothetical" architect's model is nothing at all, and imaginary analogue models will never show us how things work in the large. But theoretical models (whether treated as real or fictitious) are not literally constructed: the heart of the method consists in *talking* in a certain way.

The theoretical model need not be built; it is enough that it be *described.* But freedom to describe has its own liabilities. The inventor of a theoretical model is undistracted by accidental and irrelevant properties of the model object, which must have just the properties he assigns to it; but he is deprived of the controls enforced by the attempt at actual construction. Even the elementary demand for self-consistency may be violated in subtle ways unless independent tests are available; and what is to be meant by the reality of the model becomes mysterious.

Although the theoretical model is described but not constructed, the sense of "model" concerned is continuous with the senses previously examined. This becomes clear as soon as we list the conditions for the use of a theoretical model.

1. We have an original field of investigation in which *some* facts and regularities have been established (in any form, ranging from disconnected items and crude generalizations to precise laws, possibly organized by a relatively well-articulated theory).
2. A need is felt, either for explaining the given facts and regularities, or for understanding the basic terms applying to the original domain, or

for extending the original corpus of knowledge and conjecture, or for connecting it with hitherto disparate bodies of knowledge—in short, a need is felt for further scientific mastery of the original domain.

3. We describe some entities (objects, materials, mechanisms, systems, structures) belonging to a relatively unproblematic, more familiar, or better-organized secondary domain. The postulated properties of these entities are described in whatever detail seems likely to prove profitable.

4. Explicit or implicit rules of correlation are available for translating statements about the secondary field into corresponding statements about the original field.

5. Inferences from the assumptions made in the secondary field are translated by means of the rules of correlation and then independently checked against known or predicated data in the primary domain.

The relations between the "described model" and the original domain are like those between and analogue model and its original. Here, as in the earlier case, the key to understanding the entire transaction is the identity of structure that in favorable cases permits assertions made about the secondary domain to yield insight into the original field of interest.

NOTES

1. A good example of the experimental use of models is described in Victor P. Starr's article, "The General Circulation of the Atmosphere," *Scientific American*, CXCV (December 1956), 40–45. The atmosphere of one hemisphere is represented by water in a shallow rotating pan, dye being added to make the flow visible. When the perimeter of the pan is heated the resulting patterns confirm the predictions made by recent theories about the atmosphere.

2. "Economics as a Social Science," in *The Social Sciences at Mid-Century: Essays in Honor of Guy Stanton Ford* (Minneapolis, 1952), p. 73.

3. It is perhaps worth noting that nowadays logicians use "model" to stand for an "interpretation" or "realization" of a formal axiom system. See John G. Kemeny, "Models of Logical Systems," *Journal of Symbolic Logic*, XIII (March 1948), 16–30.

4. *The Scientific Papers of James Clerk Maxwell* (Cambridge University Press, 1890), I, 155–56.

5. Ibid., I, 159–160.

6. Ibid., II, 360.

7. Ibid., II, 322.

8. Ibid., II, 323.

9. Sir William Thomson, *Baltimore Lectures* (London, 1904), pp. 8–12.

10. *Ibid*, p. 12

Ronald Giere

Explaining Science
Models and Theories

Mechanics texts continually refer to such things as "the linear oscillator," "the free motion of a symmetrical rigid body," "the motion of a body subject only to a central gravitational force," and the like. Yet the texts themselves make clear that the paradigm examples of such systems fail to satisfy fully the equations by which they are described. No frictionless pendulum exists, nor does any body subject to no external forces whatsoever. How are we to make sense of this apparent conflict?

MODELS

I propose that we regard the simple harmonic oscillator and the like as *abstract entities* having all and only the properties ascribed to them in the standard texts. The distinguishing feature of the simple harmonic oscillator, for example, is that it satisfies the force law $F = -kx$. The simple harmonic oscillator, then, is a constructed entity. Indeed, one could say that the systems described by the various equations of motion are *socially* constructed entities. They have no reality beyond that given to them by the community of physicists.

I suggest calling the idealized systems discussed in mechanics texts "theoretical models" or, if the context is clear, simply "models." This suggestion fits well with the way scientists themselves use this (perhaps overused) term. Moreover, this terminology even overlaps nicely with the usage of logicians for whom a model of a set of axioms is an object, or a set of objects, that satisfies the axioms. As a theoretical model, the simple harmonic oscillator, for example, perfectly satisfies its equations of motion.

The relationship between some (suitably interpreted) equations and their corresponding model may be described as one of characterization, or even definition. We may even appropriately speak here of "truth." The interpreted equations are *true of* the corresponding model. But truth here has no *episte-*

From Ronald Giere, *Explaining Science: A Cognitive Approach* (Chicago: University of Chicago Press, 1988), 78–86.

mological significance. The equations truly describe the model because the model is defined as something that exactly satisfies the equations.

The statements used to characterize models come in varying degrees of abstraction. At its most abstract the linear oscillator is a system with a linear restoring force, plus any number of other, secondary forces. The simple harmonic oscillator is a linear oscillator with a linear restoring force and no others. The damped oscillator has a linear restoring force plus a damping force. And so on. Similarly, the mass-spring oscillator identifies the restoring force with the stiffness of an idealized spring. In the pendulum oscillator, the restoring force is a function of gravity and the length of the string. And so on.

"The linear oscillator," then, may best be thought of not as a single model with different specific versions, but as a *cluster* of models of varying degrees of specificity. Or, to invoke a more biological metaphor, the linear oscillator may be viewed as a family of models, or still better, a family of families of models.

HYPOTHESES

As the ordinary meaning of the word 'model' suggests, theoretical models are intended to be models *of* something, and not merely exemplars to be used in the construction of other theoretical models. I suggest that they functions as "representations" in one of the more general senses now current in cognitive psychology. Theoretical models are the means by which scientists represent the world—both to themselves and for others. They are used to represent the diverse systems found in the real world: springs and pendulums, projectiles and planets, violin strings and drum heads.

This understanding of the role of theoretical models in science immediately raises issues of "realism," which have recently much vexed both philosophers and sociologists of science. Here I wish only to attempt a little more precision in describing the relationship between a theoretical model and that of which it is a model. This requires introducing a new concept, that of a "theoretical hypothesis." Here again I intend the term to overlap with the use of the term 'hypothesis' by scientists themselves, although my usage will be more systematic.

Unlike a model, a theoretical hypothesis is, in my account, a *linguistic* entity, namely, a statement asserting some sort of relationship between a model and a designated real system (or class of real systems). A theoretical hypothesis, then, is true or false according to whether the asserted relationship holds or not. The relationship between model and real system, however, cannot be one of truth or falsity since neither is a linguistic entity. It must, therefore, be something else.

The appropriate relationship, I suggest, is *similarity*. Hypotheses, then, claim a *similarity* between models and real systems. But since anything is similar to anything else in some respects and to some degree, claims of similarity are vacuous without at least an implicit specification of relevant *respects* and *degrees*. The general form of a theoretical hypothesis is thus: Such-and-such identifiable real system is similar to a designated model in indicated respects and degrees. To take a different example:

> The positions and velocities of the earth and moon in the earth-moon system are very close to those of a two-particle Newtonian model with an inverse square central force. Here the respects are "position" and "velocity," while the degree is claimed to be "very close."

A less stilted formulation of the above hypothesis, one closer to how scientists actually talk, would be:

> The earth and moon form, to a high degree of approximation, a two-particle Newtonian gravitational system.

The latter formulation tends to blur the distinction between the theoretical model and the real system, a distinction that a theory of science should, I think, keep sharp. It also fails to distinguish respects and degrees, lumping both into the vaguer notion of "approximation." But so long as these distinctions are understood, the more relaxed formulation is generally clear enough.

That theoretical hypotheses can be true or false turns out to be of little consequence. To claim a hypothesis is true is to claim no more or less than that an indicated type and degree of similarity exists between a model and a real system. We can therefore forget about truth and focus on the details of the similarity. A "theory of truth" is not a prerequisite for an adequate theory of science.

WHAT IS A SCIENTIFIC THEORY?

Even just a brief examination of classical mechanics as presented in modern textbooks provides a basis for some substantial conclusions about the overall structure of this scientific theory as it is actually understood by the bulk of the scientific community.

What one finds in standard textbooks may be described as a cluster (of clusters of clusters) of models, or, perhaps even better, as a population of models consisting of related families of models. The various families are constructed by combining Newton's laws of motion, particularly the second law, with various force functions—linear functions, inverse square functions, and so on.

The models thus defined are then multiplied by adding other force functions to the definition. These define still further families of models. And so on.

Since it seems that theories are things one finds written down in textbooks, among other places, one is inclined to look for something "linguistic" to be the theory itself. But one cannot identify the theory of mechanics with any particular set of *sentences*. That would make the translation of a textbook from English into French a different theory, which seems absurd. Philosophers long ago introduced he more abstract entities called "statements" or "propositions," to handle just this problem. A proposition is supposed to be, roughly, what a sentence asserts. One can therefore assert the same proposition using different sentences, some in English and some in French, for example.

If one looks for statements, the obvious candidates are Newton's laws of motion and the various force laws one finds in the standard texts. But if they are understood as statements making claims directly about the world, all the laws of motion and force laws one finds written down are known to be false — a discomforting fact to say the least. That is largely why I have argued that these laws are to be interpreted as providing definitions of various models, models that are non-linguistic, though abstract, entities.

On the other hand, if understood as definitions, the laws of motion make no claims about the world. Now physicists are ambiguous about whether Newton's laws are empirical claims or definitions. But it would be a rare physicist who would agree that the whole theory consists of nothing but definitions. Newtonian mechanics, most physicists would insist, does make claims about the world, and anyone who tries to say otherwise will get short shrift.

An obvious alternative is to focus on theoretical *hypotheses*. These are appropriately linguistic, and they make claims about the world—they may be true or false. The trouble with this suggestion is that Newton's laws and the force laws turn out not to be among the statements that are part of the theory of Newtonian mechanics. Consider the most outlandishly general hypothesis possible: The whole universe is similar to a Newtonian model defined as follows. . . . This hypothesis incorporates Newton's laws, but the laws themselves fail to appear as separate statements making up the theory.

My preferred suggestion is that we understand a theory as comprising two elements: (1) a population of models, and (2) various hypotheses linking those models with systems in the real world. Thus, what one finds in textbooks is not literally the theory itself, but statements defining the models that are part of the theory. One also finds formulations of some of the hypotheses that are also part of the theory. That characterization seems to me sufficiently close to how physicists think and talk to be useful.

The links between models are not logical connections, but relations of similarity. In some cases the difference between two models is that one is an approximation of the other—again not a logical relationship. The links between models and the real world are nothing like correspondence rules linking terms with things or terms with other terms. Rather, they are again relations of similarity between a whole model and some real system. A real system is *identified* as being similar to one of the models. The *interpretation* of terms used to define the models does not appear in the picture; neither do the defining linguistic entities, such as equations.

ARE THEORIES WELL-DEFINED ENTITIES?

It is a consequence of the above interpretation that a scientific theory turns out not to be a well-defined entity. That is, no necessary and sufficient conditions determine which models or which hypotheses are part of the theory.

This is most obvious in the case of hypotheses. I suspect that most scientists would agree that general hypotheses asserting similarities between planets and Newtonian models with inverse square forces are part of the theory. Many would also agree if one of the planets is designated as the earth. But what about the claim that the pendulum on my antique clock resembles a model of a harmonically driven linear oscillator? Is that claim part of the "theory" of classical mechanics? Here it seems that one could argue the case either way. My view is that it matters not at all how, or even whether, one answers such questions.

The population of models for classical mechanics is also not well defined because there are no necessary and sufficient conditions for what constitutes an admissible force function. This raises an interesting question. What determines whether a model is to count as a proper Newtonian model?

One answer is that to be part of the theory of classical mechanics a model must bear a "family resemblance" to some family of models already in the theory. That such family resemblances among models exist is undeniable. On the other hand, nothing in the structure of any models themselves could determine that the resemblance is sufficient for membership in the family. That question, it seems, is solely a matter to be decided by the judgments of the members of the scientific community at the time. This is not to say that there is an objective resemblance to be judged correctly or not. It is to say that the collective judgments of the scientists *determine* whether the resemblance is sufficient. This is one respect in which theories are not only constructed, but socially constructed as well.

Case Studies

. . . .

Astronomy Track

PAPER 5: EINSTEIN'S GENERAL THEORY OF RELATIVITY AND THE ADVANCE OF THE PERIHELION OF MERCURY

THE CASE

While Newtonian theory was wildly successful, there was still one case that remained disturbing to astronomers concerned about the solar system. Mercury's orbit had been known since the nineteenth century to have a strange property. If one charted the trajectory, it was roughly elliptical as one would expect, but with trip around the sun, Mercury's orbit would rotate slightly. If you examined the point at which Mercury was at its farthest point from the sun, that point would itself advance around the sun by approximately forty-three arc seconds each century—an imperceptible change with each orbit but the accumulated effects of which could be found in the long history of observations.

While this quirk was interesting, it was not seen as wildly worrisome, because a similar problem had been seen with irregularities in the planet Uranus. In that case, astronomers Urbain-Jean-Joseph Leverrier and John Couch Adams independently hypothesized the existence of an additional, massive planet beyond it whose gravity was responsible for Uranus's perturbation. Calculating size and location from the deviation led to the discovery of Neptune in 1846.

In the same way, it was thought that Mercury's unusual orbit could be explained by a planet whose orbit was even closer to the sun. Such a planet would get lost in the sun's glare, which could account for the fact that no planet had been discovered despite being so much closer to Earth than Neptune. Leverrier calculated what its size and orbit would have to be and began the search for this new planet, which he dubbed Vulcan. Many searched for it—a few even convinced themselves that they had found it—but ultimately no new planet was located.

The problem persisted until 1916 when Albert Einstein proposed a view of gravitation radically different from that of Newton. Newton saw gravitation as a force between bodies in a space that remained always the same. Einstein saw gravitation as a force that warped space itself. Matter bends space, and the bends in space then tell bodies how to move. With this new approach to space, Einstein considered the advance of the perihelion of Mercury and discovered that when he plugged the data into his new theory, it perfectly predicted Mercury's behavior.

Urbain-Jean-Joseph-Leverrier (1811–1877) was a French mathematician. His work on the existence of Neptune was started after, but finished days before, that of Adams. While both men worked hard to find astronomers who would devote telescope time to finding the predicted planet, Leverrier ultimately convinced Johann Galle at the Berlin Observatory to perform the required observations, which led to the planet's discovery.

Albert Einstein (1879–1955) was born in Germany, followed his family to Italy, and attended university in Switzerland, where he became a patent clerk. In 1905, Einstein published his dissertation on the mechanics of mixing, his first paper on the special theory of relativity, his paper on the photoelectric effect for which he would win the Nobel Prize, and his paper on Brownian motion, which established the reality of atoms. After returning to Germany, Einstein produced his general theory of relativity in 1916, which overturned the Newtonian view of gravitation that had been in place for nearly three centuries. He fled Germany with the rise of Adolf Hitler, emigrating to the United States where he continued his scientific work and became an internationally influential figure.

YOUR JOB

Explain the problem with the orbit of Mercury. How did Leverrier account for it? How did Einstein account for it? Explain the semantic view of theories. How would an adherent of the semantic view account for this case? What types of models are involved here? What makes Einstein's models better than Leverrier's?

Physics Track

PAPER 5: THE STANDARD MODEL AND THE DISCOVERY OF THE TOP QUARK

THE CASE

As the quantum theory developed, questions about parts of atoms and the parts of parts of atoms continued. Experimentally, particle accelerators were used to bang pieces into each other, allowing scientists to see the resulting pieces that existed for short times thereafter. This allowed them to categorize the parts of atoms into two categories: leptons and hadrons. Leptons (for example, electrons) are particles that are uncuttable in that they lack internal structure. Scientists were able to identify six of these particles. But there were other atomic constituents, the hadrons, which were not uncuttable but could be broken down further. Their components were termed "quarks."

Scientists were able to find evidence of five of these quarks, but the theorists who developed quantum field theory, the theory that best accounted for these observations, derived what has come to be called "the standard model" of particle physics, which said there should be a sixth. The standard model accounted for the six leptons and the four nonatomic "messenger" particles for which there was evidence, but it said there should also be six quarks, not just the five that there was observable evidence for.

This missing piece, the top quark, was part of the model, and a search was initiated to find it. It was an odd particle, extremely heavy for a quark (about 190 times as heavy as a proton, which itself is made up of lighter and more stable quarks) and would likely only appear briefly and under unusual circumstances of very high energy according to the model. It would be hard to find, but it would be an important bit of confirmation of the standard model.

Then on March 2, 1995, a group of particle physicists known as the DZero team, who had been working with the particle accelerator at the U.S. government's Fermi National Accelerator Laboratory, or Fermilab, in Batavia, Illinois, announced that they had found evidence of the top quark. They were not the first to make this claim; in 1984 another group reported that they had seen evidence of it, but it turned out they had most likely been wrong. But the DZero team at Fermilab had combined the efforts of many physicists and produced evidence that was widely accepted as showing the existence of the top quark. The standard model had been corroborated. It predicted something we had never seen, and now we saw it. Decades of searching for the top quark, and it turned out to be in Illinois.

YOUR JOB

Explain the standard model. What are the parts of an atom according to the standard model? What observable evidence is there for the existence of quarks? Why did scientists believe there was a missing, yet unobserved quark? What sort of model is the standard model? Explain the syntactic view of theories. How would an advocate of the syntactic view account for this episode?

Chemistry Track

PAPER 5: KEKULÉ AND CHEMICAL STRUCTURE

THE CASE

The atomic work predating and immediately following John Dalton focused on deriving observed chemical properties from composition. Proportions of component substances, weights, and eventually affinities were used to explain reactions and properties of the resultant molecules.

Chemists in the 1830s and 1840s began to experiment with the possibility that chemical properties were derived from not just certain atoms and how many were combined in a molecule but also the geometry of the arrangement. Perhaps, they argued, how a molecule was put together was as important as what it was put together from. Unfortunately, as the atomic constituents were not observable, it seemed that consideration of internal structure as a way of understanding bonds was also doomed to be beyond the grasp of chemists.

Theories at the time held that certain atoms, usually oxygen, served as the central focus of the molecule, the key that held the larger molecule together in complex ways. Friedrich August Kekulé, following on the work of Charles Gerhardt and Edward Frankland, proposed a theory of chemical types where atoms were of type one, two, three, or four. The type indicated the number of atoms with which it would bond, what are now call valences.

Recognizing that certain atoms were of certain types, Kekulé could explain why atoms of the same type, say, oxygen and sulfur, would form compounds of similar composition. Further, structural diagrams for molecules could be designed that would explain why certain molecules were emitted in certain reactions instead of others that could be concocted using the same basic stuff. Kekulé realized that carbon atoms could bond to each other in long chains, opening significant research in organic chemistry. He also came to realize that chains were not the only sort of structure one could create when he made a closed chain model of the benzene molecule.

Following Kekulé, chemists went on to not only study the geometric structure of molecules, but to predict other possible combinations giving rise to new chemicals.

THE SCIENTIST

Friedrich August Kekulé (1829–1896) originally intended to study architecture. While he never became an architect, his contributions to chemistry hinged upon uncovering the architecture of molecules. He claimed that his most famous discovery, the structure of benzene rings, came to him in a dream in which six serpents appeared before him, each biting the tail of another, forming a six-sided ring. He awoke, the story goes, startled and sure he had figured out the problem.

YOUR JOB

Explain Kekulé's theory of atomic types and how it led to structural chemistry. How could geometric representations of molecules explain things that mere combinations of atoms could not? Explain the semantic view of theories. What are models and what are they supposed to do in science? How would a semantic theorist look at Kekulé's work and see it as an instance of the use of models? What sort of models are they? In the decades before Kekulé, chemists argued that since the constituent atoms were unobservable, it as not scientific to talk in realistic terms of molecular structure. After

Kekulé's work, is there reason to think that such models are real or should they be seen as merely useful?

Genetics Track

PAPER 5: THE HUMAN GENOME PROJECT

THE CASE

With the discovery of DNA as the operative mechanism in the transfer of genetic information, questions soon arose as to the composition of the DNA in various species. How much of the genetic coding was similar between species? How much is the same among members of a given species and how much difference accounts for the traits of individuals? What could a grasp of the coding tell us about evolutionary history and the sequence of branching species?

To answer these questions, groups of scientists began to chart the genome of various species. Starting with simpler species and working toward humans, the idea was to create maps of the genetic information that made an organism a member of its given species. Great time, effort, and money were put toward this project, culminating in the Human Genome Project, which has achieved increasingly impressive results.

With maps of the genome, scientists are better able to understand inherited illness and potentially to develop gene therapies to aid those who had little or no hope of aid before. They have a better grasp of humankind and anthropological history, now that the basis of inherited differences between groups and the relations between them can be seen. New questions arise that scientists can begin to answer, but previously had not thought to ask.

YOUR JOB

Explain what the Human Genome Project is and what its goals are. Explain the semantic view of theories. What sort of model is the human genomic map? How can we have a map of the human genome when every person is different? The mapped genome is often compared to an atlas as opposed to a street map. Why would a vague model succeed where a more specific model would fail? How would an advocate of the semantic view of theories account for this?

Evolutionary Biology Track

PAPER 5: PUNCTUATED EQUILIBRIUM

THE CASE

The standard model of evolutionary development held that mutations occurred randomly. Any given time an offspring was born, there was a chance of genetic alteration, and any roll of the genetic dice was as good as any other. Mutation would generally be harmful or meaningless, but occasionally would be advantageous. Since the appearance of these advantageous characteristics was random, they built up gradually, creating branches where two species separated from a common ancestor. In this way, linear development of a species is a continuous, long, slow, regular process.

But at the Annual Meeting of the Geological Society of America in 1971, there was a symposium called "Models in Paleontology." One of the papers presented there, "Speciation and Punctuated Equilibria: An Alternative to Phyletic Gradualism" by Stephen Jay Gould and Niles Eldredge, challenged the standard view of evolutionary development with a new model of how species came about.

According to Gould and Eldredge, instead of a slow, steady trickle of adaptations gradually allowing new species to emerge, there were sudden shocks to the system, causing rapid development of species—that is, rapid in terms of geological time. Most of the time, the model postulated, species were quite stable, showing few signs of development. But in response to changes in the environment, occasionally species would radically become altered in a much smaller number of generations than had been previously thought.

The question as to which of the two models was the dominant mode of speciation generated much controversy in the biological community. Cases and examples from the fossil record were collected by the opposing viewpoints.

THE SCIENTISTS

Stephen Jay Gould (1941–2002) was a paleontologist at Harvard University and a New York Yankee fan. A gifted scientist, he was also a great popularizer of science, writing many books designed to make science and the history of science accessible to the layperson.

Niles Eldredge (1943–) is a paleontologist and curator at the American Museum of Natural History. He was a classmate of Stephen Jay Gould at Columbia, and their long friendship led to their scientific collaboration.

YOUR JOB

Explain the difference between the gradualist and the punctuated equilibrium models of speciation. What evidence is there on each side of the discussion? Explain the semantic view of theories. How would a semantic view advocate make sense of this dispute? Are the models to be understood as giving true explanations or merely useful ones?

Geology Track

THE CASE

While there is surely still fieldwork to be done, the days of going into mines and along cliffs like John Woodward did are over. Much more contemporary work in geology takes place in front of a monitor. The field has been completely revolutionized by geographic information systems, or GIS, a complex web of maps and other geographical and geological information that can be used to help answer intricate questions.

Satellite imagery and the computing power of GIS systems, which give researchers the ability to focus on the fine details of small regions or to look at large scale global concerns, have been a boon to geological and geophysical investigations. The ability to conduct geological studies without getting your clothes dirty is a wonderful advance on many levels.

YOUR JOB

Find a contemporary example of a geologist or a geophysicist using GIS in his or her research. Explain the work. Explain the semantic view of theories. How would a semantic view advocate make sense of this research? What sort of models would they see at use? Are there any aspects of this research that do not fit well with the semantic view?

Psychology Track

PAPER 5: HARLOW'S MATERNAL BONDING AND ANIMAL MODELS

THE CASE

The popular image that comes to mind when one thinks of a psychologist is a white-coated researcher in a lab timing white rats as they run through a maze. This is a result of behaviorism, the view that rose in opposition to Freud's picture in which psychology was to probe the inner workings of the invisible subconscious. If it was invisible, some psychologists worried, then was is not the sort of thing real scientists concern themselves with. And out of this worry that psychology was deviating from a truly scientific path, behaviorists sought to banish from professional discourse any notion that was not directly observable. This limited psychologists to finding relations between measurable stimuli and observable reactions or behaviors, and this could be done not only in humans but also in animals, with the results then applied to human reactions to environmental stimuli.

One of the most famous experiments along these lines is Harry F. Harlow's work on monkeys and maternal bonding. Harlow removed newborn monkeys from their mothers and placed them in environments where their milk was provided by bottles built into model monkeys, one made of bare metal wire and the other covered with a soft terry cloth, the sort of fabric used in cushy towels. In a first trial, infants were given both models to feed from and overwhelmingly preferred to spend time gripping the softer "mother." When Harlow put milk only in the bare wire "monkey," the infants still preferred to snuggle with and seek comfort and protection from harm with the softer version, demonstrating that it was not only nourishment the young were seeking.

Harlow then raised one group with only wire "mothers" and another with only softer ones. He observed marked differences in the resulting groups, both in terms of behavior and physiological factors. He hypothesized that this was the result of emotional development which could not be reduced to mere physiology. Psychologists needed to use terms like "attachment," "care," and "love" in explaining why people do what they do.

In this way, he argued that the behaviorist restrictions designed to make psychology more scientific were overly stringent, that emotional notions needed to be added to the descriptions of stimuli and behavior.

THE SCIENTIST

Harry F. Harlow was born Harry Israel in Fairfield, Iowa, and only changed his name because of the insistence of his dissertation advisor, Lewis Terman, at Stanford. Even this well-known specialist, the man who created the Stanford-Binet IQ test, could not find a position for his pupil because of anti-Semitism, an ironic situation, given that Harlow's family was never Jewish. The renamed Harlow received a position at the University of Wisconsin, at which he proceeded to develop one of the most prestigious laboratories for the study of primate behavior in the world. It was in part because of this research and the effects on primates that society began to be concerned about the treatment of laboratory animals and why researchers now have stringent rules governing what they can and cannot do to them.

YOUR JOB

Explain Harlow's experiments. Explain what behaviorism is and how Harlow's results challenged its central tenets. Why are animal models used in psychology and how do we make inferences from the results of animal studies to hypotheses about humans? What sort of model are the animal models? Are there aspects of Harlow's work that do not fit well into the semantic approach?

Sociology Track

THE CASE

Talcott Parsons, Pitirim Sorokin's Harvard colleague, denied the existence of theory-independent sociological facts. The world of nature, what natural scientists study, is completely different from the world of action, a place with autonomous humans who make their own choices. Anything that can be thought of as a fact in the world of action requires a sociological theory to create the framework of the social world that the facts supposedly describe. Sociology does not produce theories from observation of facts; rather, sociology must start from theory or else there would be no facts.

Parson's view is termed "structural functionalism," meaning that all social facts must be understood in terms of the function performed for the larger structure, and parts of the system only have meaning in terms of their role or function in serving the interests of the larger system. It is the system that is the object of study in sociology, and all events are used to throw light upon the structure of the system as understood through the lens of the theoretical framework.

All social systems exist to maintain themselves through time. Societies are built to be self-perpetuating and so must develop mechanisms to solve problems. These problems may be external—in other words, the culture's interaction with the larger environment—or they may be internal—that is, problems of organization and cohesion within the culture. These problems will confront any society and therefore the sociologist can view the society in terms of the mechanisms it uses in solving these problems. The mechanisms can be categorized in four types: adaptation, goal attainment, integration, and latency (AGIL).

By adaptation, Parsons referred to the social system's ability to relate to its environment. The system needs to secure resources from the environment and it needs to be able to modify itself according to changes in the environment in ways that suit the needs of the system.

Goal attainment refers to the ways the system determines its goals, prioritizes them, allots resources, and develops strategies for meeting them, as well as the tools and methods used to actually try to achieve what it set out to do.

Integration refers to the mechanisms in place to grease the cogs. Social systems are complex entities with many moving parts, each trying to accomplish its own tasks for the good of the greater system. But those parts do not necessarily work harmoniously. Integration is concerned with the means used to secure cohesion among the parts of the system.

Latency, or what Parsons later termed pattern management, concerns the ways that the values and unique aspects of the culture are transmitted and maintained. How is the social system created and what does it do to make sure what defines it gets communicated and preserved?

THE SCIENTIST

Talcott Parsons (1902–1979) was born in Colorado Springs, Colorado, and received his advanced training in Europe, where he was exposed to the writings of sociologists like Max Weber, whose work had not yet reached America. One of the original sociologists at Harvard, Parsons was a president of the American Sociological Society. He worked not only in sociology but also in economics, education theory, and anthropology. He created an interdisciplinary department at Harvard following his idea of creating a grand theory to unite all the social sciences.

YOUR JOB

Explain Parson's structural functionalism and the role of AGIL. Explain the semantic view of theories. How would a semantic theorist make sense of Parson's work? What sort of model is AGIL? Are there aspects of Parsons's work that do not fit well in this approach?

Economics Track

PAPER 5: KEYNES, DEMAND-DETERMINED EQUILIBRIUM, AND MODELS

THE CASE

A major turning point in the history of macroeconomic thought was the 1936 publication of John Maynard Keynes's *General theory of Employment, Interest, and Money*. It was a reaction to the Great Depression, a state of affairs that should not have been possible under the economic theories of the time. According to classical capitalist theory, the mass unemployment and low interest rates should have made business investment desirable and sparked economic expansion. The business cycle should not have maintained such a prolonged period of contraction. In Marx's view, the capitalists should have been innovating until the workers rebelled. This sort of pause in history's dialectic march did not compute. Yet while both approaches described mechanisms of change and advancement, the Great Depression raged.

But Keynes proposed a new picture of the workings of a capitalist economy in which multiple demand-determined equilibrium points existed, that is, economies could remain stuck in different conditions, including the massive unemployment and lack of economic growth that were seen firsthand in the Depression. The classical theory, he argued, was an accurate account of a special case, but a more general theory connecting monetary and governmental policy with a broader understanding of supply and demand were needed to fully account for all possible macroeconomic situations.

The classical view, which followed Say's law (wherein as long as wages and prices were sufficiently flexible to maintain equilibrium, the economy would always be self-correcting), dictated that governmental noninterference with the workings of the economy was always the best policy. But under Keynes's approach, in which the classical notion of demand is augmented with the concept of aggregate demand (the combination of consumption, private investment, and government spending), and in which monetary policy is not held as separate from questions of supply and demand but is not seen as sufficient to temper recessions or booms, government deficit spending may be necessary to initiate economic expansion. However, he argues, it would need to be carefully managed to maintain a stable, more desirable equilibrium.

THE SCIENTIST

John Maynard Keynes (1883–1946) was born into the upper-class family of economist John Neville Keynes. While at Cambridge, he became a member of the famed Bloomsbury Group, a circle of elite intellectuals known as much for their cutting wit as for their intelligence. A brilliant thinker in several academic disciplines, he worked in the real world amassing great wealth as well as holding prestigious positions at Cambridge.

YOUR JOB

Explain Keynes's macroeconomic views and their relation to classical theory. How did he account for the situation of the Great Depression in ways that were not possible under the classical view? Explain the semantic view of theories. How would a semantic theorist make sense of Keynes's work? What kind of models did he employ? Are there aspects that do not fit in well with this approach?

CRITICAL VIEWS OF
SCIENTIFIC THEORIES

All of the approaches we have considered so far have one thing in common: they all begin with the premise that there is such a thing as the scientific method that can be sketched out. This is because they all share a presupposition that the universe is a reasonable place and that human reasoning is sufficient to give us an understanding of it. This position is called "modernism." The modernist worldview puts rationality in a special place, arguing that we can occupy an intellectually objective point of view that is superior to all others in giving us access to knowledge about the world. Science, they argue, is the very epitome of this objectivity. It doesn't matter who you are or when or where you are working; good science is good science and good science gives us good reason to believe in its results because its methods are the most rational.

But do we really have good reason to believe this? Is there actually a scientific method at all, some form of reasoning that is specific to science and underwrites its privileged place as a provider of maximally reasonable beliefs? And even if there is such a method, is it really as objective as the advertisements would have us believe? Does it really involve nothing more than observation, logic, and perhaps some ingenuity, or is science actually a social activity responsive to political pressures, religious forces, and cultural bigotry? Is there more to science than just successful rhetoric? Does the so-called scientific method really yield objectively true results shielded from the influences of biases of the times?

Paul Feyerabend argued that if you look at the actual history of science, you would not see the orderly progress depicted in textbooks with advancing rationality blazing a well-organized trail through the wilderness of the unknown. Rather, you would see intellectual chaos that produced wonderful things. There is not now and never has been a scientific method, according to Feyerabend, and all the better for science. What the enforcement of an imposed "scientific method" would do is intellectually handcuff scientists at the time they are most in need of freedom. Science needs to be radically free, anarchy needs to be the governing intellectual principle—or lack thereof—for the health and growth of science.

Feyerabend saw the notion of a scientific method as a historical fiction designed to give extra, undeserved credibility to scientific results. It comes from oversimplifying science, from creating an artificial narrative in which a caricature of real science appears as a cartoon superhero with perfect rationality charging fearlessly into uncertainty to save humanity with its emerging absolute and indubitable truth. The many mistakes and false starts made by scientists are hidden, swept under the rug of history, and ultimately forgotten; the debates and errors get papered over in favor of a made-up clean plot that preaches the inevitability of the current favored theory. But such a story, while it may seem useful to those who want to defend science as superior to other means of acquiring beliefs, in the end harms science by limiting its ability to progress. Honesty and a desire for the health of science itself, Feyerabend argues, require only one principle—anything goes.

What Feyerabend's argument introduces is a crucial element: the role of social and political rhetoric being substituted for rational methodology. Science is done by people, and people live in a society at a moment in history. All of our previous accounts of the scientific method have tried to remove the historical context and the people involved. Science exists as a logical/empirical entity, an abstraction generating absolute timeless truths. Scientific methodology justifies placing science in a special privileged place in terms of rational belief—it is in a sense super-human, its proclamations beyond challenge by all others.

But science is part of society. Governments, by funding research through grant-giving organizations, decide much of what gets researched and what doesn't; corporations, whose main concern is generating large profits, decide most of the rest. As such, the path of science is largely guided by what will get a politician reelected by a population who knows little about science and what applications will tap into to a rich enough market to make shareholders a healthy dividend.

Once the research is done, results become accepted knowledge by appearing in peer-reviewed journals. But these are controlled by certain scientific factions made up of powerful professors, almost all of whom have identical educational and socioeconomic backgrounds and similar political leanings, and are, by and large, of the same race and sex. Do these factors play any role at all in what scientific results gain acceptance as legitimate and which do not? If so, then the sterilized pictures of the scientific method we have been examining are fictions. To see how science really works, we need to examine the role of historical, social, economic, and political influences and these would have to be seen as critical challenges to the very project of sketching a scientific methodology.

Ruth Hubbard, a biologist and philosopher of biology, examined the roles of politics and gender in science. Scientific facts are created, she argues, and created by people in a very special social organization with strict rules. But these rules are not the sort of logical inferences that others might make them out to be, but rather rules by a chosen few who tightly control the making of science.

Because this group controls what gets taught, who is admitted into graduate programs to get the PhD needed to be in the group, who gets jobs, and what gets published, what becomes science is influenced deeply by the membership of this club. It turns out that the membership is overwhelmingly well-off financially, white, and male. The result of having the power over science concentrated in the hands of a homogeneous group is that science is not open and is not geared towards the sorts of questions or problems of interest to those not in the group.

Further, biases of the group will instill within science beliefs that privilege them and take a degrading or paternalistic view of others, for example, women, particularly poor or working-class women, especially those from minority groups. Science can then be used as a political tool of oppression while falsely claiming to remain objective and nonpolitical. We continue to see such encoding of social biases into scientific theory, Hubbard argues, with modern sociobiology.

Further, we see in the very methodologies employed by scientists a privileging of a certain gendered notion of what it is to do science and what facts are given the status as "scientific." This is not to say that such findings may not be true, but the claims of an author-absent objectivity is simply part of a mythology designed to hide the sociopolitical elements of the endeavor.

Bruno Latour picked up this thread and argues in favor of the postmodern view that scientific facts are social constructions. "Postmodernism" is a direct challenge to the tenets of modernism, the view that there are absolute truths of the world, facts that hold true independent of the human minds that may or may not conceive them, and that human beings are rational in such a way that by using their reason, they can have access to these truths.

The postmodernists argue that truth is not extrahuman, but of human construction. It is not "out there to be discovered," but rather "out there because that is where we built it." Think of borders. The Mason-Dixon Line runs between Maryland and Pennsylvania, but is it real? Yes and no. It was created by the surveyors Mason and Dixon; it is nothing but a human construction. Yet, we act as if it has a metaphysical reality independent of us and as a result real differences may be found in the lives of those who live on different sides of the line—how much you pay in taxes, whether volunteer or professional firefighters show up to your door when there is an emergency, and where your children can go to school are things that are determined by this artificial line. Real effects come from something that was arbitrarily constructed by us.

In the same way, those with Latour argue, scientific "truths" are also socially constructed. You cannot completely separate the facts that a culture creates from the social influences in the society. Politics is inextricably involved in the process and while science will have effects on human lives, like borders, it is an artificial construct.

Consider the status of Pluto. Just before the August 2006 meeting of the International Astronomer's union, it was a fact that Pluto was a planet. Anyone who denied it would have been factually wrong. After the meeting and the vote that was taken, anyone who asserted that Pluto was a planet would be factually wrong. During that time period, nothing happened to Pluto itself.

Scientists contend that they are mere observers collecting facts that independently exist in the world. They discover, not create. Yet, when one truly looks at the way science is actually done, Latour contends, it can be seen that scientists create their facts.

This view of "social constructivism" within the community of sociologists and historians of science loudly challenges the place and objectivity of science within the Academy and has given rise to the "Science Wars." On one side of the war are scientists arguing that the results of science exposed truths about the world itself and on the other side are those arguing that the results of science were constructed by scientists. Physicist Alan Sokal, in a sneak attack, published an article in the journal *Social Text*, a leading publication for the constructivists, which knowingly included not only false but nonsensical claims. When his activity was revealed in a subsequent article in the magazine *Lingua Franca*, he contended that the fact that *Social Text*'s editors could not tell the difference between a serious submission and a joke shows that the entire constructivist project was bankrupt.

Those rallying behind him are the Sokalists who Latour alluded to in his fictional dialogue with a working astronomer trying to illustrate the ways in which social construction fails to be the villain they make it out to be. Latour argued that the view of social constructivism created by the Sokalists is a false image designed to be easily to attack, an image supposedly created out of a desire to bring down science. Instead, Latour argued, social constructivism is an attempt to help further scientific progress by showing where the politics exist so that science can be maximally depoliticized.

Paul Feyerabend

· ·

Against Method

The following essay is written in the conviction that *anarchism,* while perhaps not the most attractive *political* philosophy, is certainly excellent medicine for *epistemology,* and for the *philosophy of science.*

The reason is not difficult to find.

'History generally, and the history of revolution in particular, is always richer in content, more varied, more many-sided, more lively and subtle than even' the best methodologist can imagine.[1] History is full of 'accidents and conjectures and curious juxtapositions of events'[2] and it demonstrates to us the 'complexity of human change and the unpredictable character of the ultimate consequences of any given act or decision of men.'[3] Are we really to believe that the naïve and simple-minded rules which methodologists take as their guide are capable of accounting for such a 'maze of interactions'?[4] And is it not clear that successful *participation* in a process of this kind is possible only for a ruthless opportunist who is not tied to any particular philosophy and who adopts whatever procedure seems to fit the occasion?

This is indeed the conclusion that has been drawn by intelligent and thoughtful observers. 'Two very important practical conclusions follow from this [character of the historical process],' writes Lenin,[5] continuing the passage from which I have just quoted. 'First, that in order to fulfill its task, the revolutionary class [i.e., the class of those who want to change either a part of society such as science, or society as a whole] must be able to master *all* forms or aspects of social activity without exception [it must be able to understand, and to apply, not only one particular methodology, but any methodology, and any variation thereof it can imagine] . . . ; second [it] must be ready to pass from one to another in the quickest and most unexpected manner.' 'The external conditions,' writes Einstein,[6] 'which are set for [the scientist] by the facts of experience do not permit him to let himself be too much restricted, in the construction of his conceptual world, by the adherence to an epistemological system. He, therefore, must appear to the systematic epistemologist as a type of unscrupulous opportunist. . . .' A complex medium containing surprising

From Paul Feyerabend, *Against Method* (New York: Verso, 1993), 9–19. Reprinted with permission of Verso.

and unforeseen developments demands complex procedures and defies analysis on the basis of rules which have been set up in advance and without regard to the ever-changing conditions of history.

Now it is, of course, possible to simplify the medium in which a scientist works by simplifying its main actors. The history of science, after all, does not just consist of facts and conclusions drawn from facts. It also contains ideas, interpretations of facts, problems created by conflicting interpretations, mistakes, and so on. On closer analysis we even find that science knows no 'bare facts' at all but the 'facts' that enter our knowledge are already viewed in a certain way and are, therefore, essentially ideational. This being the case, the history of science will be as complex, chaotic, full of mistakes, and entertaining as are the minds of those who invented them. Conversely, a little brainwashing will go a long way in making the history of science duller, simpler, more uniform, more 'objective' and more easily accessible to treatment by strict and unchangeable rules. Slop

Scientific education as we know it today has precisely this aim. It simplifies 'science' by simplifying its participants: first, a domain of research is defined. The domain is separated from the rest of history (physics, for example, is separated from metaphysics and from theology) and given a 'logic' of its own. A thorough training in such a 'logic' then conditions those working in the domain; it makes *their actions* more uniform and it freezes large parts of the *historical process* as well. Stable 'facts' arise and persevere despite the vicissitudes of history. An essential part of the training that makes such facts appear consists in the attempt to inhibit institutions that might lead to a blurring of boundaries. A person's religion, for example, or his metaphysics, or his sense of humour (his *natural* sense of humour and not the inbred and always rather nasty kind of jocularity one finds in specialized professions) must not have the slightest connection with his scientific activity. His imagination is restrained, and even his language ceases to be his own. This is again reflected in the nature of scientific 'facts' which are experienced as being independent of opinion, belief, and cultural background.

It is thus *possible* to create a tradition that is held together by strict rules, and that is also successful to some extent. But is it *desirable* to support such a tradition to the exclusion of everything else? Should we transfer to it the sole rights for dealing in knowledge, so that any result that has been obtained by other methods is at once ruled out of court? And did scientists ever remain within the boundaries of the traditions they defined in this narrow way? To these questions my answer will be a firm and resounding NO.

There are two reasons why such an answer seems to be appropriate. The first reason is that the world which we want to explore is a largely unknown entity. We must, therefore, keep our options open and we must not restrict ourselves in advance. Epistemological prescriptions may look splendid when compared with other epistemological prescriptions, or with general principles—but who can guarantee that they are the best way to discover, not just a few isolated 'facts,' but also some deep-lying secrets of nature? The second reason is that a scientific education as described above (and as practiced in our schools) cannot be reconciled with a humanitarian attitude. It is in conflict 'with the cultivation of individuality which alone produces, or can produce, well-developed human beings';[7] it 'maims by compression, like a Chinese lady's foot, every part of human nature which stands out prominently, and tends to make a person markedly different in outline'[8] from the ideals of rationality that happen to be fashionable in science, or in the philosophy of science. The attempt to increase liberty, to lead a full and rewarding life, and the corresponding attempt to discover the secrets of nature and of man, entails, therefore, the rejection of all universal standards and of all rigid traditions. (Naturally, it also entails the rejection of a large part of contemporary science.)

It is surprising to see how rarely the stultifying effect of 'the Laws of Reason' or of scientific practice is examined by professional anarchists. Professional anarchists oppose any kind of restriction and they demand that the individual be permitted to develop freely, unhampered by laws, duties or obligations. And yet they swallow without protest all the severe standards which scientists and logicians impose upon research and upon any kind of knowledge-creating and knowledge-changing activity. Occasionally, the laws of scientific method, or what are thought to be the laws of scientific method by a particular writer, are even integrated into anarchism itself. 'Anarchism is a world concept based upon a mechanical explanation of all phenomena,' writes Kropotkin.[9] Its method of investigation is that of the exact natural sciences . . . the method of induction and deduction.' 'It is not so clear,' writes a modern 'radical' professor at Columbia,[10] 'that scientific research demands an absolute freedom of speech and debate. Rather the evidence suggests that certain kinds of unfreedom place no obstacle in the way of science. . . .'

There are certainly some people to whom this is 'not so clear.' Let us therefore, start with our outline of an anarchistic methodology and a corresponding anarchistic science. There is no need to fear that the diminished concern for law and order in science and society that characterizes an anarchism of this kind will lead to chaos. The human nervous system is too well organized for that.[11] There may, of course, come a time when it will be necessary to give reason a temporary advantage and when it will be wise to defend its rules to the exclusion of everything else. I do not think we are living in such a time today.

The idea of a method that contains firm, unchanging, and absolutely binding principles for conducting the business of science meets considerable difficulty when confronted with the results of historical research. We find, then, that there is not a single rule, however plausible, and however firmly grounded in epistemology, that is not violated at some time or another. It becomes evident that such violations are not accidental events, they are not results of insufficient knowledge or of inattention which might have been avoided. On the contrary, we see that they are necessary for progress. Indeed, one of the most striking features of recent discussions in the history and philosophy of science is the realization that events and developments, such as the invention of atomism in antiquity, the Copernican Revolution, the rise of modern atomism (kinetic theory; dispersion theory; stereochemistry; quantum theory), the gradual emergence of the wave theory of light, occurred only because some thinkers either *decided* not to be bound by certain 'obvious' methodological rules, or because they *unwittingly broke* them.

This liberal practice, I repeat, is not just a *fact* of the history of science. It is both reasonable and *absolutely necessary* for the growth of science. More specifically, one can show the following: given any rule, however 'fundamental' or 'rational,' there are always circumstances when it is advisable not only to ignore the rule, but to adopt its opposite. For example, there are circumstances when it is advisable to introduce, elaborate, and defend *ad hoc* hypotheses, or hypotheses which contradict well-established and generally accepted experimental results, or hypotheses whose content is smaller than the content of the existing and empirically adequate alternative, or self-inconsistent hypotheses, and so on.[12]

There are even circumstances—and they occur rather frequently—when *argument* loses its forward-looking aspect and becomes a hindrance to progress. Nobody would claim that the teaching of *small children* is exclusively a matter of argument (though argument may enter into it, and should enter into it to a larger extent than is customary), and almost everyone now agrees that what looks like a result of reason—the mastery of a language, the existence of a richly articulated perceptual world, logical ability—is due partly to indoctrination and partly to a process of *growth* that proceeds with the force of natural law. And where arguments do seem to have an effect, this is more often due to their *physical repetition* than to their *semantic content*.

Having admitted this much, we must also concede the possibility of non-argumentative growth in the *adult* as well as in (the theoretical parts of) in-

stitutions such as science, religion, prostitution, and so on. We certainly cannot take it for granted that what is possible for a small child—to acquire new modes of behavior on the slightest provocation, to slide into them without any noticeable effort—is beyond the reach of his elders. One should rather expect that catastrophic changes in the physical environment, wars, the breakdown of encompassing systems of morality, political revolutions, will transform adult reaction patterns as well, including important patterns of argumentation. Such a transformation may lie in the fact that it increases the mental tension that preceded *and caused* the behavioral outburst.

Now, if there are events, not necessarily arguments, which *cause* us to adopt new standards, including new and more complex forms of argumentation, is it then not up to the defenders of the *status quo* to provide, not just counterarguments, but also contrary *causes?* ('Virtue without terror is ineffective,' says Robespierre.) And if the old forms of argumentation turn out to be too weak a cause, must not these defenders either give up or resort to stronger and more 'irrational' means? (It is very difficult, and perhaps entirely impossible, to combat the effects of brainwashing by argument.) Even the most puritanical rationalists will then be forced to stop reasoning and to use *propaganda and coercion,* not because some of his *reasons* have ceased to be valid, but because the *psychological conditions* which make them effective, and capable of influencing others, have disappeared. And what is the use of an argument that leaves people unmoved?

Of course, the problem never arises quite in this form. The teaching of standards and their defense never consists merely in putting them before the mind of the student and making them as *clear* as possible. The standards are supposed to have maximal *causal efficacy* as well. This makes it very difficult indeed to distinguish between the *logical force* and the *material effect* of an argument. Just as a well-trained pet will obey his master no matter how great the confusion in which he finds himself, and no matter how urgent the need to adopt new patterns of behaviour, so in the very same way a well-trained rationalist will obey the mental image of *his* master, he will conform to the standards of argumentation he has learned, he will adhere to these standards no matter how great the confusion in which he finds himself, and he will be quite incapable of realizing that what he regards as the 'voice of reason' is but a *causal after-effect* of the training he had received. He will be quite unable to discover that the appeal to reason to which he succumbs so readily is nothing but a *political manoeuvere.*

That interests, forces, propaganda and brainwashing techniques play a much greater role than is commonly believed in the growth of science, can also be seen from an analysis of the *relation between idea and action*. It is often taken for granted that a clear and distinct understanding of new ideas precedes, or should precede, their formulation and their institutional expression. *First*, we have an idea, or a problem, *then* we act, i.e., either speak, or build, or destroy. Yet this is certainly not the way in which small children develop. They use words, they combine them, they play with them, until they grasp a meaning that has so far been beyond their reach. And the initial playful activity is an essential prerequisite of the final act of understanding. There is no reason why this mechanism should cease to function in the adult. We must expect, for example, that the *idea* of liberty could be made clear only by means of the very same actions, which were supposed to *create* liberty. Creation of a *thing*, and creation plus full understanding of a *correct idea* of the thing, *are very often parts of one and the same indivisible process* and cannot be separated without bringing the process to a stop. The process itself is not guided by a well-defined programme, and cannot be guided by such a programme, for it contains the conditions for the realization of all possible programmes. It is guided by a vague urge, by a 'passion' (Kierkegaard). The passion gives rise to specific behaviour which in turn creates the circumstances and the ideas necessary for analyzing and explaining the process, for making it 'rational.'

The development of the Copernican point of view from Galileo to the 20th century is a perfect example of the situation I want to describe. We start with a strong belief that runs counter to contemporary reason and contemporary experience. The belief spreads and finds support in other beliefs that are equally unreasonable, if not more so (law of inertia; the telescope). Research now gets deflected in new directions, new kinds of instruments are built, 'evidence' is related to theories in new ways until there arises an ideology that is rich enough to provide independent arguments for any particular part of it and mobile enough to find such arguments whenever they seem to be required. We can say today that Galileo was on the right track, for his persistent pursuit of what once seemed to be a silly cosmology has by now created the material needed to defend it against all those who will accept a view only if it is told in a certain way and who will trust it only if it contains certain magical phrases, called 'observation reports.' And this is not an exception—it is the normal case: theories become clear and 'reasonable' only *after* incoherent parts of them have been used for a long time. Such unreasonable, nonsensical, unmethodical foreplay thus turns out to be an unavoidable precondition of clarity and of empirical success.

Now, when we attempt to describe and to understand developments of this kind in a general way, we are, of course, obliged to appeal to the existing forms of speech which do not take them into account and which must be distorted, misused, beaten into new patterns in order to fit unforeseen situations (without a constant misuse of language there cannot be any discovery, any progress). 'Moreover, since the traditional categories are the gospel of everyday thinking (including ordinary scientific thinking) and of everyday practice, [such an understanding] in effect presents rules and forms of false thinking and action—false, that is, from the standpoint of (scientific) common sense.'[13] This is how *dialectical thinking* arises as a form of thought that 'dissolves into nothing the detailed determinations of the understanding,'[14] formal logic included.

(Incidentally, it should be pointed out that my frequent use of such words as 'progress,' 'advance,' 'improvement,' etc., does not mean that I claim to possess special knowledge about what is good and what is bad in the sciences and that I want to impose this knowledge upon my readers. *Everyone can read the terms in his own way* and in accordance with the tradition to which he belongs. Thus for an empiricist, 'progress' will mean transition to a theory that provides direct empirical tests for most of its basic assumptions. Some people believe the quantum theory to be a theory of this kind. For others, 'progress' may mean unification and harmony, perhaps even at the expense of empirical adequacy. This is how Einstein viewed the general theory of relativity. *And my thesis is that anarchism helps to achieve progress in any one of the senses one cares to choose.* Even a law-and-order science will succeed only if anarchistic moves are occasionally allowed to take place.)

It is clear, then, that the idea of a fixed method, or of a fixed theory of rationality, rests on too naïve a view of man and his social surroundings. To those who look at the rich material provided by history, and who are not intent on impoverishing it in order to please their lower instincts, their craving for intellectual security in the form of clarity, precision, 'objectivity,' 'truth,' it will become clear that there is only one principle that can be defended under *all* circumstances and in all stages of human development. It is the principle: *anything goes.*

NOTES

1. 'History generally, and the history of revolution in particular, is always richer in content, more varied, more multiform, more lively and ingenuous than is imagined by even the best parties, the most conscious vanguards of the most advanced classes' (V.I. Lenin, 'Left-Wing Communism—An Infantile Disorder,' *Selected Works*, Vol. 3, London, 1967, p. 401). Lenin is addressing parties and revolutionary vanguards rather than scientists and methodologists; the lesson, however, is the same. Cf. footnote 5.

2. Herbert Butterfield, *The Whig Interpretation of History,* New York, 1965, p.66.

3. Ibid., p. 21.

4. Ibid., p. 25, cf. Hegel, *Philosophie der Geschichte, Werke,* Vol. 9, ed. Edward Gans, Berlin, 1837, p. 9: 'But what experience and history teach us is this, that nations and governments have never learned anything from history, or acted according to rules that might have derived from it. Every period has such peculiar circumstances, is in such an individual state, that decisions will have to be made, and decisions *can* only be made, in and out of it.'—'Very clever'; 'shrewd and very clever'; 'NB' writes Lenin in his marginal notes to this passage. (*Collected Works,* Vol. 38, London, 1961, p. 307.)

5. ibid., We see here very clearly how a few substitutions can turn a political lesson into a lesson for *methodology.* This is not at all surprising. Methodology and politics are both means for moving from one historical stage to another. We also see how an individual, such as Lenin, who is not intimidated by traditional boundaries and whose thought is not tied to the ideology of a particular profession, can give useful advice to everyone, philosophers of science included. In the 19th century the idea of an elastic and historically informed methodology was a matter of course. Thus Ernst Mach wrote in his book *Erkenntnis und Irrtum,* Neudruck, Wissenschaftliche Buchgesellschaft, Darmstadt, 1980, p. 200: 'It is often said that research cannot be taught. That is quite correct, in a certain sense. The schemata of *formal* logic and of *inductive* logic are of little use, for the intellectual situations are never exactly the same. But the examples of great scientists are very suggestive.' They are not suggestive because we can abstract rules from them and subject future research to their jurisdiction; they are suggestive because they make the mind nimble and capable of intervening entirely new research traditions. For a more detailed account of Mach's philosophy see my essay *Farewell to Reason,* London 1987, Chapter 7, as well as Vol. 2, Chapters 5 and 6 of my *Philosophical Papers,* Cambridge, 1981.

6. Albert Einstein, *Albert Einstein: Philosopher-Scientist,* ed. P.A. Schilpp, New York, 1951, pp. 683f.

7. John Stuart Mill, 'On Liberty,' in *the Philosophy of John Stuart Mill,* ed. Marshall Cohen, New York, 1961, p. 258.

8. Ibid., p. 265.

9. Peter Alexivich Kropotkin. 'Modern Science and Anarchism,' *Kropotkin's Revolutionary Pamphlets,* ed. R.W. Baldwin, New York, 1970, pp. 150–2. 'It is one of Ibsen's great distinctions that nothing was valid for him but science.' B. Shaw, *Back to Methuselah,* New York, 1921, p. xcvii. Commenting on these and similar phenomena Strindberg writes (*Antibarbarus*): 'A generation that had the courage to get rid of God, to crush the state and church, and to overthrow society and morality, still bowed before Science. And in Science, where freedom ought to reign, the order of the day was "believe in the authorities or off with your head."'

10. R.P. Wolff, *The Poverty of Liberalism,* Boston, 1968, p. 15. For a criticism of Wolff see footnote 52 of my essay 'Against Method,' in *Minnesota Studies in the Philosophy of Science,* Vol. 4, Minneapolis, 1970.

11. Even in undetermined and ambiguous situations, uniformity of action is soon achieved and adhered to tenaciously. See Muzafer Sherif, *The Psychology of Social Norms,* New York, 1964.

12. One of the few thinkers to understand this feature of the development of knowledge was Niels Bohr: '. . . he would never try to outline any finished picture, but would patiently go

through all the phases of the development of a problem, starting from some apparent paradox, and gradually leading to its elucidation. In fact, he never regarded achieved results in any other light than as starting points for further exploration. In speculating about the prospects of some line of investigation, he would dismiss the usual consideration of simplicity, elegance or even consistency with the remark that such properties can only be properly judged *after* [my italics] the event. . . .' L. Rosenfeld in *Niels Bohr. His Life and Work as Seen by his Friends and Colleagues,* S. Rosenthal (ed.), New York, 1967, p. 117. Now science is never a completed process, therefore it is always 'before' the event. Hence simplicity, elegance or consistency are *never* necessary conditions of scientific practice.

Considerations such as these are usually criticized by the childish remark that a contradiction 'entails' everything. But contradictions d not 'entail' anything unless people use them in certain ways. And people will use them as entailing everything only if they accept some rather simple-minded rules of derivation. Scientists proposing theories with logical faults and obtaining interesting results with their help (for example: the results of early forms of the calculus; of a geometry where lines consist of points, planes of lines and volumes of planes; the predictions of the older quantum theory and of early forms of the quantum theory of radiation—and so on) evidently proceed according to different rules. The criticism therefore falls back on its authors unless it can be shown that a logically decontaminated science has better results. Such a demonstration is impossible. Logically perfect versions (if such versions exist) usually arrive only long after the imperfect versions have enriched science by their contributions. For example, wave mechanics was not a 'logical reconstruction' of preceding theories; it was an attempt to preserve their achievements and to solve the physical problems that had arisen from their use. Both the achievements and the problems were produced in a way very different from the ways of those who want to subject everything to the tyranny of 'logic.'

13. Herbert Marcuse, *Reason and Revolution,* London, 1941, p. 130.

14. Hegel, *Wissenschaft der Logik,* Vol. 1, Hamburg, 1965, p.6.

Ruth Hubbard

· ·

"Science, Facts, and Feminism"

Feminists acknowledge that making science is a social process and that scientific laws and the "facts" of science reflect the interests of the university-educated, economically privileged, predominately white men who have produced them. We also recognize that knowledge about nature is created by an interplay between objectivity and subjectivity, but we often do not credit sufficiently the way women's traditional activities in the home, garden, and sickroom have contributed to understanding nature.

THE FACTS OF SCIENCE

The Brazilian educator, Paulo Freire, has pointed out that people who want to understand the role of politics in shaping education must "see the reasons behind the facts."[1] I want to begin by exploring some of the reasons behind a particular kind of facts, the facts of natural science. After all, facts aren't just out there. Every fact has a factor, a maker. The interesting question is: as people move through the world, how do we sort those aspects of it that we permit to become facts from those that we relegate to being fictions—untrue, imagined, imaginary, or figments of the imagination—and from those that, worse yet, we do not even notice and that therefore do not become fact, fiction, or figment? In other words, what criteria and mechanisms of selection do scientists use in the making of facts?

One thing is clear: making facts is a social enterprise. Individuals cannot just go off by themselves and dream up facts. When people do that, and the rest of us do not agree to accept or share the facts they offer us, we consider them schizophrenic, crazy. If we do agree, either because their facts sufficiently resemble ours or because they have the power to force us to accept their facts as real and true—to make us see the emperor's new clothes—then the new facts become part of our shared reality and their making, part of the fact-making enterprise.

Making science is such an enterprise. As scientists, our job is to generate facts that help people understand nature. But in doing this, we must follow

Ruth Hubbard, "Science, Facts, and Feminism," *Hypatia* 3 (1988): 5–17. Reprinted with the permission of John Wiley and Sons.

rules of membership in the scientific community and go about our task of fact-making in professionally sanctioned ways. We must submit new facts to review by our colleagues and be willing to share them with qualified strangers by writing and speaking about them (unless we work for private companies with proprietary interests, in which case we still must share our facts, but only with particular people). If we follow proper procedures, we become accredited fact-makers. In that case our facts come to be accepted on faith and large numbers of people believe them even though they are in no position to say why what we put out are facts rather than fiction. After all, a lot of scientific facts are counterintuitive, such as that the earth moves around the sun or that if you drop a pound of feathers and a pound of rocks, they will fall at the same rate.[2]

What are the social or group characteristics of those of us who are allowed to make scientific facts? Above all, we must have a particular kind of education that includes graduate, and post-graduate training. That means that in addition to whatever subject matter we may learn, we have been socialized to think in particular ways and have familiarized ourselves with that narrow slice of human history and culture that deals primarily with the experiences of western European and North American upper class men during the past century or two. It also means that we must not deviate too far from accepted rules of individual and social behavior and must talk and think in ways that let us earn the academic degrees required of a scientist.

Until the last decade or two, mainly upper-middle and upper class youngsters, most of them male and white, have had access to that kind of education. Lately, more white women and people of color (women and men) have been able to get it, but the class origins of scientists have not changed appreciably. The scientific professions still draw their members overwhelmingly from the upper-middle and upper classes.

How about other kinds of people? Have they no role in the making of science? Quite the contrary. In the ivory (that is, white) towers in which science gets made, lots of people are from working-class and lower-middle class backgrounds, but they are the technicians, secretaries, and clean-up personnel. Decisions about who gets to be a faculty-level fact-maker are made by professors, deans, and university presidents who call on scientists from other, similar institutions to recommend candidates who they think will conform to the standards prescribed by universities and the scientific professions. At the larger, systemic level, decisions are made by government and private funding agencies which operate by what is called peer review. What that means is that small groups of people with similar personal and academic backgrounds decide whether a particular fact-making proposal has enough merit to be financed. Scientists who work in the same, or related, fields mutually sit on each other's decision making panels and whereas criteria for access are supposedly objective and meritocratic, orthodoxy and conformity count for a lot. Someone whose ideas and/or personality are out of line is less likely to succeed than "one of the boys"—and these days some of us girls are allowed to join the boys, particularly if we play by their rules.

Thus, science is made, by and large, by a self-perpetuating, self-reflexive group by the chosen for the chosen. The assumption is that if the science is "good," in a professional sense, it will also be good for society. But no one and no group are responsible for looking at whether it is. Public accountability is not built into the system.

What are the alternatives? How could we have a science that is more open and accessible, a science *for* the people? And to what extent could—or should—it also be a science *by* the people? After all, divisions of labor are not necessarily bad. There is no reason and, indeed, no possibility that in a complicated society like ours, everyone is able to do everything. Inequalities which are bad, come not from the fact that different people do different things, but from the fact that different tasks are valued differently and carry with them different amounts of prestige and power.

For historical reasons, this society values mental labor more highly than manual labor. We often pay more for it and think that it requires more specifically human qualities and is therefore superior. This is a mistake, especially in the context of a scientific laboratory, because it means that the laboratory chief—the person "with ideas"—often gets the credit, whereas the laboratory workers—the people who work with their hands (as well as, often, their imaginations)—are the ones who perform the operations and make the observations that generate new hypotheses and that permit hunches, ideas, and hypotheses to become facts.

But it is not only because of the way natural science is done that head and hand, mental and manual work, are often closely linked. Natural science requires a conjunction of head and hand because it is an understanding of nature *for use*. To understand nature is not enough. Natural science and technology are inextricable, because we can judge that our understanding of nature is true only to the extent that they can be applied and used as technology. The science/technology distinction, which was introduced one to two centuries ago, does not hold up in the real world of economic, political and social practices.

As I said before, to be believed, scientific facts must fit the world-view of the times. Therefore, at times of tension and upheaval, such as the last two decades, some researchers always try to "prove" that differences in the political, social, and economic status of women and men, blacks and whites, or poor people and rich people, are inevitable because they are the results of people's inborn qualities and traits. Such scientists have tried to "prove" that blacks are innately less intelligent than whites, or that women are innately weaker, more nurturing, less good at math than men. If, for the purposes of this discussion, we focus on sex differences, it is clear that the ideology of women's nature can differ drastically from the realities of women's lives and indeed be antithetical to them. In fact, the ideology functions, at least in part, to obscure the ways in which women live and to make people look away from the realities or ask misleading questions about them. So, for example, the ideology that labels women as the natural reproducers of the species, and men as producers of goods, has not been used to exempt women from also producing goods and services, but to shunt us out of higher paying jobs, the professions, and other kinds of work that require continuity and provide a measure of power over one's own, and at times, other people's lives. Most women who work for pay do so in job categories, such as secretary or nurse, which often involve a great deal of concealed responsibility, but are underpaid. This is one reason why insisting on equal pay *within* job categories cannot remedy women's economic disadvantage. Women will continue to be underpaid as long as women's jobs are less well paid than men's jobs and as long as access to traditional men's jobs is limited by social pressures, career counseling, training and hiring practices, trade union policies, and various other subtle and not so subtle societal mechanisms, such as research that "proves" that girls are not as good as boys at spatial perception, mathematics, and science. An entire range of discriminatory practices is justified by the claim that they follow from the limits that biology places on women's capacity to work. Though exceptions are made during wars and other emergencies, they are forgotten as soon as life resumes its normal course. Then women are expected to return to their subordinate roles, not because the quality of their work during the emergencies had been inferior, but because these roles are seen as natural.

A few years ago, a number of women employees in the American chemical and automotive industries were actually forced to choose between working at relatively well-paying jobs that had previously been done by men or remaining fertile. In one instance, five women were required to submit to sterilization *by hysterectomy* in order to avoid being transferred from work in the lead pigment department at the American Cyanamid plant in Willow Island, West Virginia to janitorial work at considerably lower wages and benefits.[3] Even though none of these women was pregnant or planning a pregnancy in the near future (indeed, the husband of one had had a vasectomy), they were considered "potentially pregnant" unless they could prove they were sterile. This goes on despite the fact that exposure to lead can damage sperm as well as eggs and can affect the health of workers (male and female) as well as a "potential fetus." It is as though fertile women are at all times potential parents, men, never. But it is important to notice that this vicious choice is being forced only on women who have recently entered relatively well-paid, traditionally male jobs. Women whose work routinely involves reproductive hazards because it exposes them to chemical or radiation hazards, but who have traditionally female jobs such as nurses, X-ray technologists, laboratory technicians, cleaning women in surgical operating rooms, scientific laboratories or the chemical and biotechnology industries, beauticians, secretaries, workers in the ceramics industry, and domestic workers are not warned about the chemical or physical hazards of their work to their health or that of a fetus, should they be pregnant. In other words, scientific knowledge about fetal susceptibility to noxious chemicals and radiation is used to keep women out of better paid job categories from which they had previously been excluded by discriminatory employment practices, but in general, women (or, indeed, men) are not protected against health endangering work.

The ideology of women's nature that is invoked at these times would have us believe that a woman's capacity to become pregnant leaves her always physically disabled by comparison with men. The scientific underpinnings for these ideas were elaborated in the nineteenth century by the white, university-educated, mainly upper class men who made up the bulk of the new profession of obstetrics and gynecology, biology, psychology, sociology and anthropology. These professionals used their theories of women's innate frailty to disqualify the girls and women of their own race and class who would have competing with them for education and professional status. They also realized that they might lose the kinds of personal attention they were accustomed to get from mothers, wives, and sisters if women of their own class gained access to the professions. They did not invoke women's weakness when it came to poor women spending long hours working in homes and factories belonging to members of the upper classes, nor against the ways black slave women were made to work on the plantations and in the homes of their masters and mistresses.

Nineteenth century biologists and physicians claimed that women's brains were smaller than men's and that women's ovaries and uteruses required much energy and rest in order to function properly. They "proved" that therefore young girls must be kept away from schools and colleges once they begin to menstruate and warned that without this kind of care, women's uteruses and ovaries will shrivel up and the human race will die out. Yet again, this analysis was not carried over to poor women, who were not only required to work hard, but often were said to reproduce *too* much. Indeed, scientists interpreted the fact that poor women could work hard and yet bear many children as a sign that they were more animal-like and less highly evolved than upper class women.

During the past decade, feminists have uncovered this history. We have analyzed the self-serving theories and documented the absurdity of the claims as well as their class and race biases and their glaringly political intent.[4] But this kind of scientific mythmaking is not past history. Just as in the nineteenth century, medical men and biologists fought women's political organizing for equality by claiming that our reproductive organs made us unfit for anything other than childbearing and childrearing, just as Freud declared women to be intrinsically less stable, intellectually inventive and productive than men, so beginning in the 1970's, there has been a renaissance in sex differences research that has claimed to prove scientifically that women are innately better than men at home care and mothering while men are innately better fitted than women for the competitive life of the marketplace.

Questionable experimental results obtained with animals (primarily that prototypic human, the white laboratory rat) are treated as though they can be applied equally well to people. On this basis, some scientists are now claiming that the secretion of different amounts of so-called male hormones (androgens) by male and female fetuses produced life-long differences in men's and women's brains. They claim not only that these (unproved) differences in fetal hormone levels exist, but that imply (without evidence) that they predispose women and men *as groups* to exhibit innate differences in our abilities to localize objects in space, in our verbal and mathematical aptitudes, in aggressiveness and competitiveness, nurturing ability, and so on.[5] Sociobiologists claim that some of the sex differences in social behavior that exist in Western, capitalist societies (such as, aggressiveness, competitiveness, and dominance among men, coyness, nurturance, and submissiveness among women) are human universals that have existed in all times and cultures. Because these traits are said to be ever-present, sociobiologists deduce that they must have evolved through Darwinian natural selection and are now part of our genetic inheritance.[6]

Sociobiologists have tried to prove that women's disproportionate contributions to child- and homecare are biologically programmed because women have a greater biological "investment" in our children than men have. They offer the following rationale: an organism's biological fitness, in the Darwinian sense, depends on producing the greatest possible number of offspring, who themselves survive long enough to reproduce, because this is what determines the frequency with which an individual's genes will be represented in successive generations. The calculus goes as follows. Eggs are larger than sperm and women can produce many fewer of them than men can sperm. Therefore each egg that develops into a child represents a much larger fraction of the total number of children a woman can produce, hence her "reproductive fitness," than a sperm that becomes a child does of a man's "fitness." In addition, women "invest" the nine months of pregnancy in each child. Women must therefore be more careful than men to acquire well-endowed sex partners who will be good providers to make sure that their few investments (read, children) mature. Thus, from seemingly innocent biological asymmetries between sperm and eggs flow such major social consequences as female fidelity, male promiscuity, women's disproportional contribution to caring for home and children, and the unequal distribution of labor by sex. As sociobiologist, David Barash, says, "mother nature is sexist," so don't blame her human sons.[7]

In devising these explanations, sociobiologists ignore the fact that human societies do not operate with a few superstuds, nor do stronger or more powerful men as a rule have more children than weaker ones. Men, in theory, could have many more children than women can, but in most societies equal numbers of men and women engage in producing children, though not in caring for them. These kinds of absurdities are useful to people who have a stake in maintaining present inequalities. They mystify procreation, yet have a superficial ring of plausibility and thus offer naturalistic justification for discriminatory practices.

As the new scholarship on women has grown, a few anthropologists and biologists have tried to mitigate the male bias that underlies these kinds of theories by describing how females contribute to social life and species survival in important ways that are overlooked by scientists who think of females only in relation to reproduction and look to males for everything else.[8] But, unless scientists challenge the basic premises that underlie the standard, male-centered descriptions and analyses, such revisions do not offer radically different formulations and insights.[9]

I want to come back to Paulo Freire, who says, "Reality is never just simply the objective datum, the concrete fact, but is also people's [and I would say certain people's] perception of it." And he speaks of "the indispensable unity between subjectivity and objectivity in the act of knowing."[10]

The recognition of this "indisputable unity" is what feminist methodology is about. It is especially necessary for a feminist methodology in science because the scientific method rests on a particular definition of objectivity, that we feminists must call into question. Feminists and others who draw attention to the devices that the dominant group has used to deny other people access to power—be it political power or the power to make facts—have come to understand how that definition of objectivity functions in the processes of exclusion I discussed at the beginning.

Natural scientists attain their objectivity by looking upon nature (including other people) in small chunks and as isolated objects. They usually deny, or at least do not acknowledge, their relationship to the "objects" they study. In other words, natural scientists describe their activities as though they existed in a vacuum. The way language is used in scientific writing reinforces this illusion because it implicitly denies the relevance of time, place, social context, authorship, and personal responsibility. When I report a discovery, I do not write, "One sunny Monday after a restful weekend, I came into the laboratory, set up my experiment and shortly noticed that . . ." No, proper style dictates, "It has been observed that . . ." This removes the relevance of time and place, and implies that the observation did not originate in the head of a human observer, specifically my head, but out there in the world. By deleting the scientist-agent as well as her or his participation as observer, people are left with the concept of science as a thing in itself, that truly reflects nature and that can be treated as though it were as real as, and indeed equivalent to, nature.

A particularly blatant example of the kind of context-stripping that is commonly called objectivity is the way E. O. Wilson opens the final chapter of his *Sociobiology: The New Synthesis*.[11] He writes, "Let us now consider man in the free spirit of natural history, as though we were zoologists from another planet completing a catalog of social species on earth." That statement epitomizes the fallacy we need to get rid of. There is no "free spirit of natural history," only a set of descriptions put forward by the mostly white, educated, Euro-American men who have been practicing a particular kind of science during the past two hundred years. Nor do we have any idea what "zoologists from another planet" would have to say about "man" (which, I guess is supposed to mean "people") or about other "social species on earth," since that would depend on how these "zoologists" were used to living on their own planet and by what experiences they would therefore judge us. Feminists must insist that subjectivity and context cannot be stripped away, that they must be acknowledged if we want to use science as a way to understand nature and society and to use the knowledge we gain constructively.

For a different kind of example, take the economic concept of unemployment which in the United States has become "chronic unemployment" or even "the normal rate of unemployment." Such pseudo-objective phrases obscure a wealth of political and economic relationships which are subject to social action and change. By turning the activities of certain people who have the power to hire or not hire people into depersonalized descriptions of economic fact, by turning activities of scientists into "factual" statements about nature or society, scientific language helps to mystify and intimidate the "lay public," those anonymous others, as well as scientists, and make them feel powerless.

Another example of the absurdity of pretended objectivity, is a study that was described in the *New York Times* in which scientists suggested that they had identified eight characteristics in young children that were predictive of the likelihood that the children would later develop schizophrenia. The scientists were proposing a longitudinal study of such children as they grow up to assess the accuracy of these predictions. This is absurd because such experiments cannot be done. How do you find a "control" group for parents who have been told that their child exhibits five out of the eight characteristics, or worse yet, all eight characteristics thought to be predictive of schizophrenia? Do you tell some parents that this is so although it isn't? Do you not tell some parents whose children have been so identified? Even if psychiatrists agreed on the diagnosis of schizophrenia—which they do not—this kind of research cannot be done objectively. And certainly cannot be done ethically, that is, without harming people.

The problem is that the context-stripping that worked reasonably well for the classical physics of falling bodies has become the model for how to do every kind of science. And this even though physicists since the beginning of this century have recognized that the experimenter is part of the experiment and influences its outcome. That insight produced Heisenberg's uncertainty principle in physics: the recognition that the operations the experimenter performs disturb the system so that it is impossible to specify simultaneously the position and the momentum of atoms and elementary particles. So, how about standing the situation on its head and using the social sciences, where context-stripping is clearly impossible, as a model and do all science in a way the experimenter as a self-conscious subject who lives, and does science, within the context in which the phenomena she or he observes occur? Anthropologists often try to take extensive field notes about a new culture as quickly as possible after they enter it, before they incorporate the perspective and expectations of the culture, because they realize that once they know the foreign culture well and feel at home in it, they will begin to take some of its most significant aspects for granted and they will stop seeing them. Yet they realize at the same time that they must also acknowledge the limitations their own personal and social backgrounds impose on the way they perceive the foreign society. Awareness of our subjectivity and context must be part of doing science because there is no way we can eliminate them. We come to the objects we study with our particular personal and social backgrounds and with inevitable interests. Once we acknowledge those, we can try to understand the world, so to speak, from inside instead of pretending to be objective outsiders looking in.

The social structure of the laboratory in which scientists work and the community of inter-personal relationships in which they live are also part of the subjective reality and context of doing science. Yet, we usually ignore them when we speak of a scientist's scientific work despite the fact that natural scientists work in highly organized social systems. Obviously, the sociology of laboratory life is structured by class, sex, and race, as is the rest of society. We saw before that to understand what goes on in the laboratory we must ask questions about who does what kinds of work. What does the lab chief—the person whose name appears on the stationary or the door—contribute? How are decisions made about what work gets done and in what order? What role do women, whatever our class and race, or men of color and men from working class backgrounds play in this performance?

Note that women have played a very large role in the production of science—as wives, sisters, secretaries, technicians, and students of "great men"—though usually not as accredited scientists. One of our jobs as feminists must be to acknowledge that role. If feminists are to make a difference in the ways science is done and understood, we must not just try to become scientists who occupy the traditional structures, follow established patterns of behavior, and accept prevailing systems of explanation, we must understand and describe accurately the roles women have played all along in the process of making science. But we must also ask why certain ways of systematically interacting with nature and of using the knowledge so gained are acknowledged as science whereas others are not.

I am talking of the distinction between the laboratory and that other, quite differently structured place of discovery and fact-making, the household, where women use a different brand of botany, chemistry, and hygiene to work in our gardens, kitchens, nurseries, and sick rooms. Much of the knowledge women have acquired in those places is systematic and effective and has been handed on by word of mouth and in writing. But just as our society downgrades manual labor, it also downgrades knowledge that is produced in other than professional settings, however systematic it may be. It downgrades the orally transmitted knowledge and the unpaid observations, experimentation and teaching that happen in the household. Yet here is an unvalidated (in fact, devalued and invalidated) by the institutions that catalog and describe, and thus define, what is to be called knowledge. Men's exploration of nature also began at home, but later were institutionalized and professionalized. Women's explorations have stayed close to home and their value has not been acknowledged.

What I am proposing is the opposite of the project the domestic science movement put forward at the turn of the century. That movement tried to make women's domestic work more "scientific" in the traditional sense of the word.[12] I am suggesting that we acknowledge the scientific value of many of the facts and knowledge that women have accumulated and passed on in our homes and in volunteer organizations.

I doubt that women as gendered beings have something new or different to contribute to science, but women as political beings do. One of the most important things we must do is to insist on the political content of science and on its political role. The pretense that science is objective, apolitical and value-neutral is profoundly political because it obscures the political role that science and technology play in underwriting the existing distribution of power in society. Science and technology always operate in somebody's interest and serve someone or some group of people. To the extent that scientists are "neutral," that merely means that they support the existing distribution of interests and power.

If we want to integrate feminist politics into our science, we must insist on the political nature and content of scientific work and of the way science is taught and otherwise communicated to the public. We must broaden the base of experience and knowledge on which scientists draw by making it possible for a wider range of people to do science, and to do it in different ways. We must also provide kinds of understanding that are useful and usable by a broad range of people. For this, science would have to be different from the way it is now. The important questions would have to be generated by a different social process. A wider range of people would have to have access to making scientific facts and to understanding and using them. Also, the process of validation would have to be under more public scrutiny, so that research topics and facts that benefit only a small elite while oppressing large segments of the population would not be acceptable.

Our present science, which supposedly exists to explain nature and let us live more comfortably in it, has in fact mystified nature. As Virginia Woolf's Orlando says as she enters a department store elevator,

> The very fabric of life now. . . . is magic. In the eighteenth century, we knew how everything was done, but here I rise through the air, I listen to voices in America, I see men flying—but how it's done, I can't even begin to wonder.[13]

OTHER WAYS TO DO SCIENCE?

The most concrete examples of a different kind of science that I can think of come from the women's health movement and the process by which the Boston's Women's Healthbook Collective's *The New Our Bodies, Ourselves*[14] or the Federation of Feminist Women's Health Center's *A New View of a Woman's Body*[15] have been generated. These groups have consciously tried to involve a range of women in setting the agenda, as well as in asking and answering the relevant questions. But there is probably no single way in which to change present-day science, and there shouldn't be. After all, one of the problems with science as it exists now, is that scientists narrowly circumscribe the allowed ways to learn about nature and reject deviations as deviance.

Of course it is difficult for feminists who, as women, are just gaining a toehold in science, to try to make fundamental changes in the ways scientists perceive science and do it. This is why many scientists who are feminists live double-lives and conform to the pretenses of an apolitical, value-free meritocratic science in our working lives while living our politics elsewhere. Meanwhile, many of us who want to integrate our politics with our work, analyze and critique the standard science, but no longer do it. Here again, feminist health centers and counseling groups come to mind as efforts to integrate feminist inquiry and political praxis. It would be important for feminists, who are trying to reconceptualize reality and recognize knowledge and its uses in areas other than health, to create environments ("outstitutes") in which we can work together and communicate with other individuals and groups, so that people with different backgrounds and agendas can exchange questions, answers, and expertise.

NOTES

1. *The Politics of Education.* Paolo Freire. (South Hadley: Bergin and Garvey, 1985). p.2.

2. Recently some physicists have hypothesized that a pound of feathers falls more *rapidly* than a pound of rocks—an even more counterintuitive "fact" than what I learned in high school physics.

3. Stellman, J.M. and M.S. Henifin. "No Fertile Women Need Apply: Employment Discrimination and Reproductive Hazards in the Workplace." in *Biological Woman—The Convenient Myth,* (eds.) R. Hubbard, M.S. Henifin, and B. Fried (Cambridge: Schenkman, 1982). pp. 117–45.

4. See *Genes and Gender II: Pitfalls in Research on Sex and Gender.* (eds.)R. Hubbard and M. Lowe. (Staten Island: Gordian Press, 1979), *Women's Nature: Rationalizations of Inequality.* (eds.) M. Lowe and R. Hubbard (New York: Pergamon, 1983), *Science and Gender.* R. Bleier. (New York: Pergammon, 1984), *Myths of Gender.* A. Fausto-Sterling. (New York: Basic Books, 1985).

5. See *Man & Woman, Boy & Girl.* J. Money and A.A. Ehrhardt. (Baltimore: Johns Hopkins University Press, 1972), *Sexual Differentiation of the Brain.* R. W. Goy and B.S. McEwen. (Cambridge: MIT Press, 1980), *Science* 1981, vol. 211. pp. 1263–1324.

6. *Sociobiology: The New Synthesis.* E.O. Wilson. (Cambridge: Harvard University Press, 1975).

7. See *The Selfish Gene.* R. Dawkins. (New York: Oxford University Press, 1976) and *The Whispering Within.* D. Barash. (New York: Harper & Row, 1979), esp. pp. 46–90.

8. See *Primate Behavior and the Emergence of Human Culture.* J.B. Lancaster. (New York: Holt, Rinehart and Winston, 1975), *The Woman that Never Evolved.* S.B. Hrdy. (Cambridge: Harvard University, 1981), "Empathy, Polyandry, and the Myth of the Coy Female" in *Feminist Approaches to Science,* (ed.) R. Bleier. (New York: Pergamon, 1986). pp. 119–146, *Females of the Species.* B. Kevles. (Cambridge: Harvard University Press, 1986).

9. For examples of more fundamental criticisms of evolutionary thinking and sociobiology, see "Sociobiology and Biopsychology: Can Science Prove the Biological Basis of Sex Differences in Behavior?" Lowe, M. and R. Hubbard in *Genes and Gender II: Pitfalls in Research on Sex and Gender.* (eds.) R. Hubbard and M. Lowe. (Staten Island: Gordian Press, 1979). pp. 91–112, "Have Only Men Evolved?" R. Hubbard. in *Biological pp.* 17–46, *Not in Our Genes.* R.C. Lewontin, S. Rose and L.J. Kamin (New York: Pantheon, 1984).

10. Freire 1985, p. 81.

11. Wilson 1975, p. 547.

12. *Men's Ideas/Women's Realities: Popular Science, 1870–1915.* (ed.) L.M. Newman. (New York: Pergamon, 1985). pp. 156–191.

13. *Orlando.* V. Woolf. (New York: Harcourt Brace Jovanovich, Harvest Paperback Edition, 1928). p. 300.

14. (New York: Simon and Schuster, 1984).

15. (New York: Simon and Schuster, 1981).

Bruno Latour

"The Science Wars: A Dialogue"

She: So you're a sociologist and you do research on scientists? Well, then you can explain something to me. People in my lab are forever talking about the "Science Wars." What's all the fuss about?

He: If only I knew! I'd know what front to fight on, what equipment to carry, and what camouflage to wear. As things are, people are firing in all directions. It isn't easy to know what's going on.

She: I've heard the main thing is to avoid relativism. But I'm a physicist, and that presents a real difficulty. Without relativity there'd be no possibility of making measurements and we'd each be prisoners, to all eternity, in some single point of view. In my discipline, we need the relativity of frames of reference in order even to begin work. I have a special need for relativity because I work on events close to the Big Bang. You don't need relativity, too?

He (sighing): Yes, of course, but relativism is one of the victims of this war; it's a refugee. For you, the word means relativity. But in the humanities and in ethics, it's an insult implying: "you think that all points of view are equally valid, that all cultures are equal, that truth and error are on the same plane, that Rembrandt and graffiti have the same value, and that we can't distinguish between creationists and evolutionists because everything's valid and anything goes."

She: But you really think all of that! I'm appalled. My lab colleagues were right: "Never date a sociologist . . ."

He: But of course I don't think so. I told you that was an insult, not a concept. The relativist is always the other guy, the guy to accuse of not respecting axiology, of not distinguishing between a mad and a sane scientist, between a cardinal and a Galileo, between a Holocaust denier and a genuine historian.

She: Well, do you see the difference? Because if not, you're a relativist for sure.

From Bruno Latour, "The Science Wars: A Dialogue," *Common Knowledge* 8, no. 1 (2002): 71–79. Copyright © 2002, Duke University Press. All rights reserved. Reprinted by permission of the publisher.

He: Of course I see the difference. What do you take me for? The difference between the departments of geology or geoscience and the curio cabinets of the creationists (I've visited some in San Diego—the "creationist research centers"!) is so huge that I don't see the point of adding an even more absolute distinction between true and false. On the one hand, there are those who, for the last two centuries, have constructed the history of a world several billion years old, and on the other there are those obsessed by the Bible and at war with abortion. There's no connection between the two. They live in incommensurable worlds.

She: So if I understand you correctly, you reject the accusation of relativism but claim there's no need for an *absolute* distinction between true and false in order to distinguish between this case and that. In my field, if you reject absolute frames of reference, you're a relativist. But for us, that's a positive designation, and relativity's the only means of achieving commensurability.

He: Very well, if you wish: I'm a relativist in the sense that I, like you, reject an absolute point of reference. I agree that this rejection permits me to establish relations and distinctions, and to measure the gaps between points of view. For me, being a relativist means being able to establish relations between frames of reference, and so, being able to pass from one framework to another in converting measurements (or, at least explanations and descriptions). It's a positive term, I agree, to the extent that the opposite of *relativist* is *absolutist*.

She: If what you say is true, why do my colleagues so attack you? Are you keeping something from me? You're a wolf in sheep's clothing, *n'cest-ce pas?*

He: Forgive me, but your colleagues aren't simply physicists, they're politicians too, and it's for political reasons that they call me every name under the sun. They're wolves pretending to be sheep under attack by wolves.

She: Hardly! And it's you they accuse of playing politics. They say you confuse matters of scientific truth with questions of value and that, for your disciples, everything is politics. To determine if quasars are really there, in the constellation of Betelgeuse, and if they date from a billion years after the Big Bang, all we'd have to do is gather the people in my lab and let them vote—after which, abracadabra, by consensus, the four quasars in question would appear in the sky at just the right time. As easy as adding a regulation to the traffic code or passing a law on compensation for acts of God.

He *(sighing again)*: Easy! Only because you think politics consists of meetings and voting. Decisions are made and new things come into existence—abracadabra, as you say—all by themselves! But politics is a little more complicated than that.

She: Of course, yes of course—politics also consists of interests, passions, values, questions of ethics. But ultimately, is it Yes or No? Are you claiming that I can arbitrarily modify the number of quasars in the constellation of Betelgeuse, that my scientific papers aren't subject to any constraint from celestial phenomena, that science is simply a "language game" (I hear that the à la mode expression)—are you claiming that I can just say anything that occurs to me?

He: Yes, you can say just anything that occurs to you. The question you've just asked is proof of that!

She *(heatedly)*: Instead of insulting me, you'd do better to explain how a quasar is a social construction made out of whole cloth by my colleagues and me. You've written, it appears, some awful things about the "social construction of reality." And to think I'm the one accused of saying just anything that occurs to me!

He: You see, this is what the Science Wars amount to: two intelligent academics posing stupid questions to each other. First of all, "social construction" doesn't mean a thing. And second, I'm not the one who uses the term—some of my colleagues do. At any rate, it's not the term that's the problem, it's your perversity and your scandalous double standards.

She: Now you're really over the top—you stand publicly accused of imposture and you permit yourself not only to insult me but also to claim that I'm a fraud?

He: But you *are* a fraud! Maybe the word's a little violent, but your colleagues insulted me first! Look, when you use a radio telescope, when you do simulations on your computers, when you print your maps in "false colors," when you calculate the redshift, when you apply the theories of particle physicists—do these instruments, theories, methodologies play a role or not in the conclusions you reach?

She: That's self-evident. Of course. We couldn't say a thing without them. The existence of quasars could never have been proven if–

He: Then what would you place in the debit column?

She: I don't know. Whatever prevents me from talking about quasars: poor instruments, confused data, disputes among theoreticians—above all, inadequate budget. We can't transform the planet into an immense radio

telescope, we can't coordinate our efforts to accomplish it, unless—
which is incredible, because if we could coordinate our machinery, we
could achieve . . . incidentally, at the last meeting of the International
Association, I was elected to organize the second phase of the Sloan
Digital Sky Survey, which would interest you because . . .

He: If you don't mind, let's not get lost. Your business interests me, but I'd
like to finish with this little matter of accounting. So—it would never
occur to you to say, "I've come to posit the existence of quasars *despite* the
existence of radio telescopes and the panoply of equipment and theories
that are connected to them"?

She: Certainly not. Because, as I've just told you, I've been elected a member
of the council entrusted with coordinating all the telescopes on Earth to
make one huge antenna by 2005. And you haven't been listening to me.

He: Not so. I'm listening with great satisfaction as you entangle yourself in
contradictions.

She *(piqued):* How am I contradicting myself? I'd like to know.

He: Because you're sweating blood to get new machinery in the credit
column of your lab accounts. The more powerful your machines are, the
more—so you claim—you can say exact things about your quasars. . . .

She: Naturally. That's how we work. What could be wrong?

He: What's wrong, my dear physicist, is that you change your accounts
ledger depending on your audience—whether it's me or the general
public. You always have two columns, one for credit and one for debit.
But on the credit side, you now place the quasars, as if they're beyond
discussion, and on the debit side you place your instruments, your
budgets, theories, papers, colleagues—and you whine: "If only I didn't
have these machines and impediments, I could at last talk plainly and
without obfuscation about my quasars."

She *(coldly):* I said ex-act-ly the opposite. I said that without radio telescopes
we couldn't speak about quasars.

He: Why, then, did you pretend, in making fun of me, that there's a choice
to be made between politics and reality? *Either* you play politics and
arbitrarily decide, abracadabra, by consensus at a meeting of your lab
colleagues, on the existence of the four quasars for the constellation of
Betelgeuse *or else* the quasars determine what you say about them in
print. You were the one who imposed this awkward choice on me, this
choice of "language game" versus "reality." There are indeed two columns
here: a debit column and a credit column; a column of language games,

social construction, and discourse, and a column of reality, truth, and exactitude. You have two languages, and your tongue is as forked as a viper's. When it suits you, when you're asking for money, you say, "The instruments permit quasars to speak." And on the other hand, when it suits you, you say, "We must choose between social constructions and reality." Personally, I think that's the epitome of fraud . . .

She (*slightly embarrassed*): Hmm, perhaps I haven't been clear. It was my colleagues who said that you force a choice between social construction and external reality. And they said that, if you had free rein, there would be no way to distinguish between the sciences and all the absurdities of pataphysics, numerology, and astrology. They went to a talk that Alan Sokal gave and I was shocked by their report of what he said. According to them, you pose a matter of life or death for scientists. We can't let this happen.

He: But what is "that"? So as far as I'm concerned, what we can't let happen is for the "Sokalists" to perpetuate this fraud, this intellectual imposture, this accounting racket whereby, on the one hand, reality and social construction are synonymous (the better the instruments are, the better reality can be grasped), and on the other hand, social construction and reality are in opposition. I'm sorry, but I think that there's the real scandal. If we were talking about the mafia, we'd say they were laundering dirty money . . . and what's more, it's anti-science. The Sokalist imposture renders the defense of scientific activity impossible.

She: So now you're interested in defending scientific activity, Mr. Sociologist—since when have you posed as a friend of the sciences?

He (*amused*): Oh, for some thirty years. I find the sciences interesting, rich, cultivated, civilized, useful, passionately engaging; and I can't understand how so many scientists comply in making them cold, stupid, uncouth, contradictory, antisocial, useless, and boring.

She: I'm completely lost. I also find the sciences passionately engaging. I devote my life to them, they *are* my passion. Then why are we in opposing camps? If you're right, we should be allies.

He (*somewhat tenderly*): But we are, my dear physicist, of course we're allies. It's the battle cries of the science warriors and *that alone* that forces us to believe in opposing camps, to rally and align ourselves as if there were a battle, But there *isn't* a battle . . .

She (*once again distrustful*): No, if that were the problem, the Science Wars would not be so intense. My colleagues were foaming at the mouth when they came back from Sokal's seminar. The danger you represent must be more real than that of a dispute over accounting practices or the limits of constructivism.

He: Of course we present a danger. We're the Sokalists' *political* adversaries.

She: So you admit, after all, that you want to *politicize* the sciences.

He: No, I attest I want to *depoliticize* the sciences so that they can't be used in this unsavory way as a tool for silencing political discussion.

She: Okay, then: the Sokalists, as you call them, are the ones who play politics. That's all there is to it?

He: There's more. While emphasizing the link between their language and reality, their constructions and truth, their instruments and the external world, they still act as if they and they alone had *unmediated* access to reality, truth, and the external world. They act as if they possessed a magical machine that speaks the truth and pays no price for it in controversy, in construction in the laboratory, in arduous historical labor.

She: They don't say that, though. They're too reasonable to say so.

He: Oh yes, they do say so. But they have cooked their books, their crooked ledger, and can have it all both ways. When it suits them, they point to the link between instruments and truth. And when it suits them, they act as if the laws of physics fell from heaven, and as if those who point up the role of instruments and language games are madmen or criminals.

She *(ironically):* Funny to hear you saying that, because, so I've been told, it's you who they accuse of double-dealing. On occasion you say that you're a social constructionist and, when it suits you, that you're the most loyal friend of the sciences and a born-again realist. And in this way you give both your publics (those against science and those for it) what they want to hear without blemishing your reputation.

He: With the Sokalists, of course. I have to speak two languages because they don't understand what I am saying. I speak of wave-particles and they say one has to choose: either it's a wave or a particle.

She: You're not going to start doing physics, surely.

He: I'm using an image to show you the extent of their incomprehension. They haven't even begun to pose the question that we're trying to resolve in the history, sociology, and anthropology of science: how human beings can speak truly about events, about the irruption of new objects into the world. For the science warriors, there simply isn't a problem. They think that I'm playing the fiend, that I'm avoiding difficulties. Whereas I'm actually studying what they're scrupulously avoiding with their fraudulent accounts—and that is: how human beings imbue and fill the world with language. How do you yourself, my dear, set about to speak the truth about quasars, which are scarcely a billion years younger than the Big Bang itself? But instead of listening, understanding, and reconstructing the difficulty involved, the science warriors deny the difficulty altogether. They arrive in the middle of the discussion in their clumsy clogs and shout, "The question shall not be posed! Over here we have the quasars of Betelgeuse and over there is Mme. X, the physicist. Those who wish to complicate this matter are dangerous relativists." For my part, I say, "Let us do our work. You go do your dirty business elsewhere. If you don't understand the problem we're posing, don't disturb those of us who do."

She *(softened completely):* But this problem I do understand! It even fascinates me, it occupies me night and day. How can one speak the truth?—You're right, the question can't just be put aside. . . . Is that the kind of research you do?

He *(moved a little):* Yes, that's my quasar, my Betelgeuse, that's what occupies my nights and days.

She: You, too, are a researcher . . . I thought that sociologists . . . *[fading sarcastically yet tenderly].* In fact, you do have a proper job, then.

He: I believe so, yes. I hope so. Only by modifying the concept of science can we prevent the political use that your physics friends make of it, and it is this attempt, at bottom, that they can't forgive us. The controversy doesn't directly concern a problem of research.

She: I still don't understand what's political about their attitude.

He: But obviously, in insisting ceaselessly on the existence of an external world beyond discussion, directly known without mediation, without controversy, without history, they render all political will impotent. Public life is reduced to a rump of itself.

She: But if I've followed you, you also believe in an external reality, or haven't I understood, after all?

He: Oh, I ought to kiss you. Sign a certificate for me, "Mme. X, physicist, certifies on her honor she has proof, Mr. Y, sociologist, believes in external reality." It is the phrase *beyond discussion* that is at issue. For my part, external realities are what make me speak; they augment and complicate, they enlarge discussion.

She: Oh, for me too. You can't imagine the difficulties I've had in convincing my colleagues that there are four and not three quasars in this corner of the universe and that one of them is the oldest object ever discovered.

He: But they, the science warriors, equate external reality with what's beyond discussion, with silence, with what permits miserable human beings *to be* silenced. Those who speak without saying anything, the politicians. . . .

She: Politicians do perhaps speak without saying much, but what about me? What if they tried to silence me with their reality-beyond-discussion? Actually, Professor _____, a real macho jerk that one, did try to shut me up on the pretext that I'd made a mistake in the calculation of the redshift. I certainly told him what's what. You're right! We have to fight against those who want to shut our mouths. If that's what the Science Wars are, then I'm ready to fight beside you.

He: Beside *me?* But we're in opposing camps, according to you. And those who want to close discussion by confusing reality with silence are your colleagues, my dear friend—your dear colleagues, those who you said. . . .

She: Oh my, that's possible too. I don't know any longer where I stand. These Science Wars are so obscure. . . .

He: That's what I told you from the beginning. Why not talk, quite simply, about peace?

She: Yes, let's speak about something more interesting than wars. I could explain the business about the antenna as large as the planet . . . I'm sure that would passionately interest you. . . .

Case Studies

Astronomy Track

PAPER 6: THE DEMOTION OF PLUTO

THE CASE

The successful discovery of Neptune was a lesson that astronomers did not soon forget. Vulcan may not have been found, but perhaps the same line of thought could still be fruitful elsewhere. When the size and orbit of Neptune had been sufficiently determined, it was scrutinized for unexpected deviations. When tiny discrepancies were found, the race was on to calculate how big and where in the sky the next planet might be found.

Many tried for many years to no avail, but in February 1930, using photographic methods, researchers at the observatory in Flagstaff, Arizona, finally located it. The announcement led astronomers at all the major observatories to look and confirmations followed. Thus, Pluto was discovered. While the mass of Pluto turned out to be significantly smaller than predicted, in 1978 James W. Christie discovered a moon, Charon, orbiting Pluto and therefore it was generally accepted as the ninth planet.

At least, it was until the chorus of voices in the International Astronomical Union grew strong enough to remove it. In 2006 at a meeting in Prague, a new category of heavenly body, the dwarf planet, was created and Pluto was categorized off the list of major planets in the solar system by a vote of 424 professional astronomers.

Some complained that the move was not legitimate because, less than 4 percent of the attending astronomers voted, since the issue was brought to the floor late in the conference after many had left. But the decision was made according to protocol and the change in status became official. Pluto was deemed less than a full planet, making it a fact that the solar system had eight, not nine, full-fledged planets orbiting the sun.

YOUR JOB

What is the International Astronomical Association? Why do they determine what is and what is not a planet? Discuss Hubbard's and Latour's accounts of the roles of power in science and how they shape what gets counted as real science. How can the events in Prague be understood through their lens? Was the solar system different after the vote? What does the move mean for our understanding of the solar system? For our understanding of astronomy?

Physics Track

THE CASE

One of the biggest concerns in physics has become the integration of Einstein's theory of general relativity (which is used to explain the big and the fast) with quantum theory (which covers the small). It is a problem that was identified in the 1930s, and even with the advances to modern quantum field theory, it has remained in the background. A complete physical theory ought to unify the field of physics and the physical fields it describes.

A promising possibility to do just that has been string theory, in which space-time itself is thought to be made up of something that can be pictured as loops of string and the various particles described by the standard model are thought of as modes of vibration of these tiny strings. The rules that govern the strings' motions are able to account for both relativistic and quantum effects, thereby allowing the desired unification. Each independent mode of vibration is equivalent to a dimension; therefore, string theory requires a space with eleven dimensions. There is great hope that string theory, or some derived extension of it, could be the next major revolutionary step in the advancement of physics, allowing for the grand unification that has been sought for decades.

But not everyone is buying into the project. Physicist Lee Smolin, for example, is not only not sold on string theory but is convinced that those pushing string theory are stifling research on competing ideas. In "How Do You Fight Sociology?," the sixteenth chapter of his book *The Trouble with Physics: The Rise of String Theory, the Fall of a Science, and What Comes Next,* he argues that the string theorists have been able to mount a successful public relations campaign and have created an insular group that actively excludes other opportunities. The push behind string theory, he argues, has harmed the progress of particle physics.

THE SCIENTIST

Lee Smolin (1955–) studied both physics and philosophy at Hampshire College before becoming a theoretical physicist. A founder of the theory of loop quantum gravity, he works at the Perimeter Institute for Theoretical Physics at the University of Waterloo in Ontario.

YOUR JOB

What is string theory designed to do? What are its opponents' main arguments against it? Are they arguments from physics or philosophy? What is Smolin's argument about why string theory is retarding progress in particle physics? In the article, " 'Theory of

Everything' Tying Researchers Up in Knots" in the March 14, 2005, edition of *The San Francisco Chronicle,* Nobel laureate Robert Laughlin is quoted as saying, "I think string theory is textbook 'post-modernism' fueled by irresponsible expenditures of money." Explain Latour's postmodern account of science. How is string theory postmodern? How would the debate over string theory as discussed by Smolin be accounted for by a postmodernist like Latour?

Chemistry Track

PAPER 6: THE "DISCOVERY" AND RETRACTION OF ELEMENTS 116 AND 118

THE CASE

In 1999, researchers at the U.S. Department of Energy's Lawrence Berkeley National Laboratory published a paper in the extremely prestigious journal *Physical Review Letters* claiming to have discovered two new elements, ununhexium (atomic number 116) and ununoctium (atomic number 118). Based upon calculations by theoretical chemist Robert Smolańczuk, it was predicted that by focusing a beam of high-energy krypton ions on bits of lead, the reaction would have what was referred to as "islands of stability." This stability meant that for a millisecond, an atom with 118 protons and 293 neutrons would exist before decaying into an atom with 116 protons and 173 neutrons, which itself would last only for a millisecond before decaying into a smaller atom itself.

The researchers set up the experiment using the lab's large cyclotron and analyzed the data using a powerful computer. Based on that data interpretation, they claimed to have observed the two "islands of stability" in the subsequent reaction. The computer program's analysis gave reason to think that the effects of such states of matter could be discerned from the raw data.

Three years later, however, the team retracted that paper and repudiated its result. They now claimed that they had never observed the superheavy elements after all. According to their press release, "A technical committee of experts from the Laboratory's physics, supercomputing, and nuclear science division is reviewing the data and methodology from that 1999 result. Subsequent re-analyses of the original data with different software codes have been unsuccessful in observing atomic decay patterns, or chains, which would confirm the existence of element 118."

YOUR JOB

Explain the circumstances surrounding the "discovery" and subsequent rejection of the discovery of elements 116 and 118. Explain what Feyerabend, Hubbard, or Latour says about the role of social power in science. Were the elements really observed? If so, how could scientists be wrong about having observed it? If a computer program is needed for the observation, is it really an observation or is it an interpretation? Do different software programs observe different realities? Are software programs like telescopes and microscopes? Why was a team of prestigious researchers needed? What would these writers say about the social conditions that led the scientists to retract their observation?

Genetics Track

PAPER 6: PUBLIC AND CORPORATE GENOMIC RESEARCH

THE CASE

With the ability to decode the genome of organisms, new questions about scientific knowledge have arisen. Knowing a portion of the human genome in detail, for example, will allow one to formulate new treatments for heritable diseases, treatments that would be not only save many lives, but generate great profits for whoever discovers the genetic cause and formulates the genetic cure.

New tools like the polymerase chain reaction allow us to insert portions of code into the DNA of organisms and create new and improved versions created to solve human problems. Living organisms now are to the experimental biologist what a block of marble is to the sculptor. The biologists who create a new strain of wheat that is cheaper to grow, yields greater harvests, or is resistant to pests and disease would not only profit humanity but also would stand to reap substantial profits themselves.

The claim is that such profits would cause science to advance at a much greater pace. The incentive of monetary reward will drive scientists onward in a way that mere curiosity would not. But in order for this incentive to work, scientists must be guaranteed that their finds are protected as property. Knowledge about nature—indeed, facts about our very own bodies—therefore, is not something that we can have, but is owned by someone else.

Much research is done with financing from government agencies and this knowledge, coming from public funds, belongs to the public. But more and more is done by biotech firms with the idea that it is done for profit.

YOUR JOB

Explain why some argue that private funding of science is imperative for the advancement of science. Why would this mean that genes should be patented? How would Latour or Hubbard understand this attempt to own scientific knowledge? What concern does this lead to? What would be the argument that genes should not be patented? Who is right?

Evolutionary Biology Track

PAPER 6: INTELLIGENT DESIGN AND SCIENCE CURRICULA

THE CASE

In *No Free Lunch*, William Dembski argues that intelligent design theory is a legitimate scientific theory. Intelligent design is an alternative to Darwinian evolution in which random mutations are not considered to be responsible for speciation, but rather, species are part of the predetermined design of an intelligent creator. Nature's complexity is to be understood not as the result of natural and sexual selection determined by the pressures of the environment, but rather as something set up in advance according to a plan in the mind of a being that made it.

The primary means by which Dembski argues for its scientific status is an appeal to complexity theory. The scientists working on the SETI (Search for Extraterrestrial Intelligence) project, for example, examine signals collected by a large array of radio telescopes and analyze them for patterns that would indicate an intelligently designed message. Mathematical means are used to differentiate random noise from intentional signals. There is an empirically determinable difference between designed and random phenomena. If it is scientific to use such methods in one case, Dembski contends, then applying the same methods to a different subject matter, speciation, should be equally scientific.

The reason that intelligent design is considered nonscientific, he argues, is not because of its merits, but because those who advocate evolutionary theory control all of the keys to the scientific kingdom. It is a threat to their power and place, and so evolutionary theory advocates work hard to make sure that intelligent design is maligned and kept out of science classrooms. Instead of allowing students to make up their own minds about the two theories, intelligent design advocates argue, Darwinian advocates use their power and dominance to keep the conversation from even happening.

THE SCIENTIST

William Dembski (1960–) holds two PhDs, one in mathematics from the University of Chicago and another in philosophy from the University of Illinois at Chicago, in addition to a master's degree in divinity from Princeton Theological Seminary. He is a research professor in philosophy at Southwest Baptist Theological Seminary and a senior fellow at the Center for Science and Culture at the Discovery Institute, a think-tank dedicated to intelligent design.

YOUR JOB

Explain the intelligent design theory. What is complexity theory and how is it supposed to make intelligent design scientific? Explain Latour's or Hubbard's sociological view of science. How would they see this controversy? Should intelligent design be treated as a scientific view on this view? Is this correct? Why or why not?

Geology Track

PAPER 6: THE GRAND CANYON, POLITICS, AND THE AGE OF THE EARTH

THE CASE

The Grand Canyon is a living laboratory for those with an interest in geology. Whether one looks from the rim or hikes among the strata, the rocks cannot help but affect a visitor, and the mind immediately begins to wonder about the forces that gave rise to such a marvel. To help answer such questions, visitor center's gift shop features many volumes about the geology of the region.

In 2003, a new volume appeared among the books, entitled *The Grand Canyon: A New View.* Written by Tom Vail, a young earth creationist and tour guide who leads raft trips of the canyon, the book argues that the scientific accounts are flawed and gives a new version designed to be compatible with a literalist interpretation of the Bible in which the canyon is no more than a couple of thousand years old.

Park rangers working in the Grand Canyon wrote a rebuttal to the book, trying to help park staff better communicate the scientific view to visitors, but the publication was blocked by the government.

In the article "At Grand Canyon Park, a Rift Over Creationist Book" appearing in the *Washington Post,* Vail is quoted as saying, "None of it is science. Science has to be repeatable and measurable. What they call science is theory just as what is in my book is theory."

YOUR JOB

Explain the controversy surrounding *The Grand Canyon: A New View.* Explain Feyerabend's, Hubbard's, or Latour's view of science. Are science and politics really ever separate on this view? How would an adherent of your selected view respond to the Grand Canyon case? How should they respond?

Psychology Track

PAPER 6: THE CLASSIFICATION OF HOMOSEXUALITY AND THE SOCIAL CONSTRUCTION OF KNOWLEDGE

THE CASE

The *Diagnostic and Statistical Manual* (DSM), first published in 1952, is used by psychiatrists to diagnose mental illness. The first edition included homosexuality as a "sociopathic personality disorder." In the second edition, put out in 1968, homosexuality was listed as a "sexual deviation." Freudians, for example, had long considered homosexuality to be a "perverse orientation" that showed some subconscious problems that did not allow for the normal functioning of the individual.

In 1973, at the annual meeting of the American Psychiatric Association in San Francisco, a movement arose to have homosexuality removed as a disorder. Led by gay rights activists and a subset of psychiatrists, homosexuality was removed in the seventh printing of the second edition of the manual, the *DSM* II.

When the *DSM* III appeared, homosexuality was not considered by itself to be a disorder, but a new diagnosis, gender identity disorder, appeared to label young people who believed themselves to be homosexual or who displayed cross-gender interests or behaviors. Questions about the etiology, classification, and treatment criteria that must be specified to consider a condition to be a disorder have been challenged on several grounds from within and without the psychiatric community.

YOUR JOB

Explain the debate that occurred in 1973 around the categorization of homosexuality as a disorder. Explain one of the contemporary critiques of gender identity disorder. Select either Latour's or Hubbard's critique of scientific methodology and explain the view in detail. How would these episodes be viewed through the lens of this critique? Are there aspects that are not well-accounted for in this view?

Sociology Track

PAPER 6: BURAWOY, PUBLIC SOCIOLOGY, AND POLITICS

THE CASE

In his presidential address in 2004 at the annual meeting of the American Sociological Association, Michael Burawoy argued that the profession needed to focus on public sociology as opposed to merely academic sociology. By his account, public sociology "defines, promotes and informs public debates about class and racial inequalities, new gender regimes, environmental degradation, multiculturalism, technological revolutions, market fundamentalism, and state and non-state violence."

The knowledge that accompanies sociological scholarship brings with it a moral sense to play the role of the expert in social discourse and politics. This does not mean acquiring power to become the sociological version of Plato's philosopher-kings, but rather, that whenever one has a detailed understanding of an issue with significant human ramifications, one also acquires the responsibility to use that understanding to further human welfare.

Michael Burawoy (1947–) studies industrial workplaces around the world. While doing his research for his master's degree at the University of Zambia, he became a steelworker in that country. He is currently a professor of sociology at the University of California at Berkeley.

YOUR JOB

Explain what Burawoy means by public sociology. What is his argument for the necessity of public sociology? Explain either Latour's or Hubbard's critical analysis of science. How would they view Burawoy's call? What is the relationship between science and politics? Does becoming a public sociologist damage the objectivity of the researcher's work?

Economics Track

PAPER 6: THE WORLD BANK, INTERNATIONAL MONETARY FUND, AND THE POLITICS OF MACROECONOMICS

THE CASE

The lessons of the Great Depression and the instability caused by the bankrupted German state after World War I, coupled with the new appreciation for macroeconomic planning, led to a meeting at Breton Woods, New Hampshire, in 1944. At that meeting, the International Monetary Fund and the World Bank were launched.

The International Monetary Fund (IMF) is designed to be a short-term lender to countries needing to make payments on debts. By keeping countries from defaulting, confidence in the borrowing country's economy would keep investment funds coming in instead of going out, thereby allowing the economy a chance to grow out of its problems and insuring stability. The World Bank was designed to provide more substantive loans to nations with the expressed mission of reducing global poverty through development projects and post-conflict reconstruction of infrastructure.

The hope is that by creating supernational institutions, the local politics that could stand in the way of sound macroeconomic decision making could be avoided and a safety net provided to keep nations' economies from collapsing as a result of war, natural disaster, or gross mismanagement.

But the World Bank and IMF cannot be entirely apolitical institutions. Powerful and wealthy nations hold significant power in determining which loans are approved and the conditions, termed "structural adjustment policies," that must be met by the country to qualify for the needed loan. These structural adjustments often include the privatization of government-owned and -run industries, as well as austerity measures that force cut backs in health care, public assistance, education funding, and opening markets to foreign competition. Additionally, the appointments to the IMF's and World Bank's most powerful positions are affected by the distribution of global political power.

Trade can create wealth and jobs that in turn can help alleviate poverty. Development of infrastructure is a necessary precondition for a growing economy that can help the people of the country live better, healthier, more fulfilled lives. The job of the IMF and World Bank is to help foster this, and that requires macroeconomic judgments. But these judgments are made by members of the IMF and World Bank who are nominated by politicians from certain countries.

YOUR JOB
Explain what the World Bank and IMF are supposed to do. Explain some of the critics' arguments that the World Bank and IMF are not succeeding in their mission to eliminate poverty. Explain the critical view of Hubbard or Latour. How do politics influence science on this view? How would someone of this position understand the case of the World Bank and IMF? Are there aspects that are not well accounted for by this view?

Closing Remarks

The idea of "the scientific method" seems so easy, straightforward, and un-problematic at first glance. Of course, when one moves beyond that first glance to a deep stare, the questions become far more interesting. I have intentionally avoided using the word "conclusion" in naming this section because I do not intend to conclude anything. Philosophy is, at root, an ongoing conversation, one I hope your work has opened to you. This part of the discussion can focus the issues but surely cannot resolve them.

By no means is this meant to imply that nothing within philosophy is ever settled. Surely some statements are true, others false, and thereby, some views are problematic and rightly rejected while others rationally maintained. But several of the debates we have examined remain open with smart people framing quite different, competing views.

Is there a logic of scientific discovery? The positions of the hard-core deductivists and inductivists clearly overstate the case in favor of one. There cannot be a "turn the crank" mechanism for generating novel purported laws of nature. William Whewell must have been right that doing science, to some measure, is a creative act. The great geniuses who produced historic advances had insight beyond what most scientists considered. And yet, is it truly an intellectual context free of structure? Are there no logical constraints or patterns of reasoning governing the way scientific proposals are derived?

What of the logic of confirmation? Is inductive evidence a part of the process? Could Karl Popper have been correct that David Hume's problem of induction makes it impossible to use such evidence or is it simply a place where the philosophical game is so removed from the real practice of science we worry about things that no working scientist would find the slightest bit disconcerting? Are the paradoxes of evidence actual problems for science or just problems for philosophers thinking about science from a viewpoint removed from any real contact with the actual endeavor?

Even if we did allow for inductive evidence to be meaningful, what is it that is being confirmed? Are the parts of scientific theories independently testable or only as a whole? Could Thomas Kuhn have been correct that the explanation for theory choice is largely historical instead of rational? Were Paul Feyerabend, Ruth Hubbard, and Bruno Latour correct that sociological factors are always and necessarily part of the discussion and that to look at the question of scientific progress from a purely logical position oversimplifies the real life process to the point of naiveté?

Is this confirmation the mark of truth or mere utility? Imre Lakatos famously maintained that all scientific theories are born refuted, in other words, in due time all theories will be shown to be false, to have counterinstances. Does this mean that we should view science as providing us with nothing but useful tools, not actual mirrors of the relations among real things and their properties? What is the relation, if any, between our best theories and the underlying nature they seemingly strive to describe? If we think of theories as models, is there anything actually being modeled, and do our models actually bring our minds closer to it or are we just giving ourselves heuristic means by which to derive good predictions?

It should be clear that just like science itself, philosophical thought about science also progresses. Are the two evolutions interrelated? Is there one correct view that is uniquely descriptive of the way science is, has been, and must always be done? Or perhaps has scientific methodology developed along with the advancements in science itself, changing as the questions of science have changed? Does the scientific method respond only to advances in scientific research or does it also reflect the state of politics, religion, or the distribution of wealth and power as well?

These are questions that philosophers continue to discuss in various terms along various lines. Hopefully they are questions that now engage you in a way that may not have seemed relevant before beginning this investigation. Hopefully you now have educated, insightful thoughts and opinions on these questions that draw on your historical understanding of your science of interest. Hopefully you can now enter into a passionate discussion about them. Indeed, hopefully you realize that you just have. That passionate discussion *is* the philosophy of science.

Deductivism Case Study Readings

ASTRONOMY TRACK

Reading 1: Aristotle—*On the Heavens*

BOOK II

CHAPTER I

That the heavens as a whole neither came into being nor admits of destruction, as some assert, but is one and eternal, with no end or beginning of its total duration, containing and embracing in itself the infinity of time, we may convince ourselves not only by the arguments already set forth but also by a consideration of the views of those who differ from us in providing for its generation. If our view is a possible one, and the manner of generation which they assert is impossible, this fact will have great weight in convincing us of the immortality and eternity of the world. Hence it is well to persuade oneself of the truth of the ancient and truly traditional theories, that there is some immortal and divine thing which possesses movement, but movement such as has no limit and is rather itself the limit of all other movement.

CHAPTER III

. . . Everything which has a function exists for its function. The activity of God is immortality, i.e. eternal life. Therefore the movement of that which is divine must be eternal. But such is the heaven, viz. a divine body, and for that reason to it is given the circular body whose nature it is to move always in a circle. Why, then, is not the whole body of the heaven of the same character as that part? Because there must be something at rest at the centre of the revolving body; and of that body no part can be at rest, either elsewhere or at the centre. It could do so only if the body's natural movement were towards the centre. But the circular movement is natural, since otherwise it could not be eternal: for nothing unnatural is eternal. The unnatural is subsequent to the natural, being a derangement of the natural which occurs in the course of its generation. Earth then has to exist; for it is earth which is at rest at the centre. (At present we may take this for granted: it shall be explained later.) But if earth must exist, so must fire. For, if one of a pair of contraries naturally exists, the other, if it is really contrary, exists also naturally. In some form it must be present, since the matter of contraries is

From Aristotle, "De Caelo," *On the Heavens*, in *The Oxford Translation of Aristotle*, vol. 2, ed. W. D. Ross, trans. J. L. Stocks (Oxford: Oxford University Press, 1930), 283, 286, 288–89. Reprinted by permission of Oxford University Press.

the same. Also, the positive is prior to its privation (warm, for instance, to cold), and rest and heaviness stand for the privation of lightness and movement. But further, if fire and earth exist, the intermediate bodies must exist also: each element stands in a contrary relation to every other. (This, again, we will here take for granted and try later to explain.) these four elements generation clearly is involved, since none of them can be eternal: for contraries interact with one another and destroy one another. Further, it is inconceivable that a movable body should be eternal, if its movement cannot be regarded as naturally eternal: and these bodies we know to possess movement. Thus we see that generation is necessarily involved. But if so, there must be at least one other circular motion: for a single movement of the whole heaven would necessitate an identical relation of the elements of bodies to one another. This matter also shall be cleared up in what follows: but for the present so much is clear, that the reason why there is more than one circular body is the necessity of generation, which follows on the presence of fire, which, with that of the other bodies, follows on that of earth; and earth is required because eternal movement in one body necessitates eternal rest in another.

CHAPTER VI

We have next to show that the movement of the heaven is regular and not irregular. This applies only to the first heaven and the first movement; for the lower spheres exhibit a composition of several movements into one. If the movement is uneven, clearly there will be acceleration, maximum speed, and retardation, since these appear in all irregular motions. The maximum may occur either at the starting-point or at the goal or between the two; and we expect natural motion to reach its maximum at the goal, unnatural motion at the starting-point, and missiles midway between the two. But circular movement, having no beginning or limit or middle in the direct sense of the words, has neither whence nor whither nor middle: for in time it is eternal, and in length it returns upon itself without a break. If then its movement has no maximum, it can have no irregularity, since irregularity is produced by retardation and acceleration. Further, since everything that is moved is moved by something, the cause of the irregularity of movement must lie either in the mover or in the moved or both. For if the mover moved not always with the same force, or if the moved were altered and did not remain the same, or if both were to change, the result might well be an irregular movement in the moved. But none of these possibilities can be conceived as actual in the case of the heavens. As to that which is moved, we have shown that it is primary and simple and ungenerated and indestructible and generally unchanging; and the mover has an even better right to these attributes. It is the primary that moves the primary, the simple the simple, the indestructible and ungenerated that which is indestructible and ungenerated. Since then that which is moved, being a body, is nevertheless unchanging, how should the mover, which is incorporeal, be changed?

It follows then, further, that the motion cannot be irregular. For if irregularity occurs, there must be change either in the movement as a whole, from fast to slow and slow to fast, or in its parts. That there is no irregularity in the parts is obvious, since, if there were, some divergence of the stars would have taken place before now in the infinity of time, as one moved slower and another faster: but no alteration of their intervals is ever observed. Nor again is a change in the movement as a whole admissible. Retardation is always due to incapacity, and incapacity is unnatural. The incapacities of animals, age, decay, and the like, are all unnatural, due, it seems, to the fact that the whole animal complex is made up of materials which differ in respect of their proper places, and no single part occupies its own place. If therefore that which is primary contains nothing unnatural, being simple and unmixed and in its proper place and having no contrary, then it has no place for incapacity, nor, consequently, for retardation or (since acceleration involves retardation) for acceleration. Again, it is inconceivable that the mover should first show incapacity for an infinite time, and capacity afterwards for another infinity. For clearly nothing which, like incapacity, unnatural ever continues for an infinity of time; nor does the unnatural endure as long as the natural, or any form of incapacity as long as the capacity. But if the movement is retarded it must necessarily be retarded for an infinite time. Equally impossible is perpetual acceleration or perpetual retardation. For such movement would be infinite and indefinite, but every movement, in our view, proceeds from one point to another and is definite in character. Again, suppose one assumes a minimum time in less than which the heaven could not complete its movement. For, as a given walk or a given exercise on the harp cannot take any and every time, but every performance has its definite minimum time which is unsurpassable, so, one might suppose, the movement of the heaven could not be completed in any and every time. But in that case perpetual acceleration is impossible (and, equally, perpetual retardation: for the argument holds of both and each), if we may take acceleration to proceed by identical or increasing additions of speed and for an infinite time. The remaining alternative is to say that the movement exhibits an alternation of slower and faster: but this is a mere fiction and quite inconceivable. Further, irregularity of this kind would be particularly unlikely to pass unobserved, since contrast makes observation easy.

That there is one heaven, then, only, and that it is ungenerated and eternal, and further that its movement is regular, has now been sufficiently explained.

PHYSICS TRACK

Reading 1: Epicurus—*Letter to Herodotus*

EPICURUS TO HERODOTUS, GREETING.

For those who are unable to study carefully all my physical writings or to go into the longer treatises at all, I have myself prepared an epitome of the whole system, Herodotus, to preserve in the memory enough of the principle doctrines, to the end that on every occasion they may be able to aid themselves on the most important points, so far as they take up the study of Physics. Those who have made some advance in the survey of the entire system ought to fix in their minds under the principle headings an elementary outline of the whole treatment of the subject. For a comprehensive view is often required, the details but seldom.

To the former, then—the main heads—we must continually return, and must memorize them so far as to get a valid conception of the facts, as well as the means of discovering all the details exactly when once the general outlines are rightly understood and remembered; since it is the privilege of the mature student to make a ready use of his conceptions by referring every one of them to elementary facts and simple terms. For it is impossible to gather up the results of continuous diligent study of the entirety of things, unless we can embrace in short formulas and hold in mind all that might have been accurately expressed even to the minutest detail.

Hence, since such a course is of service to all who take up natural science, I, who devote to the subject my continuous energy and reap the calm enjoyment of a life like this, have prepared for you just such an epitome and manual of the doctrines as a whole.

In the first place, Herodotus, you must understand what it is that words denote, in order that by reference to this we may be in a position to test opinions, inquiries, or problems, so that our proofs may not run on untested *ad infinitum*, nor the terms we use be empty of meaning. For the primary signification of every term employed must be clearly seen, and ought to need no proving; this being necessary, if we are to have something to which the point at issue or the problem or the opinion before us can be referred.

Next, we must by all means stick to our sensations, that is, simply to the present impressions whether by the mind or of any criterion whatever, and similarly to our actual feelings, in order that we may have the means of determining that which needs confirmation and that which is obscure.

When this is clearly understood, it is time to consider generally things which are obscure. To begin with, nothing comes into being out of what is non-existent. For in that case anything would have arisen out of anything, standing as it would o its proper germs. And if that which disappears had been destroyed and become non-existent, everything would have perished, that into which the things were dissolved being non-existent. Moreover, the sum total of things was always such as it is now, and such it will ever remain. For there is nothing into which it can change. For outside the sum of things there is nothing which could enter into it and bring about this change.

Further, the whole of being consists of bodies and space. For the existence of bodies is everywhere attested by sense itself, and it is upon sensation that reason must rely when it attempts to infer the unknown from the known. And if there were no space (which we call also void and place and intangible nature), bodies would have nothing in which to be and through which to move, as they are plainly seen to move. Beyond bodies and space there is nothing which by mental apprehension or on its analogy can conceive to exist. When we speak of bodies and space, both are regarded as wholes or separate things, not as the properties or accidents of separate things.

Again, of bodies, some are composite, others the elements of which these composite bodies are made. These elements are indivisible and unchangeable, and necessarily so, if things are not all to be destroyed and pass into non-existence, but are to be strong to endure when the composite bodies are broken up, because they possess a solid nature and are incapable of being anywhere or anyhow dissolved. It follows that the first beginnings must be indivisible, corporeal entities.

Furthermore, the atoms, which have no void in them—out of which composite bodies arise and into which they are dissolved—vary indefinitely in their shapes; for so many variety of things as we see could never have arisen out of a recurrence of a definite number of the same shapes. The like atoms of each shape are absolutely infinite; but the variety of shapes, though indefinitely large, is not absolutely infinite.

The atoms are in continual motion through all eternity. Some of them rebound to a considerable distance from each other, while others merely oscillate in one place when they chance to have got entangled or to be enclose by a mass of other atoms shaped for entangling.

This is because each atom is separated from the rest by void, which is incapable of offering any resistance to the rebound; while it is the solidity of the atom which makes it rebound after a collision, however short the distance to which it rebounds, when it finds itself imprisoned in a mass of entangling atoms. Of all this there is no beginning, since both atoms and void exist from everlasting.

The repetition at such length of all that we are now recalling to mind furnishes an adequate outline for our conception of the nature of things.

CHEMISTRY TRACK

Reading 1: Paracelsus—*Hermetic and Alchemal Writings*

I praise alchemy, which compounds secret medicines, whereby all hopeless maladies are cured. They who are ignorant of this deserve neither to be called chemists nor physicians. For these remedies lie either in the power of the alchemists or in that of the physicians. If they reside with the latter, the former are ignorant of them. If with the former, the latter have not learned them. How, therefore, do those men deserve any praise? I, for my part, have rather judged that such a man shall be highly extolled who is able to bring Nature to such a point that she will lend help, that is, who shall know how after the extraction of the health-giving parts, what is useless is to be rejected; who is also acquainted with the efficacy, for he must see that it is impossible that the preparation and the science—in other words, the chemia and the medicine—can be separated from one another, because should anyone attempt to separate them he will introduce more obscurities into medicine, and the result will be absolute folly. I do not think I need labour very hard in order that you may recognize the certainty of my reasons.

. . . The separation of those things which grow out of the earth and are combustible, such as fruits, herbs, flowers, leaves, grasses, roots, woods, etc., is also arranged in many ways. By distillation is separated from them first the phlegma, afterwards the Mercury, after this the oil, Fourthly the sulphur, lastly their salt. When all these separations are made according to Spagyric Art, remarkable and excellent medicaments are the result, both for internal and external use.

As to the manner in which God created the world, take the following account. He originally reduced it to one body, while the elements were developing. This body He made up of three ingredients, Mercury, Sulphur, and Salt, so that these three should constitute one body. Of these three are composed all the things which are, or are produced, in the four elements. These three have in themselves the force and the power of all perishable things. In them lie hidden the mineral, day, night, heat, cold, the stone, the fruit, and everything else, even while not yet formed. It is even as with wood which is thrown away and is only wood, yet in it are hidden all forms of animals, of plants, of

From Paracelsus, *Hermetic and Alchemical Writings*, ed. A. E. Waite (London: James Elliot and Co., 1894), 167, 204–5, 208–9, 257.

instruments, which anyone who can carve what else would be useless, invents and produces. So the body of Iliaster [primitive chaos] was a mere trunk, but in it lay hidden all
herbs, waters, gems, minerals, stones, and chaos itself, which things the Creator alone
carved and fashioned most subtly, having removed and cast away all that was extrane-
ous. First of all, He produced and separated the air. This being formed, from the remainder issued forth the other three elements, fire, water, earth. From these He afterward took away the fire, while the other two remained, and so on in due succession.

Now, as to the philosophy of the three prime elements, it must be seen how these
flourish in the element of air. Mercury, Sulphur, and Salt are so prepared as the element of air that they constitute the air, and make up that element. Originally, the sky is
nothing but white Sulphur coagulated with the spirit of Salt and clarified by Mercury,
and the hardness of this element is in this pellicle and shell thus formed from it. Then
secondly, from the three primal parts it is changed into two—one part being air and
the other chaos—in the following way. The Sulphur resolves itself by the spirit of Salt
in the liquor of Mercury, which of itself is a liquid distributed from heaven to earth,
and is the albumen of the heaven, and the mid space. It is clear, a chaos, subtle, and
diaphanous. All density, dryness, and all its subtle nature, are resolved, nor is it any
longer the same as it was before. Such is the air. The third remnant of the three pri-
mals has passed into air, thus; If wood is burnt it passes into smoke. So this passes
into air, remains in the air to the end of its elements, and becomes Sulphur, Mercury,
and Salt, which are substantially consumed and turned into air, just as the wood which
becomes smoke. It is, in fact, nothing but the smoke of the three primal elements of
the air. So, then, nothing further arises from the elements of air beyond what has been
mentioned.

CONCERNING SALT AND SUBSTANCES COMPREHENDED UNDER SALT
God has driven and reduced man to such a pitch of necessity and want that he is unable in any way to live without Salt, but has most urgent need thereof for his food and
eatables. This is man's need and condition of compulsion. The causes of this compulsion I will briefly explain.

Man consists of three things: Sulphur, Mercury, and Salt. Of these consists also
whatever anywhere exists, and of neither more nor fewer constituents. These are the
body of every single thing, whether endowed with sense or deprived thereof. Now,
since man is divided into species, he is therefore subject to decay, nor can he escape
it except in so far as God has endowed him with a congenital balsam which also itself consists of three ingredients. This is Salt, preserving man form decay, so also Salt
naturally diffused into us by God preserves our body from putrification. Let that theory
stand, then, that man consists of three bodies, and that one of these is Salt, as the con-

servative element which prevents the body born with it from decaying. As, therefore, all created things, all substances, consist of these three, it is necessary that they should be sustained and conserved by their nutriments according to this kind.

GENETICS TRACK

Reading 1: Aristotle—*On the Generation of Animals*

BOOK 1

CHAPTER 18

Now the offspring comes *from* the semen and it is plainly in one of the two following senses that it does so—either the semen is the material from which it is made, or it is the efficient cause. For assuredly it is not in the sense of A being *after* B, as the voyage comes *from*, i.e., after, the Panathenaea; nor yet as contraries come from contraries, for then one of the two contraries ceases to be, and a third substance must exist as an immediate underlying basis from which the new thing comes into being. We must discover, then, in which of the two other classes semen is to be put, whether it is to be regarded as matter, and therefore acted upon by something else, or as a form, and therefore acting upon something else, or as both at once. For perhaps at the same time we shall see clearly also how all the products of semen come into being from contraries, since coming into being from contraries is also a natural process, for some animals do so, i.e., from male and female, others from only one parent, as is the case with plants and all those animals in which male and female are not separately differentiated. Now that which comes from the generating parents is called the seminal fluid, being that which first has in it a principle of generation, in the case of all animals whose nature is to unite; semen is that which has in it the principles from *both* united parents, as the first mixture which arises from the union of the male and female, be it a foetus or an ovum, for these already have in them that which comes from both.

CHAPTER 20

That, then, the female does not contribute semen to generation, but does contribute something, and that this is the matter of the catamenia, or that which is analogous to it in bloodless animals, is clear from what has been said, and also from a general and abstract survey of the question. For there must needs be that which generates and that

From Aristotle, "De Generation Animalium," *Parts of Animals*, in *The Oxford Translation of Aristotle*, vol. 5, ed. W. D. Ross, trans. Arthur Platt (Oxford: Oxford University Press, 1912), 724, 729, 730. Reprinted by permission of Oxford University Press.

from which it generates; even if these be one, still they must be distinct in form and their essence must be different; and in those animals that have these powers separate in two sexes the body and nature of the active and passive sex must also differ. If, then, the male stands for the effective and active, and the female, considered as female, for the passive, it follows that what the female would contribute to the semen of the male would not be semen but material for the semen to work upon. This is just what we find to be the case, for the catamenia have in their nature an affinity to the primitive matter.

CHAPTER 21

So much for the discussion of this question. At the same time, the answer to the next question we have to investigate is clear from these considerations, I mean how it is that the male contributes to generation and how it is that the semen from the male is the cause of the offspring. Does it exist in the body of the embryo as a part of it from the first, mingling with the material which comes from the female? Or does the semen communicate nothing to the material body of the embryo but only to the power and movement in it? For this power is that which acts and makes, while that which is made and receives the form is the residue of the secretion in the female. Now the latter alternative appears to be the right one both *a priori* and in view of the facts. For, if we consider the question on general grounds, we find that, whenever one thing is made from two of which one is active and the other is passive, the active agent does not exist in that which is made; and, still more generally, the same applies when one thing moves and another is moved; the moving thing does not exist in that which is moved. But the female, as female, is passive, and the male, as male, is active, and the principle of the movement comes from him. Therefore, if we take the highest genera under which they each fall, the one being active and motive and the other passive and moved, that one thing which is produced comes from them only in the sense in which a bed comes into being from the carpenter and wood, or in which a ball comes into being from the wax and the form. It is plain that it is not necessary that anything at all should come away from the male, and if anything does come away it does not follow that this gives rise to the embryo as being in the embryo, but only imparts the motion and as the form; so the medical art cures the patient.

From what has been said it is plain . . . that the contribution of the female to the generative product is not the same as that of the male, but the male contributes the principle of movement and the female the material. This is why the female does not produce offspring by herself, for she needs a principle, i.e., something to begin the movement in the embryo and to define the form it is to assume. Yet, in some animals, as birds, the nature of the female unassisted can generate to a certain extent, for they do form something, only it is incomplete; I mean the so-called wind eggs.

CHAPTER 22

For the same reason the development of the embryo takes place in the female; neither the male himself nor the female emits semen into the male, but the female receives within herself the share contributed by both, because in the female is the material from which is made the resulting product. Not only must the mass of material exist there from which the embryo is formed in the first instance, but further material must constantly be added that it may increase in size. Therefore the birth must take place in the female. For the carpenter must keep in close connexion with his timber and the potter with his clay, and generally all workmanship and the ultimate movement imparted to matter must be connected with the material concerned, as, for instance, architecture is *in* the buildings it makes.

From these considerations we may also gather how it is that the male contributes to generation. The male does not emit semen at all in some animals, and where he does this is no part of the resulting embryo; just so no material part comes from the carpenter to the material, i.e., the wood in which he works, nor does any part of the carpenters art exist within what he makes, but the shape and the form are imparted from him to the material by means of the motion he sets up. It is his hands that move his tools, his tools that move the material; it is his knowledge of his art, and his soul, in which is the form, that move his hands or any other part of him with a motion of some definite kind, a motion varying with the varying nature of the object made. In like manner, in the male of those animals which emit semen, Nature uses the semen as a tool and as possessing motion in actuality, just as tools are used in the products of any art, for in them lies, in a certain sense the motion of the art. Such, then, is the way in which these males contribute to generation.

EVOLUTIONARY BIOLOGY TRACK

Reading 1: Aristotle—*On the Generation of Animals*

BOOK II

CHAPTER 1

That the male and the female are the principles of generation has been previously stated, as also what is their power and their essence. But why is it that one thing becomes and is male, another female? It is the business of our discussion as it proceeds to try and point out (1) that the sexes arise from Necessity and the first efficient cause,

From Aristotle, "De Generation Animalium," *Parts of Animals*, in *The Oxford Translation of Aristotle*, vol. 5, ed. W. D. Ross, trans. Arthur Platt (Oxford: Oxford University Press, 1912), 731–32, 733, 734–35. Reprinted by permission of Oxford University Press.

(2) from what sort of material they are formed. That (3) they exist because it is better and on account of the final cause, takes us back to a principle still further remote.

Now (1) some existing things are eternal and divine whilst others admit of both existence and non-existence. But (2) that which is noble and divine is always, in virtue of its own nature, the cause of the better in such things as admit of being better or worse, and what is not eternal does admit of existence and non-existence, and can partake in the better and the worse. And (3) soul is better than body, and living, having soul, is thereby better than the lifeless which has none, and being is better than not being, living than not living. These, then, are the reasons of the generation of animals. For since it is impossible that such a class of things as animals should be of an eternal nature, therefore that which comes into being is eternal in the only way possible. Now it is impossible for it to be eternal as an individual (though of course the real essence of things is in the individual)—were it such it would be eternal—but it is possible for it as a species. This is why there is always a class of men and animals and plants. But since the male and female essences are the first principles of these, they will exist in the existing individuals for the sake of generation. Again, as the first efficient or moving cause, to which belong the definition and the form, is better and more divine in its nature than the material on which it works, it is better that the superior principle should be separated from the inferior. Therefore, wherever it is possible and so far as it is possible, the male is separated from the female. For the first principle of the movement, or efficient cause, whereby that which comes into being is male, is better and more divine than the material whereby it is female. The male, however, comes together and mingles with the female for the work of generation, because this is common to both.

A thing lives, then, in virtue of participating in the male and female principles, wherefore even plants have some kind of life; but the class of animals exists in virtue of sense-perception. The sexes are divided in nearly all of these that can move about, for the reasons already stated, and some of them, as said before, emit semen in copulation, others not. The reason of this is that the higher animals are more independent in their nature, so that they have greater size, and this cannot exist without vital heat; for the greater body requires more force to move it, and heat is a motive force. Therefore, taking a general view, we may say that sanguinea are of greater size than bloodless animals, and those which move about than those which remain fixed. And these are just the animals which emit semen on account of their heat and size. . . .

Some animals then, as said before, do not come into being from semen, but all the sanguinea do so which are generated by copulation, the male emitting semen into the female when this has entered into her the young are formed and assume their peculiar character, some within the animals themselves when they are viviparous, others in eggs.

There is a considerable difficulty in understanding how the plant is formed out of the seed or any animal out of the semen. Everything that comes into being or is made must (1) be made out of something, (2) be made by the agency of something, and (3) must become something. Now that out of which it is made is the material; this some animals have in its first form within themselves, taking it from the female parent, as all those which are not born alive but produced as a scolex or an egg; others receive it from the mother for a long time by sucking, as the young of all those which are not only externally but also internally viviparous. Such, then, is the material out of which things come into being, but we now are inquiring not out of what the parts of an animal are made, but by what agency. Either it is something external which makes them, or else something existing in the seminal fluid and the semen; and this must either be soul or a part of soul, or something containing soul.

Now it would appear irrational to suppose that any of either the internal organs or the other parts is made by something external, since one thing cannot set up a motion in another without touching it, nor can a thing be affected in any way by another if it does not set up a motion in it. Something then of the sort we require exists in the embryo itself, being either a part of it or separate from it. To suppose that it should be something else separate from it is irrational. For after the animal has been produced does this something perish or does it remain in it? But nothing of the kind appears to be in it, nothing which is not a part of the whole plant or animal. Yet, on the other hand, it is absurd to say that it perishes after making either all the parts or only some of them. If it makes some of the parts and then perishes, what is to make the rest of them? Suppose this something makes the heart and then perishes, and the heart makes another organ, by the same argument either all the parts must perish or all must remain. Therefore it is preserved and does not perish. Therefore it is a part of the embryo itself which exists in the semen from the beginning; and if indeed there is no part of the soul which does not exist in some part of the body, it would also be a part containing soul in it from the beginning. . . .

Has the semen soul, or not? The same argument applies here as in the question concerning the parts. As no part, if it participate not in soul, will be a part except in an equivocal sense (as the eye of a dead man is still called an 'eye'), so no soul will exist in anything except that of which it is soul; it is plain therefore that semen both has soul, and is soul, potentially.

But a thing existing potentially may be nearer or further from its realization in actuality, as e.g. a mathematician when asleep is further from his realization in actuality as engaged in mathematics than when he is awake, and when awake again but not studying mathematics he is further removed than when he is so studying. Accordingly it is not any part that is the cause of the soul's coming into being, but it is the first moving cause from outside. (For nothing generates itself, though when it has come into being it thenceforward increases itself.) Hence it is that only one part comes into being first and not all of them together. But that must first come into being which has a principle of increase (for this nutritive power exists in all alike, whether animals or plants, and this is the same as the power that enables an animal or plant to generate another like itself, that being the function of them all if naturally perfect). And this is necessary for the reason that whenever a living thing is produced it must grow. It is produced, then, by something else of the same name, as e.g. man is produced by man, but it is increased by means of itself. There is, then, something which increases it. If this is a single part, this must come into being first. Therefore if the heart is first made in some animals, and what is analogous to the heart in the others which have no heart, it is from this or its analogue that the first principle of movement would arise.

We have thus discussed the difficulties previously raised on the question what is the efficient cause of generation in each case, as the first moving and formative power.

GEOLOGY TRACK

Reading 1: John Woodward—*An Essay towards a Natural History of the Earth*

My principle Intention indeed was to get as compleat and satisfactory *Information* of the whole *Mineral Kingdom* as I possibly could. To which End, I made strict *Enquiry* whenever I came, and laid out for Intelligence of *all Places* where the *Entrails* of the *Earth* were *laid open*, either by Nature, (if I may so say) or by Art, and humane industry. And wheresoever I had Notice of any considerable *natural Spelunca* or *Grotto*: any sinking of *Wells*: or digging for *Earths, Clays, Marle, Sand, Gravel, Chalk, Cole, Stone, Marble, Ores of Metalls*, or the like, I forthwith had recourse thereunto; where taking a just Account of every observable *Circumstance* of the Earth, Stone, Metall, or other Matter, from the *Surface* quite down to the *Bottom* of the *Pit*, I enter'd it carefully into a *Journal*, which I carry'd along with me for that Purpose. And so passing on from Place to Place, I *noted* whatever I found *memorable* in each particular *Pit, Quarry*, or *Mine*: and 'tis out of these *Notes* that my *Observations* are compil'd. . .

I likewise drew up a *List* of *Queries* upon this Subject; which I dispatch'd into all Parts of the *World*, far and near, wherever I my self, or any of my Acquaintance, had any Friend resident to transmit those *Queries* unto.

The Result was, that in time I was abundantly assured, that the *Circumstances* of these *Things* in *remoter Countries* were much the same with those of ours here: that the Stone, and other *terrestrial Matter*, in France, Flanders, Holland, Spain, Italy, Germany,

John Woodward, *An Essay towards a Natural History of the Earth* (London: F. Crooke, 1661), 9–11.

Denmark, Norway, and *Sweden*, was distinguished into *Strata*, or *Layers*, as it is in *England*: that those *Strata* were divided by *parallel Fissures*: that there were enclosed in the *Stone*, and all the other denser kinds of *terrestrial Matter*, great Numbers of *Shells*, and other *Productions of the Sea*; in the same Manner as in that of this *Island*. To be short, by the same Means I got sufficient Intelligence that *these Things* were found in like Manner in *Barbary*, in *Egypt*, in *Guiney*, and other Parts of *Africa*: in *Arabia*, *Syria*, *Persia*, *Malabar*, *China*, and other *Asiatick* Provinces: in *Jamaica*, *Barbadoes*, *Virginia*, *New-England*, *Brasil*, *Peru*, and other Parts of *America*. . . .

I shall distribute them into two general *Classes* or *Sections*, whereof the *former* will comprehend my Observations upon all the *Terrestrial Matter* that is naturally disposed into *Layers*, or *Strata*; such as our common *Sand-Stone*, *Marble*, *Cole*, *Chalk*, all Sorts of *Earth*, *Marle*, *Clay*, *Sand* with some others.

Of this various Matter, thus formed into *Strata*, the far greatest Part of the *Terrestrial Globe* consists, from its *Surface* downwards to the greatest *Depth* we ever dig or mine. And it is upon my Observations on this that I have grounded all my *general Conclusions* concerning the *Earth*: all that relate to its *Form*: all that relate to the *Universal* and other *Deluges*: in a Word, all that relate to the several *Vicissitudes* and *Alterations* that it hath yet undergone. . . .

For upon the particular Observations of the said *Metallick* and *Mineral Bodies*, I have not founded any thing but what purely and immediately concerns the *Natural History of those Bodies*.

A DISSERTATION CONCERN*ing* Shells, and *other* marine Bodies, *found at* Land: *Proving that they originally generated and formed at* Sea: *that they are the real Spoils of once living Animals: and not Stones, or natural Minerals, as some late Learned Men have thought.*

There being, I say, beside *these*, such vast *Multitudes* of *Shells* contained in *Stone*, &c. which are *entire*, *fair*, and absolutely free from such *Mineral Contagion*: which are to be match'd by others at this Day found upon our *Shores*, and which do not *differ* in any *Respect* from them; being of the *same Size* that those are of, and the *same Shape* precisely: of the *same Substance* and *Texture*; as consisting of the *same peculiar Matter*, and this constituted and disposed in the *same Manner*, as is that of their respective *Fellow-kinds at Sea*: the *Tendency of the Fibres* and *Striae* the same: the *Composition* of the *Lamallae*, constituted by these *Fibres*, alike in both: the same *Vestigia* of *Tendons* (by Means whereof the Animal is fastened and joined to the *Shell*) in each: the same *Papilae*: the same *Sutures*, and every thing else, whether *within* or *without* the *Shell* in its *Cavity*, or upon its *Convexity*, in the *Substance*, or upon the *Surface* of it. Besides, these *Fossil Shells* are attended with the ordinary *Accidents* of the *marine* ones, *ex-gr.* They sometimes grow to one another, the *lesser Shells* being fixed to the *larger*: they have *Balani*, *Tubuli vermiculares*, *Pearls*, and the like, still *actualy growing* upon them. And, which is very considerable, they are most exactly the same *Specifick Gravity* with their Fellow-kinds now upon the *Shores*. Nay farther, they answer all *Chymical Tryals* in like Manner as the *Sea-Shells* do: their *Parts* when dissolved have the same *Appearance* to View, the same *Smell* and *Taste*: they have the same *Vires* and *Effects* in *Medicine*, when inwardly administered to Animal Bodies: *Aqua fortis*, Oyl of *Vitriol*, and other like *Menustra*, have the very same *Effects* upon both. . . .

When therefore I shall proved more at large, that *those* which we find at *Land*, that are not matchable with any upon our *Shores*, are many of them of *those very Kinds* which the forecited *Relations* particularly assure us are found no where but in the *deeper Parts* of the *Sea*: and that as well those which we can match, as those we *cannot*, are all *Remains* of the *Universal Deluge*, when the *Water* of the *Ocean*, being boisterously turned out upon the *Earth*, bore along with it *Fishes* of all Sorts, *Shells*, and the like moveable Bodies, which it *left behind* at its *Return* back again to its *Channel*; it will not, I presume, be thought *strange*, that amongst the *rest*, it left some of the *Pelagiae*, or those Kinds of *Shells* which naturally have their Abode at *Main-Sea*, and which therefore are now never flung upon the *Shores*. And it may very reasonably be concluded, that all these *strange* Shells, which we cannot so *match*, are of these *Pelagiae*: that the several Kinds of them are at *this Day* living in the *huge Busom* of the *Ocean*: and that there is not any one entire *Species of Shell-fish*, formerly in Being, now *perish'd* or *lost*.

So that I shall only proceed to make *Inferences* from them; which *Inferences*, in this *Part*, are all *affirmative*. Of these, the first is,

That these *Marine Bodyes* were born forth of the *Sea* by the *Universal Deluge*: and that, upon the *Return* of the *Water* back again from off the *Earth*, they were left behind at *Land*.

That during the *Time* of the *Deluge*, whilst the *Water* was out upon, and covered the *Terrestrial Globe*, all the *Stone* and *Marble* of the *Antediluvian Earth*: all the *Metalls* of it: all *Mineral Concretions*: and, in a Word, all [*Minerals*] whatever, that had before obtain'd any *Solidity*, were totally dissolved, and their constituent *Corpuscles* all *disjoyned*, their *Cobaesion* perfectly ceasing.

That at length all the *Mass*, that was thus born up in the *Water*, was again *precipitated*, and *subsided* towards the Bottom. That this *Subsidence* happened generally, and as near as possibly could be expected in so great a *Confusion*, according to the *Laws of Gravity*.

That the *Matter*, subsiding thus, *formed the Strata of Stone*, of *Marble*, of *Cole*, of *Earth*, and the rest; of which *Strata*, lying one upon another, the *Terrestrial Globe*, or at least as much of it as is ever displayed to view, doth mainly consist.

Reading 1: Hippocrates—*The Nature of Man* and *The Sacred Disease*

THE NATURE OF MAN

2.

I propose to show that the substances I believe compose the body are, both nominally and essentially, always the same and unchanging; in youth as well as in old age, in cold weather as well as in warm. I shall produce proofs and demonstrate the causes both of the growth and decline of each of the constituents of the body.

4.

The human body contains blood, phlegm, yellow bile, and black bile. These are the things that make up its constitution and cause its pain and health. Health is primarily that state in which these constituent substances are in the correct proportion to each other, both in strength and quantity, and are well mixed. Pain occurs when one of the substances presents either a deficiency or excess, or is separated in the body and not mixed with the others. It is inevitable that when one of these is separated from the rest and stands by itself, not only the part from which it has come, but also that where it collects and is present in excess, should become diseased, and because it contains too much of the particular substance, causes pain and distress. Whenever there is more than slight discharge of one of these humours outside the body, then its loss is accompanied by pain. If however, the loss change or separation from the other humours is internal, then it inevitably causes twice the pain, as I have said, for pain is produced both in the part whence it is derived and in the part where it accumulates.

5.

Now I said that I would demonstrate that my proposed constituents of the human body were always constant, both nominally and essentially. I hold that these constituents are blood, phlegm, yellow bile, and black bile. They have specific and different names because there are essential differences in their appearance. Phlegm is not like blood, nor is blood like bile, nor bile like phlegm. Indeed, how could they be alike when there is no similarity in appearance and when they are different to the sense of touch. They are dissimilar in their qualities of heat, cold, dryness, and moisture. It follows then that substances so unlike in appearance and characteristics cannot basically be identical. As evidence of the fact that they are dissimilar, each possessing its own qualities and nature, consider the following case. If you give a man medicine which brings

From Hippocrates, *The Medical Works of Hippocrates*, ed. John Chadwick and W. N. Mann (Oxford: Blackwell, 1950), 203–5, 190–91.

up phlegm, you will find his vomit is phlegm; if you give him one which brings up bile, he will vomit bile. Similarly, black bile can be eliminated by administering a medicine which brings it up, or, if you cut the body so as to form an open wound, it bleeds. These things will take place just the same every day and every night, winter and summer, so long as the subject can draw breath and expel it again, or until he is deprived of any of these congenital elements. For they must be congenital, firstly because it is obvious that they are present at every age so long as life is present and, secondly, because they were procreated by a human being who had them all and mothered in a human being similarly endowed wih all the elements which I have indicated and demonstrated.

THE SACRED DISEASE

17.

It ought to be generally known that the source of our pleasure, merriment, laughter and amusement, as of our grief, pain, anxiety and tears, is none other than the brain. It is specially the organ which enables us to think, see and hear, and to distinguish the ugly and the beautiful, the bad and the good, pleasant and unpleasant. Sometimes we judge according to convention; at other times according to the perceptions of expediency. It is the brain too which is the seat of madness and delirium, of the fears and frights which assail us, often by night, but sometimes even by day; it is there where lies the cause of insomnia and sleep-walking, of thoughts that will not come, forgotten duties and eccentricities. All such things result from an unhealthy condition of the brain; it may be warmer than it should be, or it may be colder, or moister or drier, or in any other abnormal state. Moistness is the cause of madness for when the brain is abnormally moist it is necessarily agitated and this agitation prevents sight or hearing being steady. Because of this, varying visual and acoustic sensations are produced, while the tongue can only describe things as they appear and sound. So long as the brain is still, a man is in his right mind.

18.

The brain may be attacked by phlegm and by bile and the two types of disorder which result may be distinguished thus: those whose madness results from phlegm are quiet and neither shout nor make a disturbance; those whose madness results from bile shout, play tricks and will not keep still but are always up to some mischief. Such are the causes of continued madness, but fears and frights may be caused by changes in the brain. Such a change occurs when it is warmed and that is the effect bile has when, flowing from the rest of the body, it courses to the brain along the blood-vessels. Fright continues until the bile runs away again into the blood-vessels and into the body. Feelings of pain and nausea result from inopportune cooling and abnormal consolidation of the brain and this is the effect of phlegm. The same condition is responsible for loss of memory. Those of a bilious constitution are liable to shout and cry out during the night when the brain is suddenly heated; those of phlegmatic constitution do not suffer in this way. Warming of the brain also takes place when a plethora of blood finds is way to the brain and boils. It courses along the blood-vessels I have described in great quantity when a man is having a nightmare and is in a state of terror. He reacts in sleep in the same way that he would if he were awake; his face burns, his eyes are blood-shot as they are when scared or when the mind is intent upon the commission of a crime. All this ceases as soon as the man wakes and the blood is dispersed again into the blood vessels.

SOCIOLOGY TRACK

Reading 1: Thomas Hobbes—*Leviathan*

NATURE hath made men so equall, in the faculties of body and mind; as that, though there bee found one man sometimes manifestly stronger in body, or of quicker mind then another, yet when all is reckoned together, the difference between man, and man, is not so considerable, as that one man can thereupon claim to himselfe any benefit to which another may not pretend as well as he. For as to the strength of body, the weakest has strength enough to kill the strongest, either by secret machination or by confederacy with others that are in the same danger with himselfe.

From this equality of ability ariseth equality of hope in the attaining of our Ends. And therefore if any two men desire the same thing, which nevertheless they cannot both enjoy, they become enemies; and in the way to their End (which is principally their own conservation, and sometimes their delectation only) endeavour to destroy or subdue one another. And from hence it comes to pass that where an Invader hath no more to fear than another man's single power, if one plant, sow, build, or possess a convenient seat, others may probably be expected to come prepared with forces united to dispossess and deprive him, not only of the fruit of his labour, but also of his life or liberty. And the Invader again is in the like danger of another.

And from this diffidence of one another, there is no way for any man to secure himselfe so reasonable as anticipation; that is, by force, or wiles, to master the persons of all men he can so long till he see no other power great enough to endanger him: and this is no more than his own conservation requireth, and is generally allowed. Also, because there be some that, taking pleasure in contemplating their own power in the acts of conquest, which they pursue farther than their security requires, if others, that otherwise would be glad to be at ease within modest bounds, should not by invasion

From Thomas Hobbes, *Leviathan* (Oxford: Clarendon, 1909), 86–92, 100–1.

increase their power, they would not be able, long time, by standing only on their defence, to subsist. And by consequence, such augmentation of dominion over men being necessary to a man's conservation, it ought to be allowed him.

Hereby it is manifest that during the time men live without a common Power to keep them all in awe, they are in that condition which is called Warre; and such a warre as is of every man against every man. For WARRE consisteth not in Battell only, or the act of fighting, but in a tract of time, wherein the Will to contend by Battell is sufficiently known: and therefore the notion of *Time* is to be considered in the nature of warre, as it is in the nature of Weather. For as the nature of foul weather lieth not in a shower or two of rain, but in an inclination thereto of many days together: so the nature of war consisteth not in actual fighting, but in the known disposition thereto during all the time there is no assurance to the contrary. All other time is PEACE.

Whatsoever therefore is consequent to a time of Warre, where every man is Enemy to every man, the same consequent to the time wherein men live without other security than what their own strength and their own invention shall furnish them withal. In such condition there is no place for Industry, because the fruit thereof is uncertain: and consequently no Culture of the earth; no Navigation, nor use of the commodities that may be imported by Sea; no commodious Building; no Instruments of moving and removing such things as require much force; no Knowledge of the face of the Earth; no account of Time; no Arts; no Letters; no Society; and which is worst of all, continual fear, and danger of violent death; and the life of man, solitary, poore, nasty, brutish, and short.

To this Warre of every Man against every Man, this also is consequent; that nothing can be Unjust. The notions of Right and Wrong, Justice and Injustice, have there no place. Where there is no common Power, there is no Law; where no Law, no Injustice. Force and Fraud are in war the two Cardinall vertues. Justice and Injustice are none of the Faculties neither of the Body nor Mind. If they were, they might be in a man that were alone in the world, as well as his Senses and Passions. They are Qualities that relate to men in Society, not in Solitude. It is consequent also to the same condition that there be no Propriety, no Dominion, no *Mine* and *Thine* distinct; but only that to be every man's that he can get, and for so long as he can keep it. And thus much for the ill condition which man by mere nature is actually placed in; though with a possibility to come out of it, consisting partly in the Passions, partly in his Reason.

The Passions that incline men to Peace are: Feare of Death; Desire of such things as are necessary to commodious living; and a Hope by their Industry to obtain them. And Reason suggesteth convenient Articles of Peace upon which men may be drawn to agreement. These Articles are they which otherwise are called the Lawes of Nature, whereof I shall speak more particularly in the two following Chapters.

THE RIGHT OF NATURE, which Writers commonly call *Jus Naturale,* is the Liberty each man hath to use his own power as he will himselfe for the preservation of his own Nature; that is to say, of his own Life; and consequently, of doing anything which, in his own Judgment and Reason, he shall conceive to be the aptest means thereunto.

By LIBERTY is understood, according to the proper signification of the word, the absence of external Impediments: which Impediments may oft take away part of a man's Power to do what he would; but cannot hinder him from using the power left him according as his judgment and reason shall dictate to him.

A LAW OF NATURE, (*Lex Naturalis*), is a Precept, or generall Rule, found out by Reason, by which a man is forbidden to do that which is destructive of his life, or taketh away the means of preserving the same, and to omit that by which he thinketh it may be best preserved. For though they that speak of this subject use to confound *Jus*, and *Lex, Right* and *Law*, yet they ought to be distinguished, because RIGHT consisteth in liberty to do, or to forbeare; whereas LAW determineth and bindeth to one of them: so that law and right differ as much as obligation and liberty, which in one and the same matter are inconsistent.

And because the condition of Man (as hath been declared in the precedent Chapter) is a condition of Warre of every one against every one, in which case every one is governed by his own Reason, and there is nothing he can make use of that may not be a help unto him in preserving his life against his enemies; it followeth that in such a condition every man has a Right to every thing, even to one another's body. And therefore, as long as this naturall Right of every man to every thing endureth, there can be no security to any man, how strong or wise soever he be, of living out the time which Nature ordinarily alloweth men to live. And consequently it is a precept, or generall Rule of Reason, *That every man, ought to endeavour Peace, as far as he has hope of obtaining it; and when he cannot obtain it, that he may seek and use all helps and advantages of Warre.* The first branch of which rule containeth the first and Fundamental Law of Nature, which is: *to seek peace, and follow it.* The Second, the summe of the Right of Nature; which is, *By all means we can to defend our selves.*

From this Fundamentall Law of Nature, by which men are commanded to endeavour Peace, is derived this second Law; That a man be willing, when others are so too, as farre-forth, as for Peace and defence of himselfe he shall think it necessary, to lay down this right to all things; and be contented with so much liberty against other men as he would allow other men against himselfe. For as long as every man holdeth this right, of doing anything he liketh; so long are all men in the condition of Warre. But if other men will not lay down their Right, as well as he, then there is no Reason for anyone to divest himself of his: for that were to expose himself to Prey, (which no man is bound to), rather than to dispose himself to Peace. This is that law of the Gospell: Whatsoever you require that others should do to you, that do ye to them.

FROM that law of Nature by which we are obliged to transfer to another such Rights as, being retained, hinder the peace of mankind, there followeth a Third; which is this; *That men performe their Covenants made;* without which Covenants are in vain, and but empty words; and the Right of all men to all things remaining, we are still in the condition of Warre.

But because Covenants of mutuall trust, where there is a fear of not performance on either part (as hath been said in the former chapter), are invalid, though the Originall of Justice be the making of Covenants, yet Injustice actually there can be none till the cause of such feare be taken away; which, while men are in the natural condition of Warre, cannot be done. Therefore before the names of Just and Unjust can have place, there must be some coercive Power to compel men equally to the performance of their Covenants, by the terror of some punishment greater than the benefit they expect by the breach of their Covenant, and to make good that Propriety which by mutuall Contract men acquire in recompense of the universall Right they abandon: and such power there is none before the erection of a Common-wealth. And this is also to be gathered out of the ordinary definition of Justice in the Schooles: For they say, that *Justice is the constant Will of giving to every man his own.* And therefore where there is no *Own,* that is, no Propriety, there is no Injustice; and where there is no coercive Power erected, that is, where there is no Common-wealth, there is no Propriety, all men having Right to all things: therefore where there is no Common-wealth, there nothing is unjust. So that the Nature of Justice consisteth in keeping of valid Covenants, but the Validity of Covenants begins not but with the Constitution of a Civill Power, sufficient to compell men to keep them: and then it is also that Propriety begins.

ECONOMICS TRACK

Reading 1: Aristotle—*Politics*

BOOK I

CHAPTER 4

Property is a part of the household, and the art of acquiring property is a part of the art of managing the household; for no man can live well, or indeed live at all, unless he be provided with necessaries. And as in the arts which have a definite sphere the workers must have their own proper instruments for the accomplishment of their work, so it is in the management of the household . . .

From Aristotle, "Politica," *Politics & Economics,* in *The Oxford Translation of Aristotle,* vol. 10, ed. W. D. Ross, trans. B. Jowett (Oxford: Oxford University Press, 1921), 1253, 1255, 1256–58. Reprinted by permission of Oxford University Press.

CHAPTER 8

Let us now inquire into property more generally, and into the art of getting wealth, in accordance with our usual method . . . The first question is whether the art of getting wealth is the same with the art of managing a household or a part of it, or instrumental to it; and if the last, whether in the way that the art of making shuttles is instrumental to the art of weaving, or in the way that the casting of bronze is instrumental to the art of statuary, for they are not instrumental in the same way, but the one provides the tools and the other the material; and by material I mean the substratum out of which any work is made; thus wool is the material of the weaver, bronze of the statuary. Now it is easy to see that the art of household management is not identical with the art of getting wealth, for one uses the material which the other provides. For the art which uses household stores can be no other than the art of household management . . .

Of the art of acquisition then there is one kind which by nature is a part of the management of a household, insofar as the art of household management must either find ready to hand, or itself provide, such things necessary to life, and useful for the community of the family or state, as can be stored. They are the elements of true riches; for the amount of property which is needed for a good life is not unlimited, although Solon in one of his poems says that 'No bound to riches has been fixed for man.' But there is a boundary fixed, just as there is in the other arts; for the instruments of any art are never unlimited, either in number or in size, and riches may be defined as a number of instruments to be used in a household or in a state. And so we see that there is a natural art of acquisition which is practiced by managers of households and by statesmen, and what is the reason for this.

CHAPTER 9

There is another variety of the art of acquisition which is commonly and rightly called an art of wealth-getting, and has in fact suggested the notion that riches and property have no limit. Being nearly connected with the preceding, it is often identified with it. But though they are not very different, neither are they the same. The kind already described is given by nature, the other is gained by experience and art.

Let us begin our discussion of the question with the following considerations:

Of everything we possess there are two uses: both belong to the thing as such, but not in the same manner, for one is the proper, and the other the improper or secondary use of it. For example, a shoe is used for wear, and it is used for exchange; both are uses of the shoe. He who gives a shoe in exchange for money or food to him who wants one, does indeed use the shoe as a shoe, but this is not its proper or primary purpose, for a shoe is not made to be an object of barter. The same may be said of all possessions, for the art of exchange extends to all of them, and it arises at first from what is natural, from the circumstance that some have too little, others too much.

Hence we may infer that retail trade is not a natural part of the art of getting wealth; had it been so, men would have ceased to trade when they had enough. . . .

When the use of coin had once been discovered, out of the barter of necessity articles arose the other art of wealth-getting, namely, retail trade; which was at first probably a simple matter, but became more complicated as soon as men learned by experience whence and by what exchanges the greatest profit might be made. Originating in the use of coin, the art of getting wealth is generally thought to be chiefly concerned with it, and to be the art which produces riches and wealth; having to consider how they may be accumulated. Indeed, riches is assumed by many to be only a quantity of coin. . . .

Hence some persons are led to believe that getting wealth is the object of household management, and the whole idea of their lives is that they ought either to increase their money without limit, or at any rate not to lose it. The origin of this disposition in men is that they are intent upon living only, and not upon living well; and, as their desires are unlimited, they also desire that the means of gratifying them should be without limit. Those who do aim at a good life seek the means of obtaining bodily pleasures; and, since the enjoyment of these appears to depend on property, they are absorbed in getting wealth; and so there arises a second species of wealth-getting. For as their enjoyment is in excess, they seek an art which produces the excess of enjoyment; and, if they are not able to supply their pleasures by the art of getting wealth, they try other arts, using in turn every faculty in a manner contrary to nature. . . .

Thus, then, we have considered the art of wealth-getting which is unnecessary, and why men want it; and also the necessary art of wealth-getting, which we have seen to be different from the other, and to be a part of the art of managing a household, concerned with the provision of food, not however, like the former kind, unlimited, but having a limit.

ASTRONOMY TRACK

Reading 2: Ptolemy—*Almagest*

BOOK I

1. PREFACE

The true philosophers, Syrus, were, I think, quite right to distinguish the theoretical part of philosophy from the practical. For even if practical philosophy, before it *is* practical, turns out to be theoretical, nevertheless one can see that there is a great difference between the two: in the first place, it is possible for many people to possess some of the moral virtues even without being taught, whereas it is impossible to achieve theoretical understanding of the universe without instruction; furthermore, one derives most benefit in the first case [practical philosophy] from continuous practice in actual affairs, but in the other [theoretical philosophy] from making progress in the theory. Hence we thought it fitting to guide our actions (under the impulse of our actual ideas [of what is to be done]) in such a way as never to forget, even in ordinary affairs, to strive for a noble and disciplined disposition, but to devote most of our time to intellectual matters, in order to teach theories, which are so many and beautiful, and especially those to which the epithet 'mathematical' is particularly applied. For Aristotle divides theoretical philosophy too, very fittingly, into three primary categories, physics, mathematics, and theology. For everything that exists is composed of matter, form, and motion; none of these [three] can be observed in its substratum by itself, without the others: they can only be imagined. Now the first cause of the first motion of the universe, if one considers it simply, can be thought of as an invisible and motionless deity; the division [of theoretical philosophy] concerned with investigating this [can be called] 'theology,' since this kind of activity, somewhere up in the highest reaches of the universe, can only be imagined, and is completely separated from perceptible reality. The division [of theoretical philosophy] which investigates material and ever-moving nature, and which concerns itself with 'white,' 'hot,' 'sweet,' 'soft,' and suchlike qualities one may call 'physics'; such an order of being is situated (for the most part) amongst the corruptible bodies and below the lunar sphere. That division [of theoretical philosophy] which determines the nature involved in forms and motion from place to place, and which serves to investigate shape, number, size, and place, time,

From Ptolemy, *Almagest*, trans. G. J. Toomer (London: Duckworth, 1984), 35–37, 38–40, 45–47, 426, 442–43. By permission of Gerald Duckworth & Co., Ltd.

and suchlike, one may define as 'mathematics.' Its subject-matter falls as it were in the middle between the other two, since, firstly, it can be conceived both with and without the aid of the senses, and, secondly, it is an attribute of all existing things without exception, both mortal and immortal: for those things which are perpetually changing in their inseparable form, it changes with them, while for eternal things which have an aethereal nature, it keeps their unchanging form unchanged.

From all this we concluded: that the first two divisions of theoretical philosophy should rather be called guesswork than knowledge, theology because of its invisible and ungraspable nature, physics because of the unstable and unclear nature of matter; hence there is no hope that philosophers will ever be agreed about them; and that only mathematics can provide sure and unshakable knowledge to its devotees, provided one approaches it rigorously. For this kind of proof proceeds by indisputable methods, namely arithmetic and geometry. Hence we are drawn to the investigation of that part of theoretical philosophy, as far as we are bale to the whole of it, but especially to the theory concerning divine and heavenly things. For that alone is devoted to the investigation of the eternally unchanging. For that reason it too can be eternal and unchanging (which is a proper attribute of knowledge) in its own domain, which is neither unclear nor disorderly. Furthermore it can work in the domains of the other [two divisions of theoretical philosophy] no less than they do. For this is the best science to help theology along its way, since it is the only one which can make a good guess at [the nature of] that activity which is unmoved and separated; [it can do this because] it is familiar with the attributes of those beings which are on the one hand perceptible, moving and being moved, but on the other hand eternal and unchanging, [I mean the attributes] having to do with motions and the arrangements of motions. As for physics, mathematics can make a significant contribution. For almost every peculiar attribute of material nature becomes apparent from the peculiarities of its motion from place to place. [Thus one can distinguish] the corruptible from the incorruptible by [whether it undergoes] motion in a straight line or in a circle, and heavy from light, and passive from active, by [whether it moves] towards the center or away from the center.

3. THAT THE HEAVENS MOVE LIKE A SPHERE

It is plausible to suppose that the ancients got their first notions on these topics from the following kind of observations. They saw the sun, moon, and other stars were carried from east to west along circles which were always parallel to each other, that they began to rise up from below the earth itself, as it were, gradually got up high, then kept on going round in similar fashion and getting lower, until, falling to earth, so to speak, they vanished completely, then, after remaining invisible for some time, again rose afresh and set; and [they saw] that the periods of these [motions], and also the places of rising and setting, were, on the whole, fixed and the same.

What chiefly led them to the concept of a sphere was the revolution of the ever-visible stars, which was observed to be circular, and always taking place about one centre, the same [for all]. For by necessity that point became [for them] the pole of the heavenly sphere: those stars which were closer to it revolved on smaller circles, those that were farther away described circles ever greater in proportion to their distance, until one reaches the distance of the stars which become invisible. In the case of these, too, they saw that those near the ever-visible stars remained invisible for a short time, again in proportion [to their distance]. The result was that in the beginning, they got to the aforementioned notion solely from such considerations; but from then on, in their subsequent investigation, they found that everything else accorded with it, since absolutely all phenomena are in contradiction to the alternative notions which have been propounded.

For if one were to suppose that the stars' motion takes place in a straight line towards infinity, as some people have thought, what device could one conceive of which would cause each of them to appear to begin their motion from the same starting-point each day? How could the stars turn back if their motion is towards infinity? Or, if they did turn back, how could this not be obvious? [On such a hypothesis], they must gradually diminish in size until the disappear, whereas, on the contrary, they are seen to be greater at the very moment of their disappearance, at which time they are gradually obstructed and cut off, as it were, by the earth's surface.

But to suppose that they are kindled as they rise out of the earth and are extinguished again as they fall to earth is a completely absurd hypothesis. For even if we were to concede that that the strict order in their size and number, their intervals, positions, and periods could be restored by such a random and chance process; that one whole area of the earth has a kindling nature, and another an extinguishing one, or rather that the same part [of the earth] kindles for one set of observers and extinguishes for another set; and that the same stars are already kindled or extinguished for some observers while they are not yet for others: even if, I say, we were to concede all these ridiculous consequences, what could we say about the ever visible stars, which neither rise nor set? Those stars which are kindled and extinguished ought always to be visible for observers everywhere. What cause could we assign for the fact that this is not so? We will surely not say that stars which are kindled and extinguished for some observers never undergo this process for other observers. Yet it is utterly obvious that the same stars rise and set in certain regions [of the earth] and do neither at others.

To sum up, if one assumes any motion whatever, except spherical, for the heavenly bodies, it necessarily follows that there distances, measured from the earth upwards, must vary, wherever and however, one supposes the earth itself to be situated. Hence the sizes and mutual distances of the stars must appear to vary for the same observers during the course of each revolution, since at one time they must be at a greater distance, at another at a lesser. Yet we see that no such variation occurs. For the apparent increase in their sizes at the horizons is caused, not by a decrease in their distances, but by the exhalations of moisture surrounding the earth being interposed between the place from which we observe and the heavenly bodies, just as objects placed in water appear bigger than they are, and the lower they sink, the bigger they appear.

The following considerations also lead us to the concept of the sphericity of the heavens. No other hypothesis but this can explain how sundial constructions produce correct results; furthermore, the motion of the heavenly bodies is the most unhampered and free of all motions, and freest motion belongs among plane figures to the circle and among solid shapes to the sphere; similarly, since of different shapes having an equal boundary to those with more angles are greater [in area or volume], the circle is greater than [all other] surfaces, and the sphere greater than [all other] solids, [likewise] the heavens are greater than all other bodies.

Furthermore, one can reach this kind of notion from certain physical considerations. E.g., the aether is, of all bodies, the one with constituent parts which are finest and most like each other; now bodies with parts like each other are the circular, among planes, and the spherical, among three-dimensional surfaces. And since the aether is not plane, but three-dimensional, it follows that it is spherical in shape. Similarly, nature formed all earthly and corruptible bodies out of shapes which are round but of unlike parts, but all aethereal and divine bodies out of shapes which are of like parts and spherical. For if they were flat or shaped like a discus they would not always display a circular shape to all those observing them simultaneously from different places on earth. For this reason it is plausible that the aether surrounding them, too, being of the same nature, is spherical, and because of the likeness of its parts moves in a circular and uniform fashion.

5. THAT THE EARTH IS IN THE MIDDLE OF THE HEAVENS

Once one has grasped this, if one next considers the position of the earth, one will find that the phenomena associated with it could take place only if we assume that it is in the middle of the heavens, like the centre of a sphere. For if this were not the case, the earth would have to be either

[a] not on the axis [of the universe] but equidistant from both poles, or

[b] on the axis but removed towards one of the poles, or

[c] neither on the axis nor equidistant from the poles.

. . . [I]f the earth did not lie in the middle [of the universe], the whole order of things which we observe in the increase and decrease of the lengths of daylight would be fundamentally upset. Furthermore, eclipses of the moon would not be restricted to situations where the moon is diametrically opposite the sun (whatever part of the heaven [the luminaries are in]), since the earth would often come between them when they were not diametrically opposite, but at intervals of less than a semi-circle.

BOOK III

3. ON THE HYPOTHESIS FOR UNIFORM CIRCULAR MOTION

[W]e must make the general point that the rearward displacements of the planets with respect to the heavens are, in every case, just like the motion of the universe in advance, by nature uniform and circular. That is to say, if we imagine the bodies being carried around by straight lines, in absolutely every case the straight line in question describes equal angles at the centre of its revolution in equal times. The apparent irregularity [anomaly] in their motions is the result of the position and order of those circles in the sphere of each by means of which they carry out their movements, and in reality there is in essence nothing alien to their eternal nature in the 'disorder' which the phenomena are supposed to exhibit. The reason for the appearance of irregularity can be explained by two hypotheses, which are the most basic and simple. When their motion is viewed with respect to a circle imagined to be in the plane of the ecliptic, the centre of which coincides with the centre of the universe (thus its centre can be considered to coincide with our point of view), then we can suppose, either that the uniform motion of each [body] takes place on a circle which is not concentric with the universe, or that they have such a concentric circle, but their uniform motion takes place, not actually on that circle, but on another circle, which is carried by the first circle, and [hence] is known as the 'epicycle.'

BOOK IX

5. PRELIMINARY NOTIONS [NECESSARY] FOR THE HYPOTHESES OF THE 5 PLANETS

Now that these [mean motions] have been tabulated, our next task is to discuss the anomalies which occur in connection with the longitudinal positions of the five planets . . .

There are, as we said, two types of motion which are the simplest and at the same time sufficient for our purpose, [namely] that produced by circles eccentric to [the centre of] the ecliptic, and that produced by circles concentric with the ecliptic but carrying epicycles around. There are likewise two apparent anomalies for each planet: [1] that anomaly which varies according to its position in the ecliptic, and [2] that which varies according to its position relative to the sun.

For [2] we find, from a series of different [sun-planet] configurations observed round about the same part of the ecliptic, that in the case of the five planets the time from greatest speed to mean is always greater than the time from mean speed to least. Now this feature cannot be a consequence of the eccentric hypothesis, in which exactly the opposite occurs, since the greatest speed takes place at the perigee in the eccentric hypothesis, while the arc from the perigee to the point of mean speed is less than the arc from the latter to the apogee in both [the eccentric and epicyclic] hypotheses. But it can occur as a consequence of the epicyclic hypothesis, however only when the greatest speed occurs, not at the perigee, as in the case of the moon, but at the apogee; that is to say, when the planet, starting from the apogee, moves, not as the moon does, in advance [with respect to the motion] of the universe, but instead towards the rear. Hence we use the epicyclic hypothesis to represent this kind of anomaly.

3.

It is probable that the first notions of these things came to the ancients from some such observation as this. For they kept seeing the sun and moon and other stars always moving from rising to setting in parallel circles, beginning to move upward from below as if out of the earth itself, rising little by little to the top, and then coming around again and going down in the same way until at last they would disappear as if falling into the earth. And then again they would see them, after remaining some time invisible, rising and setting as if from another beginning; and they saw that the times and also the places of rising and setting generally corresponded in an ordered and regular way.

But most of all the observed circular orbit of those stars which are always visible, and their revolution about one and the same center, led them to this spherical notion. For necessarily this point became the pole of the heavenly sphere; and the stars nearer to it were those that spun around in smaller circles, and those farther away made greater circles in their revolutions in proportion to the distance, until a sufficient distance brought one to the disappearing stars. And then they saw that those near the always-visible stars disappeared for a short time, and those farther away for a longer time proportionately. And for these reasons alone it was sufficient for them to assume this notion as a principle, and forthwith to think through also the other things consequent upon these same appearances, in accordance with the development of the science. For absolutely all the appearances contradict the other opinions.

Moreover, certain physical considerations lead to such a conjecture. For example, the fact that of all bodies the ether has the finest and most homogenous parts; but the surfaces of homogenous parts must have homogenous parts, and only the circle is such among plane figures and the sphere among solids. And since ether is not plane but solid, it can only be spherical. Likewise the fact that nature has built all earthly and corruptible bodies wholly out of rounded figures but with heterogeneous parts, and all divine bodies in the ether out of spherical figures with homogeneous parts, since if they plane or disc-like they would not appear circular to all those who see them from different parts of the earth at the same time. Therefore it would seem reasonable that the ether surrounding them and of a like nature be also spherical, and that because of the homogeneity of its parts it moves circularly and regularly.

8.

It will be sufficient for these hypotheses, which have to be assumed for the for the detailed exposition following them, to have been outlined here in such a summary way since they will finally be established and confirmed by the agreement of the consequent proofs with the appearances. In addition to those already mentioned, this general assumption would also be rightly made that there are two different prime movements in the heavens. One is that by which everything moves from east to west, always in the same way and at the same speed with revolutions in circles parallel to each other and clearly described about the poles of the regularly revolving sphere. Of these circles the greatest is called the equator, because it alone is always cut exactly in half by the horizon which is a great circle of the sphere, and because everywhere the sun's revolution about it is sensibly equinoctial. The other movement is that according to which the spheres of the stars make certain local motions in the direction opposite to that of the movement just described and around other poles than those of that first revolution. And we assume that it is so because, while from each day's observation, all the heaven bodies are seen to move generally in paths sensibly similar and parallel to the equator and rise, culminate, and set (for such is the property of the first movement), yet from subsequent and more continuous observation, even if all the other stars appear to preserve their angular distances with respect to each other and their properties as regards their place within the first movement, still, the sun and moon and planets make certain complex movements unequal to each other, but all contrary to the general movement, towards the east opposite to the movement of the fixed stars which preserve their respective angular distances and are moved as if by one sphere.

If then this movement of the planets also took place in circles parallel to the equator—that is, around the same poles as the first revolution—it would be sufficient to assume for them all one and the same revolving movement in conformity with the first. For then it would be plausible to suppose that their movement was the result of a lag and not of a contrary movement. But they always seem, at the same time they move towards the east, to deviate towards the north and south poles without any uniform magnitude's being observed in this deviation, so that this seems to befall them through impulsions. But although this deviation is irregular on the hypothesis of one prime movement, it is regular when effected by a circle oblique to the equator. And so, such a circle is conceived, one and the same for, and proper to the planets, quite exactly expressed and as it were described by the motion of the sun, but traveled also by the moon and planets which ever turn about it with every deviation from it on the part of any planet either way, a deviation within a prescribed distance and governed by rule. And since this is seen to be a great circle also because of the sun's equal oscillation to the north and south of the equator, and since the eastward movement of all the planets take place on one and the same circle, it was necessary to suppose a second movement about the poles of this oblique circle or ecliptic in the direction opposite to that of the movement.

BOOK IX

5.

Now, the relation of the anomalies to the longitudinal passage of the five planets follows the exposition of these mean movements, and we have attempted a general outline of it in the following way.

For, as we said, the very simple movements together sufficient for the problem in hand are two: One effected by circles eccentric to the ecliptic, and the other by circles concentric with the ecliptic but bearing epicycles. And likewise also the apparent anomalies for each star considered singly are two: one observed with respect to parts of the zodiac and the other with respect to the configurations of the sun. In the case of the latter anomaly, we find from different configurations observed in contiguity and in the same parts of the zodiac that, for the five planets, the time from the greatest movement to the mean movement is always longer than from the mean to the least. And such a property cannot follow from the hypothesis of eccentricity, but its contrary follows, because the greatest passage is always effected at the perigee, and in both hypotheses the arc from the perigee to the point of mean passage is less than that from this point to the apogee. But it can occur in the hypothesis of epicycles when the greatest passage is not effected at the perigee as is the case of the moon, but at the apogee—that is, when the star, starting from the apogee moves not westward as the moon, but eastward in the opposite direction. And so we suppose this anomaly to be produced by epicycles.

PHYSICS TRACK

Reading 2: James Clerk Maxwell—"Molecules"

An atom is a body which cannot be cut in two. A molecule is the smallest possible portion of a particular substance. No one has ever seen or handled a single molecule.

From James Clerk Maxwell, "Molecules," *Nature* 8 (1873): 437–41.

Molecular science, therefore, is one of those branches of study which deal with things invisible and imperceptible by our senses, and which cannot be subjected to direct experiment.

The mind of man has perplexed itself with many hard questions. Is space infinite, and if so in what sense? Is the material world infinite in extent, and are all places within that extent equally full of matter? Do atoms exist, or is matter infinitely divisible?

Take any portion of matter, say a drop of water, and observe its properties. Like every other portion of matter we have ever seen, it is divisible. Divide it in two, each portion appears to retain all the properties of the original drop, and among others that of being divisible. The parts are similar to the whole in every respect except in absolute size.

Now go on repeating the process of division till the separate portions of water are so small that we can no longer perceive or handle them. Still we have no doubt that the sub-division might be carried further, if our senses were more acute and our instruments more delicate. Thus far all are agreed, but now the question arises, Can this sub-division be repeated forever?

According to Democritus and the atomic school, we must answer in the negative. After a certain number of sub-divisions, the drop would be divided into a number of parts each of which is incapable of further sub-division. We should thus, in imagination, arrive at the atom, which, as its name literally signifies, cannot be cut in two. This is the atomic doctrine of Democritus, Epicurus, and Lucretius, and, I may add, of your lecturer.

Every substance, simple or compound, has its own molecule. If this molecule be divided, its parts are molecules of a different substance or substances from that of which the whole is a molecule. An atom, if there is such a thing, must be a molecule of an elementary substance. Since, therefore, every molecule is not an atom, but every atom is a molecule, I shall use the word molecule as the more general term.

We all know that air or any other gas placed in a vessel presses against the sides of the vessel, and against the surface of any body placed within it. On the kinetic theory this pressure is entirely due to the molecules striking against these surfaces, and thereby communicating to them a series of impulses which follow each other in such rapid succession that they produce an effect which cannot be distinguished from that of a continuous pressure.

If the velocity of the molecules is given, and the number varied, then since each molecule, on an average, strikes the side of the vessel the same number of times, and with an impulse of the same magnitude, each will contribute an equal share to the whole pressure. The pressure in a vessel of given size is therefore proportional to the number of molecules in it, that is to the quantity of gas in it.

This is the complete dynamical explanation of the fact discovered by Robert Boyle, that the pressure of air is proportional to its density. It shows also that of different portions of gas forced into a vessel, each produces its own part of the pressure independently of the rest, and this whether these portions be of the same gas or not.

Let us next suppose that the velocity of the molecules is increased. Each molecule will now strike the sides of the vessel a greater number of times in a second, but besides this, the impulse of each blow will be increased in the same proportion, so that the part of the pressure due to each molecule will vary as the *square* of the velocity. Now the increase of the square of velocity corresponds, in our theory, to a rise of temperature, and in this way we can explain the effect of warming the gas, and also the law discovered by Charles that the proportional expansion of all gases between given temperatures is the same.

The dynamical theory also tells us what will happen if molecules of different masses are allowed to knock about together. The greater masses will go slower than the smaller ones, so that, on an average, every molecule, great or small, will have the same energy of motion.

The proof of this dynamical theorem, in which I claim the priority, has recently been greatly developed and improved by Dr. Ludwig Boltzmann. The most important consequence which flows from it is that a cubic centimetre of every gas at standard temperature and pressure contains the same number of molecules. This is the dynamical explanation of Gay Lussac's law of the equivalent volumes of gases. But we must now descend to particulars, and calculate the actual velocity of a molecule of hydrogen.

A cubic centimetre of hydrogen, at the temperature of melting ice and at a pressure of one atmosphere, weighs 0.00008954 grammes. We have to find at what rate this small mass must move (whether altogether or in separate molecules makes no difference) so as to produce the observed pressure on the sides of the cubic centimetre. This is the calculation which was first made by Dr. Joule, and the result is 1,859 metres per second. This is what we are accustomed to call a great velocity. It is greater than any velocity obtained in artillery practice. The velocity of other gases is less, as you will see by the table, but in all cases it is very great as compared with that of bullets.

We have now to conceive the molecules of the air in this hall flying about in all directions, at a rate of about seventeen miles in a minute.

If all these molecules were flying in the same direction, they would constitute a wind blowing at the rate of seventeen miles a minute, and the only wind which approaches this velocity is that which proceeds from the mouth of a cannon. How, then, are you and I able to stand here? Only because the molecules happen to be flying in different directions, so that those which strike against our backs enable us to support the storm which is beating against our faces. Indeed, if this molecular bombardment were to cease, even for an instant, our veins would swell, our breath would leave us, and we should, literally, expire. But it is not only against us or against the walls of the room that the molecules are striking. Consider the immense number of them, and the fact that they are flying in every possible direction, and you will see that they cannot avoid striking each other. Every time that two molecules come into collision, the paths of both are changed, and they go off in new directions. Thus each molecule is continually getting its course altered, so that in spite of its great velocity it may be a long time before it reaches any great distance from the point at which it set out.

I have here a bottle containing ammonia. Ammonia is a gas which you can recognise by its smell. Its molecules have a velocity of six hundred metres per second, so that if their course had not been interrupted by striking against the molecules of air in the hall, everyone in the most distant gallery would have smelt ammonia before I was able to pronounce the name of the gas. But instead of this, each molecule of ammonia is so jostled about by the molecules of air, that it is sometimes going one way and sometimes another. It is like a hare which is always doubling, and though it goes a great pace, it makes very little progress. Nevertheless, the smell of ammonia is now beginning to be perceptible at some distance from the bottle. The gas does diffuse itself through the air, though the process is a slow one, and if we could close up every opening of this hall so as to make it air-tight, and leave everything to itself for some weeks, the ammonia would become uniformly mixed through every part of the air in the hall.

This property of gases, that they diffuse through each other, was first remarked by Priestley. Dalton showed that it takes place quite independently of any chemical action between the inter-diffusing gases. Graham, whose researches were especially directed towards those phenomena which seem to throw light on molecular motions, made a careful study of diffusion, and obtained the first results from which the rate of diffusion can be calculated.

We have no means of marking a select number of molecules of air, so as to trace them after they have become diffused among others, but we may communicate to them some property by which we may obtain evidence of their diffusion.

For instance, if a horizontal stratum of air is moving horizontally, molecules diffusing out of this stratum into those above and below will carry their horizontal motion with them, and so tend to communicate motion to the neighbouring strata, while molecules diffusing out of the neighbouring strata into the moving one will tend to bring it to rest. The action between the strata is somewhat like that of two rough surfaces, one of which slides over the other, rubbing on it. Friction is the name given to this action between solid bodies; in the case of fluids it is called internal friction or viscosity.

It is in fact only another kind of diffusion—a lateral diffusion of momentum, and its amount can be calculated from data derived from observations of the first kind of diffusion, that of matter. The comparative values of the viscosity of different gases were determined by Graham in his researches on the transpiration of gases through long narrow tubes, and their absolute values have been deduced from experiments on the oscillation of discs by Oscar Meyer and myself.

Another way of tracing the diffusion of molecules in rough calm air is to heat the upper stratum of the air in a vessel, and so observe the rate at which this heat is communicated to the lower strata. This, in fact, is a third kind of diffusion—that of energy, and the rate at which it must take place was calculated from data derived from experiments on viscosity before any direct experiments on the conduction of heat had been made. Prof. Stefan, of Vienna, has recently, by a very delicate method, succeeded in determining the conductivity of air, and he finds it, as he tells us, in striking agreement with the value predicted by the theory.

All these three kinds of diffusion—the diffusion of matter, of momentum, and of energy—are carried on by the motion of the molecules. The greater the velocity of the molecules and the farther they travel before their paths are altered by collision with other molecules, the more rapid will be the diffusion. Now we know already the velocity of the molecules, and therefore by experiments on diffusion we can determine how far, on an average, a molecule travels without striking another. Prof. Clausius, of Bonn, who first gave us precise ideas about the motion of agitation of molecules, calls this distance the mean path of a molecule. I have calculated, from Prof. Loschmidt's diffusion experiments, the mean path of the molecules of four well-known gases. It is a very small distance, quite imperceptible to us even with our best microscopes. Roughly speaking, it is about the tenth part of the length of a wave of light, which you know is a very small quantity. Of course the time spent on so short a path by such swift molecules must be very small. I have calculated the number of collisions which each must undergo in a second. They are reckoned by thousands of millions. No wonder that the travelling power of the swiftest molecule is but small, when its course is completely changed thousands of millions of times in a second. . . .

Thus we have been led, along a strictly scientific path, very near to the point at which Science must stop. Not that Science is debarred from studying the internal mechanism of a molecule which she cannot take to pieces, any more than from investigating an organism which she cannot put together. But in tracing back the history of matter Science is arrested when she assures herself, on the one hand, that the molecule has been made, and on the other that it has not been made by any of the processes we call natural.

Science is incompetent to reason upon the creation of matter itself out of nothing.

We have reached the utmost limit of our thinking faculties when we have admitted that because matter cannot be eternal and self-existent it must have been created. It is only when we contemplate, not matter in itself, but the form in which it actually exists, that our mind finds something on which it can lay hold.

That matter, as such, should have certain fundamental properties—that it should exist in space and be capable of motion, that its motion should be persistent, and so on, are truths which may, for anything we know, be of the kind which metaphysicians call necessary. We may use our knowledge of such truths for purposes of deduction but we have no data for speculating as to their origin.

CHEMISTRY TRACK

Reading 2: Robert Boyle—*The Skeptical Chymist*

Notwithstanding the subtle reasonings I have met with in the books of the Peripateticks [Aristotelians], and the pretty experiments that have been shew'd me in the Laboratories of Chymists, I am of so diffident, or dull a Nature, as to think that if neither of them can bring more cogent arguments to evince the truth of their assertion than are wont to be brought; a Man may rationally enough retain some doubts concerning the very number of those materiall Ingredients of mixt bodies, which some would have us call Elements, are as considerable amongst the Doctrines of natural Philosophy, as the Elements themselves are amongst the bodies of the Universe, I expect to find those Opinions solidly establish'd, upon which so many others are superstructured. But when I took the pains impartially to examine the bodies themselves that are said to result from the blended Elements, and to torture them into a confession of their constituent Principles, I was quickly induc'd to think that the number of the Elements has been contended about by Philosophers with more earnestness, than success. This unsatisfiedness of mine has been much wonder'd at, by these two Gentlemen who though they differ almost as much betwixt themselves about the question we are to consider, as I do from either of them, yet they both agree very well in this, that there is a determinate number of such ingredients as I was just now speaking of, and that what that number is, I say not; may be (for what may not such as they perswade?) but is wont to be clearly enough demonstrated both by Reason and Experience. This has occasion'd our present Conference.

Robert Boyle, *The Skeptical Chymist* (London: Bettesworth and Taylor, 1723), 4–6, 9–10, 16, 23–24, 28–29, 83, 85–86.

"NEW EXPERIMENTS TOUCHING THE RELATION BETWIXT FLAME AND AIR"

EXPERIMENT I.

A WAY OF KINDLING BRIMSTONE IN VACUO BOYLIANO UNSUCCESSFULLY TRIED

We took a small earthen melting Pot, of an almost Cylindrical figure, and well glaz'd (when it was first bak'd) by the heat; and into this we put a small cylinder of Iron of about an inch in thickness, and half as much more in Diameter, made red hot in the fire; and having hastily pump'd out the Air, to prevent the breaking of the Glass; when this vessel seem'd to be well emptied, we let down, by turning a key, a piece of Paper, wherein was put a convenient quantity of flower of Brimstone, under which the iron had been carefully plac'd; so that, being let down, it might fall upon the heated metal, which as soon as it came to do, that vehement heat did, as we expected, presently *destroy* the contiguous paper; whence the included Sulphur fell immediately upon the iron, whose upper part was a little concave, that it might contain the flowers when melted. But all the heat of the iron, though it made the Paper and Sulphur smoke, would not actually kindle either of them that we could perceive.

EXPERIMENT II.

AN INEFFECTUAL ATTEMPT TO KINDLE SULPHUR IN OUR VACUUM ANOTHER WAY.

Another way I thought of to examine the inflammability of Sulphur without Air; which, though it may prove somewhat hazardous to put it in practice, I resolved to try, and did so after the following manner:

Into a glass-bubble of a convenient size, and furnish'd with a neck fit for our purpose, we put a little flower of Brimstone (as likely to be more pure and inflammable than common Sulphur;) and having exhausted the Glass, and secured it against the return of the Air, we laid it upon burning coals, where it did not take fire, but rise all to the opposite part of the glass, in the form of a fine powder; and that part being turned downward and laid on the coals, the Brimstone, without kindling, rose again in the form of an expanded substance, which (being removed from the fire) was, for the most part, transparent, not unlike yellow varnish.

Though these unsuccessful attempts to kindle Sulphur in our exhausted Receivers, were made more discouraging by some more, that were made another way; yet judging that last way to be rational enough, we persisted somewhat obstinately in our en-

From Robert Boyle, "Tracts Written by the Honourable Robert Boyle, Containing New Experiments touching the Relation betwixt *Flame* and *Air*," in *The Philosophical works of the Honourable Robert Boyle, Esq.* (London: W. & J. Innys, J. Osborn, and T. Longman, 1725), 2:517–18.

deavours, and conjecturing that there might be some unperceived difference between Minerals, that do all of them pass, and are sold for common Sulphur, I made trial according to the way hereafter to be mentioned, with another parcel of brimstone, which differ'd not so much from the former, as to make it worth while to set down a description of it, that probably would not be useful.

EXPERIMENT III.

SHEWING THE EFFICACY OF AIR IN THE PRODUCTION OF FLAME, WITHOUT ANY ACTUALLY FLAMING OR BURNING BODY.

Having hitherto examin'd by the *presence* of the Air, what interest it has in kindling Flame; it will not be impertinent to add an Experiment or two, that we tried to shew the same interest of the Air by the effects of its *admission* into our Vacuum. For I thought, it might reasonably be supposed, that if such dispositions were introduc'd into a body, as that there should not appear any thing wanting to turn it into Flame but the presence of the Air, an actual ascension of that body might be produced by the admitted Air, without the intervention of any actual Flame, or Fire, or even heated substance; the warrantableness of which supposition may be judged by the two following Experiments.

When we had made the Experiment, ere long to be related in its due place to examine the presumption we had, that even when the Iron was not hot enough to keep the melted Brimstone in such a heat, as was requisite to make it burn without Air, or with very little, it would yet be hot enough to kindle the Sulphur, if the Air had access to it: to examine this we made two or three several Tryals, and found by them, that if some little while after the flame was extinguished, the Receiver were removed, the sulphur would Presently take fire again, and flame as vigourously as before. But I thought it might without absurdity be doubted, whether or no the agency of the Air in the production of the flame might not be somewhat less than these trials would perswade; because that, by taking off the Receiver, the Sulphur was not only exposed to fresh Air, but also advantaged with a free scope for the avolution of those fumes, which in a close Vessel might be presum'd to have been unfriendly to the Flame.

How far this doubt may, and how far it should, be admitted, we may be assisted to discern by the subjoined experiment, though made in great part for another purpose; which you will perceive by the beginning of the Memorial I made of it, that ran thus;

EXPERIMENT IV.

A DIFFERING EXPERIMENT TO THE SAME PURPOSE WITH THE FORMER.

Having a mind to try, at how great a degree of rarefaction of the Air it was possible to make Sulphur flame by the assistance of an adventitious heat, we caused such an experiment as the above mention'd to be reiterated, and the pumping to be continued for some time after the flame of the melted flowers of Brimstone appeared to be quite extinguished, and the Receiver was judged by those that managed the Pump (and that upon probable signs) to be very well exhausted. Then, without stirring the Receiver, we let in at the stop-cock very warily a little Air, upon which we could perceive, though not a constant flame, yet divers little flashes, as it were, which disclosed themselves by their bleu color to be sulphureous flames; and yet, the Air, that had suffic'd to re-kindle the Sulphur, was so little, that two exsuctions more drew it out again, and quite depriv'd us of the mentioned flashes. And when a little Air was cautiously let in again at the stop-cock, the like flashes again began to appear, which upon two executions more did again quite vanish, though upon the letting in a little fresh Air the third time, they did once more reappear.

GENETICS TRACK

Reading 2: Gregor Mendel—*Experiments in Plant Hybridization*

INTRODUCTORY REMARKS

Experience of artificial fertilisation, such as is effected with ornamental plants in order to obtain new variations in colour, has led to the experiments which will here be dis-cussed. The striking regularity with which the same hybrid forms always reappeared whenever fertilisation took place between the same species induced further experiments to be undertaken, the object of which was to follow up the developments of the hybrids in their progeny.

To this object numerous careful observers, such as Kölreuter, Gärtner, Herbert, Lecoq, Wichura, and others, have devoted a part of their lives with inexhaustible per-severance. Gärtner especially in his work *Die Bastarderzeugung im Planzenreiche,* has recorded very valuable observations; and quite recently Wichura published the results of some profound investigations into the hybrids of the willow. That, so far, no gen-erally applicable law governing the formation and development of hybrids has been successfully formulated can hardly be wondered at by anyone who is acquainted with the extent of the task, and can appreciate the difficulties with which experiments of this class have to contend. A final decision can only be arrived at when we shall have before us the results of *detailed experiments* made on plants belonging to the most di-verse orders.

Those who survey the work done in this department will arrive at the conviction that

From Gregor Mendel, *Experiments in Plant Hybridization* (Cambridge, MA: Harvard University Press, 1916), 1–2, 4–9, 11, 13–14.

among all the numerous experiments made, not one has been carried out to such an extent and in such a way as to make it possible to determine the number of different forms under which the offspring of the hybrids appear, or to arrange these forms with certainty according to their separate generations, or definitely to ascertain their statistical relations.

It requires indeed some courage to undertake a labour of such far-reaching extent; this appears, however, to be the only right way by which we can finally reach the solution of a question the importance of which cannot be overestimated in connection with the history of the evolution of organic forms.

The paper now presented records the results of such a detailed experiment. This experiment was practically confined to a small plant group, and is now, after eight years' pursuit, concluded in all essentials. Whether the plan upon which the separate experiments were conducted and carried out was the best suited to attain the desired end is left to the friendly decision of the reader.

DIVISION AND ARRANGEMENT OF THE EXPERIMENTS

If two plants which differ constantly in one or several characters be crossed, numerous experiments have demonstrated that the common characters are transmitted unchanged to the hybrids and their progeny; but each pair of differentiating characters, on the other hand, unite in the hybrid to form a new character, which in the progeny of the hybrid is usually variable. The object of the experiment was to observe these variations in the case of each pair of differentiating characters, and to deduce the law according to which they appear in successive generations. The experiment resolves itself therefore into just as many separate experiments as there are constantly differentiating characters presented in the experimental plants.

The various forms of Peas selected for crossing showed differences in length and colour of the stem; in the size and form of the leaves; in the position, colour, and size of the flower; in the length of the flower stalk; in the colour, form, and size of the pods; in the form and size of the seeds; and in the colour of the seed-coats and of the albumen. Some of the characters noted do not permit of a sharp and certain separation, since the difference is of a "more or less" nature, which is often difficult to define. Such characters could not be utilized for the separate experiments; these could only be applied to characters which stand out clearly and definitely in the plants. Lastly, the result must show whether they, in their entirety, observe a regular behavior in their hybrid unions, and whether from these facts and conclusions can be reached regarding those characters which possess a subordinate significance in the type.

The characters which were selected for experiment relate:

1. To the *difference in the form of the ripe seeds*. These are either round or roundish, the depressions, if any, occur on the surface, being always only shallow; or they are irregularly angular and deeply wrinkled.

2. To the *difference in the colour of the seed albumen*. The albumen of the ripe seeds is either pale yellow, bright yellow and orange coloured, or it possesses a more or less intense green tint. This difference of colour is easily seen in the seeds as their coats are transparent.

3. To the *difference in the colour of the seed coat*. This is either white, with which character white flowers are constantly correlated; or it is gray, gray-brown, leather-brown, with or without violet spotting, in which case the colour of the standards is violet, that of the wings purple, and the stem in the axils of the leaves a reddish tint. The gray seed-coats become dark brown in boiling water.

4. To the *difference in the form of the ripe pods*. These are either simply inflated, not contracted in places; or they are deeply constricted between the seeds and more or less wrinkled.

5. To the *difference in the colour of the unripe pods*. They are either light to dark green, or vividly yellow, in which colouring the stalks leaf-veins, and calyx participate.

6. To the *difference in the position of the flowers*. They are either axial, that is, distributed along the main stem; or they are terminal, that is, bunched at the top of the stem and arranged almost in a false umbel; in this case the upper part of the stem is more or less widened in section.

7. To the *difference in the length of the stem*. The length of the stem is very various in some forms; it is, however, a constant character for each, in so far that healthy plants, grown in the same soil, are only subject to unimportant variations in this character. The experiments with this character, in order to be able to discriminate with certainty, the long axis of 6 to 7 feet was always crossed with the short one of ¾ feet to 1 and ½ feet.

Each two of the differentiating characters enumerated above were united by cross-fertilisation. There were made for the

1st trial	60 fertilizations on	15 plants
2nd trial	58 fertilizations on	10 plants
3rd trial	35 fertilizations on	10 plants
4th trial	40 fertilizations on	10 plants
5th trial	23 fertilizations on	5 plants
6th trial	34 fertilizations on	10 plants
7th trial	37 fertilizations on	10 plants

Experiments which in previous years were made with ornamental plants have already afforded evidence that the hybrids, as a rule, are not exactly intermediate between the parental species. With some of the more striking characters, those, for instance, which relate to the form and size of the leaves, the pubescence of the several parts, etc., the intermediate, indeed, is nearly always to be seen; in other cases, however, one of the two parental characters is so preponderant that it is difficult, or quite impossible to detect the other in the hybrid.

This is precisely the case with the Pea hybrids. In the case of each of the seven crosses the hybrid-character resembles that of one of the parental forms so closely that the other either escapes observation completely or cannot be detected with certainty. This circumstance is of great importance in the determination and classification of the forms under which the offspring of the hybrids appear. Henceforth in this paper those characters which are transmitted entire, or almost unchanged in the hybridization, and therefore in themselves constitute the characters of the hybrid, are termed the *dominant,* and those which become latent in the process *recessive.* The expression "recessive" has been chosen because the characters thereby designated withdraw or entirely disappear in the hybrids, but nevertheless reappear unchanged in their progeny, as will be demonstrated later on.

Of the differentiating characters which were used in the experiments the following are dominant:

1. The round or roundish form of the seed with or without shallow depressions.
2. The yellow colouring of the seed albumen.
3. The gray, gray-brown, or leather brown colour of the seed-coat, in association with violet-red blossoms and reddish spots in the leaf axils.
4. The simply inflated form of the pod.
5. The green colouring of the unripe pod in association with the same colour of the stems, the leaf-veins and the calyx.
6. The distribution of the flowers along the stem.
7. The greater length of the stem.

THE GENERATION [BRED] FROM THE HYBRIDS

In this generation there reappear, together with the dominant characters, also the recessive ones with their peculiarities fully developed, and this occurs in the definitely expressed average proportion of 3:1, so that among each four plants of this generation three display the dominant character and one the recessive. This relates without exception to all the characters which were investigated in the experiments. The angular wrinkled form of the seed, the green colour of the albumen, the white colour of the seed-coats and the flowers, the constriction of the pods, the yellow colour of the unripe pod, of the stalk, of the calyx, and of the leaf venation, the umbel-like form of the inflorescence, and the dwarfed stem, all reappear in the numerical proportion given, without any essential alteration. *Transitional forms were not observed in any experiment.*

Since the hybrids resulting from reciprocal crosses are formed alike and present no appreciable difference in their subsequent development, consequently these results can be reckoned together in each experiment. The relative numbers which were obtained for each pair of differentiating characters are as follows:

Expt. 1: Form of seed. From 253 hybrids 7324 seeds were obtained in the second trial year. Among them were 5474 round or roundish ones and 1850 angular wrinkled ones. Therefrom the ratio 2.96:1 is deduced.

Expt. 2: Colour of albumen. 258 plants yielded 8023 seeds, 6022 yellow, and 2001 green, their ratio, therefore, is as 3.01:1.

Expt. 3: Colour of the seed-coats. Among 929 plants, 705 bore violet-red flowers and gray-brown seed-coats; 224 had white flowers and white seed-coats, giving the proportion of 3.15:1.

Expt. 4: From of the pods. Of 1181 plants, 882 had them simply inflated, and in 299 they were constricted. Resulting ratio, 2.95:1.

Expt. 5: Colour of the unripe pods. The number of trial plants was 580, of which 448 had green pods and 152 yellow ones. Consequently these stand in the ratio of 2.82:1.

Expt. 6: Position of flowers. Among 858 cases 651 had inflorescences axial and 207 terminal. Ratio, 3.14:1.

Expt. 7: Length of stem. Out of 1064 plants, in 787 cases the stem was long, and in 277 short. Hence, a mutual ratio of 2.84:1.

If now the results of the whole of the experiments be brought together, there is found, as between the number of forms with the dominant and recessive characters, an average ratio of 2.98:1, or 3:1.

THE SUBSEQUENT GENERATIONS FROM THE HYBRIDS

The proportions in which the descendants of the hybrids develop and split up in the first and second generations presumably hold good for all subsequent progeny. Experiments 1 and 2 have already been carried through six generations; 3 and 7 through five;

and 4, 5, and 6, through four; these experiments being continued from the third generation with a small number of plants, and no departure from the rule has been perceptible. The offspring of the hybrids separated in each generation in the ratio of 2:1:1 into hybrids and constant forms.

If **A** be taken as denoting one of the two constant characters, for instance the dominant, **a** the recessive, and **Aa** the hybrid form in which both are conjoined, the expression **A+2Aa+a** shows the terms in the series for the progeny of the hybrids of the two alternating characters.

The observation made by Gärtner, Kölreuter, and others, that hybrids are inclined to revert to the parental forms, is also confirmed by the experiments described. It is seen that the number of the hybrids which arise from one fertilisation, as compared with the number of forms which become constant, and their progeny from generation to generation, in continually diminishing, but that nevertheless they could not entirely disappear. If an average equality of fertility in all plants in all generations be assumed, and if, furthermore, each hybrid forms seed of which one-half yields hybrids again, while the other half is constant to both characters in equal proportions, the ratio of numbers for the offspring in each generation is seen by the following summary, in which **A** and **a** denote again the two parental characters, and **Aa** the hybrid forms. For brevity's sake it may be assumed that each plant in each generation furnishes only four seeds.

Generation	Ratios					
	A	Aa	a	A:	Aa:	a
1	1	2	1	1:	2:	1
2	6	3	6	3:	2:	3
3	28	4	28	7:	2:	7
4	120	16	120	15:	2:	15
5	496	32	496	31:	2:	31
n				2^n-1:	2:	2^n-1

In the tenth generation, for instance, $2^n-1 = 1023$. There result, therefore, in each 2048 plants which arise in this generation 1023 with the constant dominant character, 1023 with the recessive character, and only two hybrids.

EVOLUTIONARY BIOLOGY

Reading 2: Carolus Linnaeus—*Systema Naturae*

OBSERVATIONS ON THE THREE KINGDOMS OF NATURE

1. If we observe God's works, it becomes more than sufficiently evident to everybody, that each living being is propagated from an egg and that every egg produces an offspring closely resembling the parent. Hence, no new species are produced nowadays.

2. Individuals multiply by generation. Hence at present the number of individuals in each species is greater than it was at first.

3. If we count backwards this multiplication of individuals in each species, in the same way as we have multiplied forward (2), the series ends up in one single *parent*, whether that parent consists of *one single* hermaphrodite (as commonly in plants) or of a double, viz., a male and a female, (as in most animals).

4. As there are no new species (1); as like always gives birth to like (2); as one in each species was at the beginning of the progeny (3); it is necessary to attribute this progenitorial unity to some Omnipotent and Omniscient Being, namely *God,* whose work is called *Creation.* This is confirmed by the mechanism, the laws, principles, constitutions and sensations in every living individual.

5. Individuals thus procreated, lack in their prime and tender age absolutely all knowledge, and are forced to learn everything by means of their external senses. By *touch* they first of all learn the consistency of objects; by *taste* the fluid particles; by *smell* the violate ones; by *hearing* the vibration of remote bodies; and finally by *sight* the shape of visible bodies, which last sense, more than any of the others, gives the animals greatest delight.

6. If we observe the universe, three objects are conspicuous: viz., α. the very remote *coelestrial* bodies; β. the *elements* to be met everywhere; γ. the solid *natural bodies.*

7. On our earth, only two of the three mentioned above (6) are obvious; i.e. the *elements* constituting it; and the *natural* bodies constructed out of the elements, though in a way inexplicable except by creation and by the laws of procreation.

8. Natural objects (7) belong more to the field of the senses (5) than all the others (6) and are obvious to our senses anywhere. Thus I wonder why the

From Carl Linnaeus, *Systema Naturae* (1735). Facsimile of the first edition with an introduction and an English translation of the observations by M. S. J. Engel Ledeboer and H. Engel (Nieuwkoop, Holland: B. de Graff, 1964), 18–19, 26–28.

Creator put man, who is thus provided with senses (5) and intellect, on the earth globe, where nothing met his senses but natural objects, constructed by means of such an admirable and amazing mechanism. Surely for no other reason than that the observer of the wonderful work might admire and praise its Maker.

9. All that is useful to man originates from these natural objects; hence the industry of mining and metallurgy; plant-industry or agriculture and horticulture; animal husbandry, hunting and fishing. In one word, it is the foundation of every industry of building, commerce, food supply, medicine, etc. By them people are kept in a healthy state, protected against illness and recover from disease, so that their selection is highly necessary. Hence (8, 9) the necessity of natural science is self-evident.

10. The first step in wisdom is to know the things themselves; this notion consists in having a true idea of the objects; objects are distinguished and known by classifying them methodically and giving them appropriate names. Therefore, classification and name-giving will be the foundation of our science.

11. Those of our scientists, who cannot class the variations in the right species, the species in the natural genera, the genera in families, and yet constitute themselves doctors of this science, deceive others and themselves. From all those who really laid the foundation to natural science, have had to keep this in mind.

12. He may call himself a naturalist (a natural historian), who well distinguishes the parts of natural bodies by sight (5) and describes and names all these rightly in agreement with the threefold division. Such a man is a lithologist, a phytologist, or a zoologist.

13. Natural science is that classification and that name-giving (10) of the natural bodies judiciously instituted by such a naturalist (12).

14. Natural bodies are divided into *three kingdoms of nature:* viz., the mineral, vegetable, and animal kingdoms.

15. *Minerals* grow; *Plants* grow and live; *Animals* grow, live, and have feeling. Thus the limits between these kingdoms are constituted.

16. In this science of describing and picturing many have laboured for a whole life-time; how much, however, has already been observed and how much there remains to be done, the curious on-looker will easily find out for himself.

17. I have shown here a general survey of the system of natural bodies so that the curious reader with the help of this as it were geographical table knows where to direct his journey in these vast kingdoms, for to add more descriptions, space, time and opportunity lacked.

18. A new method mainly based on my own authentic observations has been used in every single part, for as I have learnt that very few people are lightly to be trusted, as far as observations go.

OBSERVATIONS ON THE ANIMAL KINGDOM

1. Zoology, that noblest part of Natural History, is much less worked up than the other two parts. If, however, we take into account either the movement, or the mechanism, or the external and internal senses, or lastly the shape of the animals, which surpasses all the others, it will be as clear as the sun to everyone, that the animals are the highest and most perfect works of the Creator.

2. If we reexamine the zoologies of the Authors we shall find for the greater part nothing but fabulous stories, a vague way of writing, pictures by the copper engravers and descriptions which are imperfect and often too extensive. There are very few indeed, who have tried to reduce zoology to genera and species according to the rules of systematics, the most noble *Willughby* and the very famous *Ray* excepted.

3. Hence I have begun to compose a kind of system of zoology by the aid of any observations I have been able to obtain with my own eyes; this I here present to you now, Illustrious Reader. First I distinguished in *Tetrapodologia* (Quadrupeds) the Orders of animals according to their teeth, in *Ornithologia* according to their bill, in *Entomologia* by their antennae, their wings, etc.

4. In *Ichthyologie* I have not made a method myself, as the greatest Ichthyologist of our time, the Very Illustrious Dr. *Petrus Artedi,* a *Swede,* has communicated his method to us, who hardly can be equaled by anyone in distinguishing the natural genera of the fishes, and the differences between the species. This I present now already to the Curious Reader in order to give him an idea of the whole work. The Illustrious Reader may soon look forward to more by the same (author), viz. *Institutiones totius Ichthyologiae.*

5. There are people who think, that *Zoology* is of less *use* than the other parts of Natural History, mainly with regard to the very small animals; but, if we consider only the noxiousness, the use and the properties of insects, which are best known so far, it easily appears of how much use and, moreover, of how great a future importance might be the *characteristics* of those which are not yet well-known to us.

6. From the following the *noxious properties* (5) of insects are more than evident: e.g., *Blatta* (cockroach) in Finland and Russia, consumes bread as well as all kinds of clothes, in such a way that the inhabitants have been forced to leave

their homes for some time in midwinter until it would perish from the cold. *Oestrum Lapponicum* (a gad-fly) destroys about one-third of the Reindeer, the cattle of the Laps, as long as there are still young ones. Of *Teredo navium* (ship-worm) it is generally known how much damage it has done to ships and (jetty-) poles. How much trouble *Culices* (mosquitoes) bring to man and cattle in the provinces bordering on Lapland I need hardly tell. What a troublesome strident noise *Gryllus domesticus* (cricket), which very familiar animals live in walls, make and how many sleepless nights they cause to those who want to sleep, is a very well-known fact. That specimens of *Muscus domesticus* (house-fly) in Norvegian Finmark, filled entire houses and left nothing intact, I have seen myself on my journey through Lapland. Everyone knows how much work and trouble *Pulex* (fleas) causes to women, and *Pediculus* (louse) to sailors and soldiers everywhere. Indeed, also quadrupeds, birds, etc. are troubled by their own lice. *Acari* (mites), the smallest animals of the insects, very often even cause a rash of the human skin. It is very well known with what enormous army *Locusta africana* devastated plants in certain areas of Europe a few years ago and in what devastating way *Eruca papilionum* (the caterpillars of the butterflies) each year eat the leaves of trees. The best gardeners know how in the early spring *Gyrinus terrestris* (flee-beetle) destroys our tiny plant germs. *Dermestes* (beetle) lacerates very precious furs and skins of quadrupeds and birds in an extraordinary way. *Oestrum bovinum* (gad-fly) greatly troubles cattle tired by the summer weather. How many people have been killed by *spiders* and *scorpions* or became insane from *tarantulas* is testified by the observations of medical people, apart from innumerable other such cases.

7. Most *useful* (5) insect products for the dyeing industry are supplied by *Cochineal, Kermes* and by *Galls* produced by gall-insects (ichneumons). The use of *Cantharides* (Spanish fly) in surgery, of *Meloë* (blister beetle) in medicine and *Bombyx* (silk-worm) in the art of weaving, of *bee*-honey in food industry, is well known.

8. The curious investigator, who wants to examine the *properties* (5) of insects, can hardly have a greater pleasure anywhere. Just examine: the rostrum of *Curculio* (snout-beetle), the horns of *Lucanus* (stag-beetle), the antennae of *Tragocerus*, the joints of *Meloë* (blister beetle), the wings of an *earwig*, the plumes of a *butterfly*, the eyes of a *Tabanus* (horse-fly), the abdomen of *Ricinus* (a tick), the sting of a *digger-wasp*, the colour of a *Spanish fly*, the elasticity of a *click-beetle*, the stridor of a *cricket*, the small of a *bug*, the smallness of a *mite*, the copulation of the *dragonflies*, the nest of an *ichneumon-fly*, the comb of the *honey-bees*, the hibernation of a *gad-fly*, the building of a *wasps*-nest, the shell of a *hermit crab*, the life of an *ephemeron*, an anthill, the trap-fall of an *ant-lion*,

a *spider*'s web, the way of swimming of *Cyclops*, the locomotion of a *whirligig*, the phosphorescence of *Lampyris* (fire-fly), the luminescence of *Scolopendria marina* (a Nereide), the sloughing of a *crab*, the spiral motion of the *caterpillar coming from a blue bottle fly*, the well-nigh indestructible life of the *aquatic maggot of the horse-fly* and the so-called metamorphoses of nearly all *Insects*.

9. The eggs of most insects are covered by a triple integument. If the first skin comes off, it (the animal) is called *eruca* (maggot or caterpillar), if the second comes off, a *propolis* (chrysalis or pupa), and lastly after losing the third one a perfect *insect*. Hence the triple hatching of the young from such eggs.

10. In the human intestine three species of animals occur, viz. Lumbrici, Ascarides, (round worms) and Taeniae (tape-worms). That the *Lambricus* of the intestine is one and the same species as the ordinary earthworm, is shown by the appearance of all its parts. That the *Ascaris* species are identical with those very small worms (Lambricus) one finds anywhere on marshy spots, becomes very clear by close inspection. *Taenia* so far has been considered a parasitic species, as it has been recovered, mostly at one time, from man, dogs, fishes, etc. and they gave a great deal of trouble to those who diligently carried out the work of investigating the generation of animals. However, in 1734, I found it on the Reuterholm trip to Dalekarlia in the presence of seven companions of mine in sour iron ochre, about which I was highly surprised, for most people try to get rid of Taenia by means of that kind of acid water. Hence it follows that worms do not take their origin from insects' eggs, flies and the like (for if that happened, they could never multiply inside the intestinal tract, and would perish during the stages of metamorphoses); but from the eggs of the worms above-mentioned, taken in with the water by drinking; from this it is evident that medicaments detrimental to insects need not necessarily kill the worms.

GEOLOGY TRACK

Reading 2: James Hutton—"System of the Earth"

The solid parts of the present land appear, in general, to have been composed of the products of the sea, and of other materials similar to those now found upon the shores. Hence we find reason to conclude,

From James Hutton, "Abstract of a Dissertation Read in the Royal Society of Edinburgh, upon the 7th of March and the 4th of April, concerning the System of the Earth, its Duration, and Stability," *The Edinburgh Magazine*, vol. 3, 184–87 (London: Sibbald, 1788).

1st, That the land on which we rest is not simple and original, but that it is a composition, and has been formed by the operation of second causes.

2dly, That, before the present land was made, there had subsisted a world composed of sea and land, in which were tides and currents, with such operations at the bottom of the sea as now take place. And,

Lastly, That, while the present land was forming at the bottom of the ocean, the former land maintained plants and animals; at least, the sea was then inhabited by animals, in a similar manner as it is at present.

Hence we are led to conclude, that the greater part of our land, if not the whole, had been produced by operations natural to this globe; but that, in order to make this land a permanent body, resisting the operations of the waters, two things had been required; 1st, The consolidation of masses formed by collections of loose or incoherent materials; *2dly*, The elevation of those consolidated masses from the bottom of the sea, the place where they were collected, to the stations in which they now remain above the level of the ocean.

Here are two different changes, which may serve mutually to throw some light upon each other; for as the same subject has been made to undergo both these changes, and as it is from the examination of this subject that we are to learn the nature of those events, the knowledge of the one may lead us to some understanding of the other.

Thus the subject is considered as naturally divided into two branches, to be separately examined: *First,* by what natural operation strata of loose materials had been formed into solid masses; *Secondly,* By what power of nature the consolidated strata at the bottom of the sea had been transformed into land.

With regard to the *first* of these, the consolidation of the strata, there are two ways in which this operation may be conceived to have been performed; first, by means of the solutions of bodies in water, and the after concretion of these dissolved substances, when separated from their solvent; *secondly,* the fusion of bodies by means of heat, and the subsequent congelation of those consolidating substances.

With regard to the operation of water, it is *first* considered, how far the power of this solvent, acting in the natural situation of those strata, might be sufficient to produce the effect; and here it is found, that water alone, without any other agent, cannot be supposed capable of inducing solidity among the materials of strata in that situation. It is, *2dly,* considered, how far, supposing water capable of consolidating the strata in that situation, it might be concluded, from examining the natural appearances, that this had actually been the case? Here again, having proceeded upon this principle, that water could only consolidate strata with such substances as it has the power to dissolve, and having found strata consolidated with every species of substance, it is concluded that strata in general have not been consolidated by means of aqueous solution.

With regard to the other probable means, heat and fusion, these are found to be perfectly competent for producing the end in view, as every kind of substance may by heat be rendered soft, or brought into fusion, and as strata are actually found consolidated with every different species of substance.

A more particular discussion is then entered into: Here, consolidating substances are considered as being classed under two different heads, viz. Siliceous and Sulphureous bodies, with a view to prove, that it could not be by means of aqueous solution that strata had been consolidated with those particular substances, but that their consolidation had been accomplished by means of heat and fusion.

Sal Gem, as a substance soluble in water, is next considered, in order to show that this body had been last in a melted state; and this example is confirmed by one of fossil alkali. The case of particular septaria of iron-stone, as well as certain crytstallized cavities in mineral bodies, are then given as examples of a similar fact; and as containing in themselves, a demonstration, that all the various mineral substances had been concreted and crystallized immediately from a state of fusion.

With regard to the second branch, in considering by what power the consolidated strata had been transformed into land, or raised above the level of the sea, it is supposed that the same power of extreme heat, by which every different mineral substance had been brought into a melted state, might be capable of producing an expansive force, sufficient for elevating the land, from the bottom of the ocean, to the place it now occupies above the surface of the sea. Here we are again referred to nature, in examining how far the strata, formed by successive sediments or accumulations deposited at the bottom of the sea, are to be found in that regular state, which would necessarily take place in their original production; or if, on the other hand, they are actually changed in their natural situation, broken, twisted, and confounded, as might be expected, from the operation of subterranean heat, and violent expansion. But, as strata are actually found in every degree of fracture, flexure, and contortion, consistent with this supposition, and with no other, we are led to conclude, that our land had been raised above the surface of the sea, in order to become a habitable world; as well as it had been consolidated by means of the same power of subterranean heat, in order to remain above the level of the sea, and to resist the violent efforts of the ocean.

This theory is next confirmed by the examination of mineral veins, those great fissures of the earth, which contain matter perfectly foreign to the strata they traverse; matter evidently derived from the mineral region, that is, from the place where the active power of fire, and the expansive force of heat, reside.

Such being considered as the operations of the mineral region, we are hence directed to look for the manifestation of this power and force, in the appearance of nature. It is here we find eruptions of ignited matter from the scattered volcano's of the globe; and these we conclude to be the effects of such a power precisely as that about which we now inquire. Volcano's are thus considered as the proper discharges of a superfluous or redundant power; not as things accidental in the course of nature, but as useful for the safety of mankind, and as forming a natural ingredient in the constitution of the globe.

PSYCHOLOGY TRACK

Reading 2: Heinrich Weber—"The Sense of Touch and the Common Feeling"

The smallest perceptible difference between two weights, which we can distinguish by the feeling of muscular exertion, appears according to my experiments to be that between weights that stand approximately in the relation of 30 to 40: that is to say, of which one is about 1/40 heavier than the other. By means of the feeling of pressure, which two weights make upon the skin, all we are able to distinguish is a difference of weights that amount to only 1/30, so that the weights accordingly stand in the relation of 29 to 30.

If we look at one line after another, anyone who possesses a very exceptional visual discrimination can according to my experiments discover a difference between two lines whose lengths are related as 50:51, or even as 100:101. Those who have a less delicate visual discrimination distinguish lines, which are separated from one another by 1/25 of their length. The smallest perceptible difference of the pitch of two tones, (which are really in unison), that a musician perceives, if he hears the tones successively, is according to Delezenne ¼ Komma (81/80) ¼. A lover of music according to him distinguishes only about ½ Komma (81/80) ½. If the tones are heard simultaneously we cannot, according to Delezenne's experiments, perceive such small tonal differences. ¼ Komma is nearly the relation of 321:322, but ½ Komma is nearly the relation 160:161.

I have shown that the result in the determination of weight is the same, whether one takes ounces or half-ounces; for it does not depend upon the number of grains that form the increment of weight, but depends on the fact that this increment makes up the thirtieth or fortieth part of the weight which we are comparing with the second weight. This likewise holds true of the comparison of the lengths of two lines and of the pitch of two tones. It makes no difference whether we compare lines that are, say, two inches or one inch long, if we examine them successively, and can see them lying parallel to each other; and yet the extent by which the one line exceeds the other is in

From Heinrich Weber, "The Sense of Touch and the Common Feeling," in *The Classical Psychologists*, ed. and trans. Benjamin Rand (Gloucester: Peter Smith, 1912), 557–60.

the former case twice as great as in the latter. To be sure, if both lines lie close together and parallel, we compare only the ends of the lines to discover how much the one line exceeds the other; and in this test, the question is only how great that length of line which overlaps the other really is, and how near the two lines lie to one another.

So too in the comparison of the pitch of two tones, it does not matter whether the two tones are seven tonal stops higher or lower, provided only they do not lie at the end of the tonal series, where the exact discrimination of small tonal differences becomes more difficult. Here again, therefore, it is not a question of the number of vibrations, by which the one tone exceeds the other, but of the relation of the numbers of the vibrations of the two tones which we are comparing. If we counted the vibrations of the two tones it would be conceivable that we should pay attention only to the number of vibrations by which one tone exceeds the other. If we fix the eyes first upon one line and afterwards upon a second, and thus permit both to be pictured successively upon the most sensitive part of the retina, we should be inclined to suppose, that we compared the traces of the impression which the first image left, with the impression which the second image made upon the same parts of the retina, and that we thereby perceived how much the second image exceeds the first, and conversely. For this is the way we compare two scale-units: we place one upon the other, so that they coincide, and thus perceive how much the one exceeds the other. From the fact that we do not employ this method which is so advantageous, it seems to follow, that we are unable to employ it, and that therefore the preceding impression left behind no such trace on the retina, or in the brain, as would permit of comparison in the manner mentioned with succeeding impressions. That it is possible for us to proceed otherwise in the comparison of the lengths of two lines appears from the fact, that we can compare two lines, which are longer than we can picture at once in their entirety on the most sensitive part of the retina. In this case we must move the eye, and only thus do we form an idea of the length of the lines. Were the impressions of visible things, which we preserve in memory, traces, which the sensuous impressions left behind in the brain, and whose spatial relations correspond to the spatial relations of the sensuous impressions, and were thus so to speak photographs of the same, it would be difficult to remember a figure, which is larger than could be pictured at once wholly upon the sensitive part of the retina. It appears to me, indeed, as if a figure, which we can survey at a single glance, impressed itself better upon our memory and our imagination, than a figure, which we can survey only successively by moving the eyes; but we can nevertheless represent also the former by means of the imagination. But in this case, the representation of the whole figure seems to be composed by us of the parts which we perceive all at once.

Reading 2: Émile Durkheim—*Suicide*

Sociological method as we practice it rests wholly on the basic principle that social facts must be studied as things, that is, as realities external to the individual. There is no principle for which we have received more criticism; but none is more fundamental. Indubitably for sociology to be possible, it must above all have an object all of its own. It must take cognizance of a reality which is not in the domain of other sciences. But if no reality exists outside of individual consciousness, it wholly lacks any material of its own. In that case, the only possible subject of observation is the mental states of the individual, since nothing else exists. That, however, is the field of psychology. From this point of view the essence of marriage, for example, or the family or religion, consists of individual needs to which these institutions supposedly correspond: paternal affection, filial love, sexual desire, the so-called religious instinct, etc. These institutions themselves, with their varied and historical forms, become negligible and of little significance.

But it seems hardly possible to us that there will not emerge, on the contrary, from every page of this book, so to speak, the impression that the individual is dominated by a moral reality greater than himself: namely, collective reality. When each people is seen to have its own suicide rate, more constant than that of general mortality, that its growth is in accordance with a coefficient of acceleration characteristic to each society; when it appears that the variations through which it passes at different times of the day, month, year, merely reflect the rhythm of social life; and that marriage, divorce the family, religious society, the army, etc., affect it in accordance with definite laws, some of which may even be numerically expressed—these states and institutions will no longer be regarded simply as characterless, ineffective ideological arrangements. Rather they will be felt to be real, living, active forces which, because of the way they determine the individual, prove their independence of him; which, if the individual enters as an element in the combination whence these forces ensue, at least control him once they are formed. Thus it will appear more clearly why sociology can and must be objective, since it deals with realties as definite and substantial as those of the psychologist or biologist.

. . . We have in fact shown that for each social group there is a specific tendency to suicide explained neither by the organic-psychic constitution of individuals nor the nature of the physical environment. Consequently, by elimination, it must necessarily depend upon social causes and be in itself a collective phenomenon; some of the facts examined, especially the geographic and seasonal variations of suicide, had definitely led us to this conclusion. We must now study this tendency more closely.

To accomplish this it would seem to be best to inquire first whether the tendency is single and indestructible or whether it does not rather consist of several different tendencies, which may be isolated by analysis and which should be separately studied. If so, we should proceed as follows. As the tendency, single or not, is observable only in its individual manifestations, we should have to begin with the latter. Thus we should observe and describe as many as possible, of course omitting those due to mental alienation. If all were found to have the same essential characteristics, they should be grouped in a single class; otherwise, which is much more likely—for they are too different not to include several varieties—a certain number of species should be determined according to their resemblances and differences. One would admit as many suicidal currents as there were distinct types, then seek to determine their causes and respective importance.

Unfortunately, no classification of the suicides of sane persons can be made in terms of their morphological types or characteristics, from almost complete lack of the necessary data.

But our aim may be achieved by another method. Let us reverse the order of study. Only in so far as the effective causes differ can their be different types of suicide. For each to have its own nature, it must also have special conditions of existence. The same antecedent or group of antecedents cannot sometimes produce one result and sometimes another, for, if so, the difference from the second from the first would itself be without cause, which would contradict the principle of causality. Every proved scientific difference between causes therefore implies a similar difference between effects. Consequently, we shall be able to determine the social types of suicide by classifying them not directly by their preliminarily described characteristics, but by the causes which produce them. Without asking why they differ from one another, we will first seek the social conditions responsible for them; then group these conditions in a number of separate classes by their resemblances and differences, and we shall be sure that a specific type of suicide will correspond to each of these classes.

But how reach these causes?

The legal establishments of fact always accompanying suicide include the motive (family trouble, physical or other pain, remorse, drunkenness, etc.), which seems to have been the determining cause, and in statistical reports of almost all countries is found a special table containing the results of these inquiries under the title: *presumptive motives of suicides*. It seems natural to profit by this already accomplished work and begin our study by a comparison of such records. They appear to show us the immediate antecedents of different suicides . . . But as Wagner long ago remarked, what are called statistics of the motives of suicides are actually statistics of the opinions concerning such motives of officials, often of lower officials, in charge of this information service . . . The value of improvised judgments, attempting to assign a definite origin for each special case from a few hastily collected bits of information is, therefore, slight.

Moreover, even if more credible, such data could not be very useful, for the motives thus attributed to the suicides, whether rightly or wrongly, are not their true causes. The proof is that the proportional number of cases assigned by statistics of each of these presumed causes remain almost identically the same, whereas the absolute figures, on the contrary, show the greatest variations. In France, from 1856–1878, suicide rises about 40 per cent, and more than 100 per cent in Saxony in the period 1854–1880 (1,171 cases in place of 547). Now, in both countries each category of motives retains the same respective importance from one period to another.

If we consider that the figures here reported are, and can be, only grossly approximate and therefore do not attach too much importance to slight differences, they will clearly appear to be practically stable. It cannot be by coincidence that all at the same time become doubly fatal. The conclusion is forced that they all depend on a more general state, which all more or less faithfully reflect. That it is this that makes them more or less productive of suicide and which is thus the truly determining cause of it.

Let us see how the different religious confessions affect suicide. If one casts a glance at the map of European suicide, it is at once clear that in purely Catholic countries like Spain, Portugal, Italy, suicide is very little developed, while it is at its maximum in Protestant countries, in Prussia, Saxony, Denmark. The following averages compiled by Morselli confirm this conclusion:

State(s)	Average number of suicides*
Protestant	190
Mixed (Protestant and Catholic)	96
Catholic	58
Greek Catholic	40

Source: Henry Morselli, *Suicide: An Essay on Comparative Moral Statistics*
(London: C. K. Paul, 1881).
*The number of suicides is the average per million inhabitants, per year.

The low proportion of the Greek Catholics cannot be surely attributed to religion; for as their civilization is very different from that of other European nations, this difference of culture may be the cause of their lesser aptitude. But this is not the case with most Catholic or Protestant societies. To be sure, they are not all on the same intellectual and moral level; yet the resemblances are sufficiently essential to make it possible to ascribe to confessional differences the marked contrast they offer in respect to suicide.

Nevertheless, this first comparison is too summary. In spite of undeniable similarities, the social environments of the inhabitants of these different countries are not identical. The civilizations of Spain and Portugal are far below that of Germany and this inferiority may conceivably be the reason for the lesser development of suicide which we have just mentioned. If one wishes to avoid this source of error and determine more definitely the influence of Catholicism and Protestantism on the suicidal tendency, the two religions must be compared in the heart of a single society.

Of all the great states of Germany, Bavaria has by far the fewest suicides. There have been barely 90 per million inhabitants yearly since 1874, while Prussia has 133 (1871–75), the duchy of Baden 156, Wurttemberg 162, Saxony 300. Now, Bavaria also has most Catholics, 713.2 to 1,000 inhabitants. On the other hand, if one compares the different provinces of Bavaria, suicides are found to be in direct proportion to the number of Protestants and in inverse proportion to the number of Catholics (See Table). Not only the proportions of averages to one another confirm the law but all the numbers of the first column are higher than those of the second and those of the second higher than those of the third without exception.

SUICIDE RATE IN BAVARIAN PROVINCES WITH CATHOLIC POPULATIONS (1867–1875)

Provinces with Catholic minority (less than 50%)	Number of suicides*	Provinces with Catholic majority (50%–90%)	Number of suicides*	Provinces more than 90% Catholic	Number of suicides*
Rhenish Palatinate	167	Lower Franconia	157	Upper Palatinate	64
Central Franconia	207	Swabia	118	Upper Bavaria	114
Upper Franconia	204			Lower Bavaria	19
Average number	192	Average number	135	Average number	75

*The number of suicides is the average per million inhabitants, per year, 1867–1875.

ECONOMICS TRACK

Reading 2: François Quesnay—"Farmers"

Those who lease country property and work it for profit. They produce the kind of wealth and resources most essential to the support of the state. Thus the work of the

From François Quesnay, "Farmers," in *Economic and Philosophical Works* [in French] (Paris: 1888), 159–60, 173–74, 180–81, 182, 189–90, 192.

farmer is of great importance to the kingdom and deserves close attention by the government.

Only vague and imperfect ideas can be derived from considering French agriculture in general terms. It is a common misconception that only regions where the land is left fallow are deficient in cultivation, and that the labor of the poor peasant is as productive as the labor of the wealthy farmer. The crops covering the land deceive us; we view them rapidly and this assures us that the land is under cultivation, but such a superficial glance does not leave us informed about the productiveness of the harvest or the condition of the crops, and it tells us even less about the profitability of cattle and other essential aspects of agriculture. Knowledge on these subjects can only be gathered by a very extensive and thorough study. The productiveness of agriculture depends on the different ways of treating the land under cultivation and on the causes of these differences. One must have a thorough knowledge of the various types of cultivation to judge the present state of agriculture in the kingdom.

In a kingdom such as France whose territory is so large and could produce much more bread grain than can be sold, this production must be confined to land with good soil. The very poor land that is presently being farmed for bread grain does not pay enough to make up for the expense of cultivation. We are not concerned here with the improvement of this land; the cost of this cannot conceivably be met in France where it is not even possible, by a large margin, to meet the cost of ordinary agriculture. But this same land can be made more profitable if it is exploited to provide food for livestock by being cultivated for feed grain or for roots, by being left as grassland or seeded as meadowland. The more the animals can be fed in the stable with this type of cultivation, the more dung they furnish to fertilize the soil, the more abundant the crops of grain and fodder become, and the greater the number of animals that can be kept. Woods and vineyards, which are of great value, can also take up much land without detriment to the cultivation of grain.

Now the claim has been made that the cultivation of the vine should be cut back so that the cultivation of bread grain could be extended. But this would mean unnecessarily depriving the kingdom of a valuable product without removing the obstacles that impede the cultivation of the land. It seems that the vine-grower finds it more to his advantage to cultivate the vine, or else he does not need as much money to keep up this cultivation as he does to prepare land for the growing of bread grain. Everyone follows his capacities; if customs that have become established for incontrovertible reasons are impeded by laws, these laws are only another hindrance to agriculture. Such legislation is all the more unnecessary where vineyards are concerned, since land is not what is lacking for the cultivation of bread grain, but rather the means for working the land profitably.

The farmer is always of greater benefit to the state, even in periods when, due to the low price of grain, he does not make any profit on his crop. At least the sum of his expenditures results in an annual increase of real wealth in the kingdom. In truth this increase in wealth cannot continue when those individuals who bear its cost receive no profit from it and in fact suffer losses that reduce their ability to continue operating. While keeping the price of grain low tends to favor the city dweller, the factory workers, and the artisans, it desolates the countryside, the wellspring of the true riches of the state. Moreover this policy has very limited success. Bread is not man's only food and to enjoy an abundance of other foods we again have to depend on agriculture, which must be encouraged.

The wealthy farmer puts the peasant to work and supports him. The peasant in turn provides the citizen who is poor with all the necessary produce. Wherever there are no farmers and the land is worked with oxen, the peasants languish in extreme poverty. The sharecropper who is himself poor cannot put them to work and they either leave the countryside or are reduced to living on oats, barley, buckwheat, potatoes, and other cheap foodstuffs which they grow themselves and which have a short growing period. The cultivation of bread grain demands too much time and effort; the peasants cannot wait two years for their first crop. The only ones who can cultivate bread grain are the farmer who can meet the costs and the sharecropper who receives help from the land-owner. The sharecropper is a feeble support for agriculture, yet he is the only one for landowners who have no farmers. The farmers themselves can draw their profit only from their superior methods of cultivation and from the good quality of the land they cultivate. The excess of their crop over their expenses constitutes their only gain. If, after making allowance for the seed grain and for his expenses, the farmer is left with one *septier* per acre, which is to his benefit since, on forty acres seeded to wheat, he then makes a profit of forty *septiers*, worth about six hundred livres. If he cultivates so well that he can retain two *septiers* per acre for himself, his profit doubles. In this case each acre must produce seven to eight *septiers;* however he can only obtain such pro-duction from good land. When some of the land he cultivates is good and some poor, his profit can only be slight.

The price of bread grain is not regulated merely by good or poor harvests. Its value is decided mainly by the degree of freedom of the grain trade. If during a period of good harvests attempts are made to restrict or impede this trade, then agricultural production is unsettled, the state is weakened, the income of the landowners is de-creased, the farmhands and day laborers, whose help is needed in agriculture, are en-couraged to be lazy and arrogant; in a word, those who cultivate the land are ruined and the countryside depopulated. To prevent the export of grain for fear of a shortage is to be unaware of the favorable situation of France, a kingdom which can produce much more than can possibly be sold abroad.

The policy of England in this respect proves, on the contrary, that the sale abroad of a part of the harvest is the surest means of supporting agriculture, maintaining abundance, and preventing famines. Since it began to favor and stimulate export, this nation has not experienced any periods in which grain was extraordinarily expensive or lost its value.

It is the farmer's wealth which renders the land fertile, multiplies the livestock, attracts the rural population and keeps it in the country. This wealth creates the strength and prosperity of the nation.

Manufacture and commerce, which are supported by the disorders of luxury, pile up men and riches in the cities, prevent the improvement of landed property, lay waste the countryside, inspire men with scorn for agriculture, lead to an excessive increase in expenditure by individuals, impair the support of the family, prevent the propagation of the human race, and weaken the state.

The downfall of an empire has often followed closely upon a flourishing commerce. Whenever a nation spends on luxuries what it gains by commerce, this leads merely to the circulation of money and brings about no real increase in wealth. It is by the sale of surplus that the subjects and the sovereign grow rich. The products of our soil must provide the raw materials for our manufactures and the commodities for our commerce. No form of commerce is secure unless it rests on this foundation. The more dazzling the commerce of a kingdom, the more it arouses the competition of neighboring nations with which the spoils must be shared. A kingdom rich in fertile land cannot be imitated in agriculture by another that is not so favored. To profit from this, however, it is necessary to eliminate the causes that bring about the flight from the countryside, which collect and keep wealth in the big cities. All the nobles, all the wealthy people, all those who have a fixed income or a pension sufficiently large to live in comfort take up residence in Paris or some other large city, and there spend almost the entire income of the kingdom. These expenditures attract a multitude of merchants, artisans, domestics, and unskilled workers. Such an uneven distribution of population and wealth is inevitable, but it goes much too far. The policy of favoring the inhabitants of the city over the rural people probably did much to start this process. Men are drawn by their self-interest and by a peaceful existence. If these advantages were provided in the countryside, it would be proportionately no less populated than the cities. Not all city-dwellers are rich or even comfortably off. The country can offer its own riches and pleasures, people leave it only to escape the harassment to which they are exposed, and the government can do away with this disadvantage. Commerce appears to flourish in the cities because they harbor many rich merchants. But what other result does this have than that almost all the money in the kingdom is used for a type of commerce that does not increase the wealth of the nation? Locke compares it to gambling where, after the gamblers have won or lost, the total sum of money remains the same as before. Internal commerce is necessary in order to satisfy needs, support luxury, and facilitate consumption; but it contributes little to the strength and prosperity of the state. If part of the immense wealth that is tied up in this commerce and yields so little to the kingdom were channeled into agriculture, this would provide much more tangible and substantial revenue. Agriculture is the patrimony of the sovereign: all its products are visible and can be properly subjected to taxation. Moneyed wealth cannot be assessed for taxes and the government is only able to obtain its share by methods costly to the state.

Nevertheless the assessment of taxes on those who work the land also presents great difficulties. Arbitrary taxes are too unjust and act too much as a deterrent, they will always be a powerful check on the revival of agriculture. Proportional assessment is scarcely possible: it does not seem feasible to regulate it by the appraisal and the tax rate of the land, since the two types of agriculture we have discussed entail many differences in the productiveness of plots of land of the same value. Thus, as long as these two types of cultivation will persist in their diversity, it will be impossible to use land as a proportional yardstick for the assessment of the taille. If land were taxed according to its present condition, the tax lists would become progressively inaccurate as large-scale farming spread. Besides, there are provinces where the profit from livestock is much more substantial than the yield from the crops, and others where the contrary is true. Furthermore, this diversity is very much subject to change. Thus it is scarcely possible to imagine an over-all plan for the proportional assessment of taxes.

If the inhabitants of the country were freed from arbitrary taxation through the taille, they would live under the same security as the inhabitants of the large cities. Many land-owners would go and work their property for profit themselves. People would no longer desert the countryside where riches and population would increase again. Thus, if all the other factors detrimental to agricultural progress are eliminated, the strength of the kingdom would gradually revive by means of the growth of population and the rise in the revenues of the state.

Bibliography

Aristotle. "De Caelo." In *The Oxford Translation of Aristotle*, ed. W. D. Ross, trans. J. L. Stocks. Vol. 2, 283, 286, 288–89. Oxford: Clarendon Press, 1930.

———. "De Generation Animalium." In *The Oxford Translation of Aristotle*, ed. W. D. Ross, trans. Arthur Platt. Vol. 5, 724, 729–35. Oxford: Clarendon Press, 1912.

———. "Physica." In *The Oxford Translation of Aristotle*, ed. W. D. Ross, trans. R. P. Hardie and R. K. Gaye. Vol. 2, 192–95. Oxford: Clarendon Press, 1930.

———. "Politica." In *The Oxford Translation of Aristotle*, ed. W. D. Ross, trans. B. Jowett. Vol. 10, 1253–58. Oxford: Clarendon Press, 1921.

———. "Analytica Posteriora." In *The Oxford Translation of Aristotle*, ed. W. D. Ross, trans. G. R. G. Mure. Vol. 1, 70–74. Oxford: Clarendon Press, 1928.

Bacon, Francis. *Novum Organum.* In *The Works of Francis Bacon*, ed. James Spedding, Robert Leslie Ellis, and Douglas Denon Heath, 59–64, 372–74. New York: Hurd and Houghton, 1877.

Black, Max. "Models and Archetypes." In *Both Human and Humane: The Humanities and Social Sciences in Graduate Education*, ed. Charles E. Boewe and Roy F. Nichols. Philadelphia: University of Pennsylvania Press, 1960.

Boyle, Robert. *The Philosophyical Works of the Honourable Robert Boyle, Esq.* London: W & J Innys, J. Osborn, and T. Longman, 1725.

———. *The Skeptical Chymist.* London: Bettesworth and Taylor, 1723.

Braithwaite, R. B. *Scientific Explanation.* Cambridge: Cambridge University Press, 1953.

Carnap, Rudolf. "Logical Foundations of the Unity of Science." In *International Encyclopedia of Unified Science*, ed. Otto Neurath, Rudolf Carnap, and Charles W. Morris. Chicago: University of Chicago Press, 1938.

Descartes, René. *Discourse on Method.* La Salle, IL: Open Court, 1952.

Duhem, Pierre. *The Aim and Structure of Physical Theory.* Princeton, NJ: Princeton University Press, 1954.

Durkheim, Émile. *Suicide: A Study in Sociology.* New York: Free Press, 1951.

Epicurus. "A Letter to Herodotus." In *Diogenes Laertius: Volume II*, trans. R. D. Hicks. Loeb Classical Library Volume 185. Cambridge, MA: Harvard University Press, 1925.

Feyerabend, Paul. *Against Method.* New York: Verso, 1993.

Giere, Ronald. *Explaining Science: A Cognitive Approach.* Chicago: University of Chicago Press, 1988.

Goodman, Nelson. *Fact, Fiction, and Forecast.* Cambridge, MA: Harvard University Press, 1979.

Hempel, Carl. *Aspects of Scientific Explanation.* New York: Free Press, 1965.

Hippocrates. *The Medical Works of Hippocrates*, ed. John Chadwick and W. N. Mann. Oxford: Blackwell, 1950.

Hobbes, Thomas. *Leviathan.* Oxford: Clarendon, 1909.

Hubbard, Ruth. "Science, Fact, and Feminism." *Hypatia* 3 (1988): 5–17.

Hume, David. *Philosophical Essays Concerning Human Understanding*. London: A. Millar, 1748.

Hutton, James. "Abstract of a Dissertation Read in the Royal Society of Edinburgh, upon the 7th of March and the 4th of April, concerning the System of the Earth, its Duration, and Stability," *The Edinburgh Magazine*, vol. 3. London: Sibbald, 1788.

Kuhn, Thomas. *The Structure of Scientific Revolutions*. Chicago: University of Chicago Press, 1962.

Lakatos, Imre. *The Methodology of Scientific Research Programmes*. Cambridge: Cambridge University Press, 1978.

Latour, Bruno. "The Science Wars: A Dialogue." *Common Knowledge* 8 (2002): 71–79.

Linnaeus, Carol. *Systema Naturae*. 1735. Facsimile of 1st ed., with introduction and English translation of the observations by M. S. J. Engel Ledeboer and H. Engel (Nieuwkoop, Holland: B. de Graaf, 1964), 18–19, 26–28.

Maxwell, James Clerk. "Molecules." *Nature* 8 (1873): 437–41.

Mendel, Gregor. *Experiments in Plant Hybridization*. Cambridge, MA: Harvard University Press, 1916.

Mill, John Stuart. *A System of Logic*. London: Longman, 1906.

Newton, Isaac. *Mathematical Principles of Natural Philosophy*. Berkeley: University of California Press, 1934.

Paracelsus. *Hermetic and Alchemical Writings*, ed. Arthur Edward Waite. London: James Elliot & Co., 1894.

Popper, Karl. *The Logic of Scientific Discovery*. New York: Routledge, 1992.

Ptolemy. *Almagest*, trans. G. J. Toomer. London: Duckworth, 1984.

Quesnay, François. "Fermiers" [Farmers]. In *Oeuvres Èconomiques et philosophiques* [Economic and Philosophical Works]. Paris: Francfort s/m, 1888.

Spector, Marschall. "Models and Theories." *British Journal for the History of Science* 62 (1965): 121–42.

Weber, Heinrich. *The Sense of Touch and the Common Feeling*. In *The Classical Psychologists*, ed. Benjamin Rand. Gloucester, MA: Peter Smith, 1912.

Whewell, William. *Novum Organum Renovatum*. London: J.W. Parker & Son, 1858.

Woodward, John. *An Essay Towards a Natural History of the Earth*. London: R. Wilkin, 1695.

Index

. . . .

Bacon, Francis (*continued*)
and, 112; hypothetico-deductivism
and, 93; inductivism of, 44; Kuhn's
references to, 184, 187; Mill's reference
to, 61; natural histories and, 184; *Novum
Organum*, 44, 46–52, 180; Popper's
references to, 152, 153
Barash, David, 300
behaviorism, 276, 277
belief: critical views of, 281, 282, 283, 290;
Hume's problem of induction and, 112–
13; Kuhn on, 184, 185, 188; Reichenbach's
degrees of, 138; scepticism about, 199
Bernard, Claude, 212n11
Bernoulli, Daniel, 78
Biot, Jean-Baptiste, 178, 180
Black, Joseph, 158
Black, Max, 233, 256–64
blackbody radiation, 189, 216
bleen, 129
Bohr, Niels: on development of knowledge,
292n12; life and work, 217
Bohr model, 216–18; Black on, 256, 263;
Pauli and, 213n23; Spector on, 242,
249–50
Boltzmann, Ludwig, 78, 359
Bombastus von Hohenheim. *See* Paracelsus
Boulding, Kenneth, 260
Boyle, Robert, 77, 79–80, 158, 362–65
Boyle's law, 79–80, 359
Brahe, Tycho, 214, 215
Braithwaite, R. B., 92, 106–11, 232; Spector
on models of, 237–47, 248, 250, 251,
253nn6–7, 254n17
Brownian motion, 254n21, 271
Bruno, Giordano, 155
Buffon, Georges-Louis Leclerc de, 162,
163
Burawoy, Michael, 321–22

Callippus, 75
Camerarius, Rudolf Jacob, 83
capitalism: Keynes on, 279–80; Marx on,
229–30; sociobiology and, 299; Weber
on religion and, 166–68
Carnap, Rudolf: excerpt from writings of,
105; hypothetico-deductivism of, 92;
logic of discovery and, 250; paranormal
phenomena and, 141; theoretical terms
and, 232, 235, 237, 244, 251, 252
case studies, xiv, xv–xvi; paper 1
(deductivism), 30–42, 327–49; paper
2 (inductivism), 75–90, 350–87;
paper 3 (hypothetico-deductivism
and falsificationism), 155–69; paper
4 (holistic view), 214–30; paper 5
(semantic view), 270–80; paper
6 (critical views), 315–23. *See also*
astronomy track; chemistry track;
economics track; evolutionary biology
track; genetics track; geology track;
physics track; psychology track;
sociology track
catastrophism, 37–38, 339–41
Catholic Church, and Galileo, 155–56
Catholicism: capitalism and, 166–68;
suicide and, 382–83
cause and effect: Hume on, 117–22, 123–26,
138; Kant on, 138
causes: Aristotle on, 2–3, 6–7, 8, 15–16;
Bacon on, 49–50; Descartes on first
causes, 28–29; Mill's methods and, 45,
57–74; Newton on, 53
change: Aristotle on, 11, 12, 14, 15, 16; as
central question in ancient thought,
30, 34, 36, 41; Mill on, 66–69; Whewell
on forces and, 97; Zeno on illusion
of, 31
Chargaff, Erwin, 221
Charles, Jacques, 77, 359
chemistry: Kuhn on crisis in, 193; Whewell
on theories of, 103–4. *See also* elements
chemistry track: paper 1 (alchemy), 32–34,
332–34; paper 2 (Boyle's law), 79–80,
362–65; paper 3 (Priestley and oxygen),
158–60; paper 4 (Dalton), 218–20; paper

5 (Kekulé), 272–74; paper 6 (elements 116 and 118), 317–18
Christie, James W., 315
chromosomes, 160–61
Clairaut, Alexis-Claude, 192
class bias, 294, 299, 303, 321
Clausius, Rudolf, 78, 361
coincidents, Aristotle on, 9
colligation of facts, 91, 94, 95, 100
commensurately universal attributes, 8–10
commonwealth, Hobbes's theory of, 40
communism, 230
complexity theory, and intelligent design, 319
computer analysis of data, 317, 318
computer models, 234
Comte, Auguste, 87
conditioned reflexes, 165–66
Conditioned Reflexes (Pavlov), 166
confirmation of hypotheses: Braithwaite on, 110; Goodman on, 127–30; Hempel on logic of, 131–37; paradoxes of, 135–36, 137n10; syntactic view of theories and, 171. *See also* evidence; hypotheses
confirmation of theories, 325–26; Carnap on, 105; Lakatos on, 205–6. *See also* theories
conjectures: Lakatos on, 200; Popper on, 151, 152–53, 207; Whewell on, 100
consilience of inductions, 94
context of discovery, ix, 171; Popper's view of, 142; Whewell's view of, 91–92
context of justification, 92
continental drift, 224–25
contradiction, Aristotle on, 7
Copernicanism: Feyerabend on, 288, 290; Galileo and, 155–56, 212n12; Kuhn on, 182, 189–90, 193, 196, 215; Lakatos on, 203, 212n12, 215; Whewell on, 98
Copernicus, Nicholas, 155, 190
corporate research, 282, 318
correspondence rules, 237, 244, 248, 249, 251, 269

corroboration: Lakatos on, 204, 205–6, 207, 210, 212n17, 213n24; Popper on, 143, 146, 151, 153
Coulomb, Charles-Augustin de, 186
creationism, 307, 308, 319, 320
creative recurrence, in societies, 228
creativity of the scientist, 91, 93, 325
Crick, Francis, 220–22, 234
crisis, 189, 190–94, 196, 206, 210
critical views of scientific theories, 281–84; case studies on, 315–23; of Feyerabend, 281–82, 285–93, 325–26; of Hubbard, 283, 294–306, 325–26; of Latour, 283–84, 307–14, 325–26
crucial experiment: Duhem on, 179–81; Lakatos on, 174, 205; Popper on, 151–52, 154n6
currency, Aristotle on, 41–42, 349. *See also* wealth

Dalton, John, 218–20, 360
Darwin, Charles, 160, 164, 222–24
De Broglie, Louis, 196
deductive logic, 1–2, 4; of Braithwaite's system of hypotheses, 108–11; Carnap on role of, 105; Hempel on role of, 134–35, 137n8; Hume on limitations of, 112; in hypothetico-deductive method, 92, 93; Lakatos on role of, 198; *modus tollens* in, 92, 93, 143, 207–8; Popper on role of, 142–43, 145–46, 148, 151, 154n5. *See also* axioms; demonstration, deductive; logic; mathematics; syllogism
Deductive Method of Mill, 69–74, 142
deductivism, 1–4, 43, 171; case studies in, 30–42; case study readings on, 327–49; hypothetico-deductivism and, 91, 93; Popper's departure from, 141–42. *See also* hypothetico-deductivism
definition: Aristotle on, 8; operational, 232; Whewell on induction and, 99–100

Engels, Friedrich, 230
*Enquiry Concerning Human Understanding,
   An* (Hume), 116–26
Epicurean theory of light, 183
Epicurus: atomism of, 32, 78, 358; letter to
   Herodotus, 32, 330–32; life and thought
   of, 32
epicycles, 75, 76, 354–55, 357; Copernicus's
   use of, 155; Kepler's abandonment of,
   214; system without, 155; Whewell on
   induction and, 98, 99, 102, 103
epilepsy, Hippocrates on, 39
epistēmē, Popper on, 152, 153
epistemology, Feyerabend's anarchism
   about, 285, 287, 288
equants, 76, 155, 212n12
equivalence condition, 114, 134–36
*Essay towards a Natural History of the Earth,
   An* (Woodward), 38, 339–41
essential attributes, Aristotle on, 8–10
ether: in Aristotle's physics, 155; light
   propagation in, 180, 193, 241–42, 262;
   models of, 241–42, 254nn14–15, 262–63;
   Ptolemy on, 353, 355–56
Euclid, 3, 116, 179
Eudoxus, 75
evidence: Braithwaite on, 110; Hempel
   on, 131; Hume on, 116, 117; probabilism
   and, 199; problem of demarcation
   and, 141; Whewell on, 97. *See also*
   confirmation of hypotheses; paradoxes
   of evidence
evolutionary biology track: paper 1
   (Aristotle), 36–37, 336–39; paper 2
   (Linnaeus), 82–83, 371–75; paper 3
   (Lamarck), 162–63; paper 4 (Darwin),
   222–24; paper 5 (punctuated
   equilibrium), 274–75; paper 6
   (intelligent design), 319–20. *See also*
   natural selection; species
exchange, Aristotle on, 41–42, 348–49
existential quantification, 133, 136n3
experience: Hume on, 117–26; inductivism

and, 43; Popper on, 146–47, 148, 150;
   syntactic view of theories and, 171.
   See also observation
experiments: Bacon on, 51–52; Boyle's
   use of, 79, 80, 362–65; Carnap on, 105;
   Descartes on, 28–29; Duhem on,
   175–81; in hypothetico-deductive
   method, 93; inductivism and, 43;
   Lakatos on, 200, 201, 205, 206; Mill's
   methods and, 57–69; Newton on,
   53–55; of Pavlov, 165–66; Popper on,
   143, 145–46, 151–52, 154n6, 212n14;
   syntactic view of theories and, 171; in
   Weber's psychology, 86–87, 165, 378–79;
   Whewell on, 102. *See also* confirmation
   of hypotheses; observation
*Experiments and Observations on Different
   Kinds of Air* (Priestley), 160
Experiments in Plant Hybridization
   (Mendel), 81–82, 365–70
Explaining Science (Giere), 265–69
explanation: Hempel on, 134–35, 137n8;
   Lakatos on, 204; mathematical models
   and, 260–61; models and, 232, 262–63;
   Popper on, 154n5

Fact, Fiction, and Forecast (Goodman),
   127–30
facts: colligation of, 91, 94, 95, 100;
   feminist view of, 294–96, 301; social
   construction of, 283, 284
falsification: Hempel on, 131; Kuhn on,
   191
falsificationism, 141–44, 171; case studies
   on, 155–69; dogmatic (naturalistic),
   200–2, 211n3; Lakatos on, 173–74,
   198–206, 210–11, 212nn16–17; of
   Popper, 142–44, 146, 147–50, 152, 173;
   sophisticated versus naïve, 198, 202–6,
   210–11, 212nn16–17
Faraday, Michael, 78, 262
"Farmers" (Quesnay), 90, 383–87
Fechner, Gustav, 86, 165

feminism, 294, 299, 305

feminist methodology, 301, 302, 303

Feyerabend, Paul, 281–82, 325–26; case studies relating to, 317–18, 320; excerpt from writings of, 285–93

final cause, 3, 50

forces, Whewell on, 96–97

form: Aristotle on, 12–13, 14, 15, 16; Aristotle's germ theory and, 35, 335, 336; Bacon on, 50, 51; Descartes on, 17

formal cause, 3, 50

forms, Platonic, 2

Foucault, Jean, 178, 180

four elements (earth, air, fire, water), 32–33; Aristotle on, 11, 31, 328; Boyle's departure from, 79, 80, 362; Hippocrates' four humours and, 39; Paracelsus on, 33, 34, 332, 333

Frankland, Edward, 273

Franklin, Benjamin, 159, 167, 182

Franklin, Rosalind, 220–22

Freire, Paulo, 294, 301

Fresnel, Augustin-Jean, 180, 183

Freud, Sigmund, 141, 225–27, 299. See also psychoanalysis

fruit flies, 160–61

funding of research, 282, 318

Galen, 33, 86

Galileo: Aristotle's theory of motion and, 189; case study on heliocentrism of, 155–57; falling bodies and, 109–10, 111nn1–2; Feyerabend on, 290; Hobbes's meeting with, 41; life of, 156; medieval elements in thought of, 197n6; simplicity of Copernicanism and, 212n12

Galle, Johann, 271

Gamow, George, 217

gasses: Boyle's law, 79–80, 359; Priestley's work on, 158–60, 360. See also ideal gas law; kinetic theory of gases

Gay-Lussac, Joseph Louis, 77, 359

Geiger, Hans, 157

gender: Aristotle on differences of, 34–35, 334–37; scientific method and, 282, 283. See also feminism

gender identity disorder, 321

General Theory of Employment, Interest, and Money (Keynes), 279

general theory of relativity, 192, 196, 201, 270–71; Feyerabend on, 291; quantum theory and, 316

genetics track: paper 1 (Aristotle), 34–35, 334–36; paper 2 (Mendel), 81–82, 365–70; paper 3 (Morgan), 160–62; paper 4 (Crick and Watson), 220–22; paper 5 (Human Genome Project), 274; paper 6 (public and corporate funding), 318

geographic information systems (GIS), 276

geology track: paper 1 (Woodward and catastrophism), 37–38, 339–41; paper 2 (Hutton and Plutonism), 84–85, 375–78; paper 3 (Lyell and uniformitarianism), 163–64; paper 4 (Wegener and plate tectonics), 224–25; paper 5 (geographic information systems), 276; paper 6 (Grand Canyon), 320

geometry: Braithwaitean models and, 239, 243–44; Descartes on, 18, 23; of Euclid, 3, 179; as Greek model for knowledge, 1, 2; Hume on uses of, 120; non-Euclidean, 199, 234. See also mathematics

Gerhardt, Charles, 273

germ theory, of Aristotle, 34–35, 160, 334–39

Giere, Ronald, 234, 265–69

God: Aristotle on, 327; Descartes on, 4, 22–26, 28–29; Great Chain of Being and, 82; Linnaeus on, 371–72; Paracelsus on, 332, 333

Gödel, Kurt, 141

Goodman, Nelson, 113–14, 115, 127–30; responses to, 138, 139

Gould, Stephen J., 275
government deficit spending, 280
government funding of research, 282, 318
Graham, Thomas, 360, 361
Grand Canyon, and age of the earth, 320
gravitation: Descartes' theory of, 201, 207, 212n10; Mill on, 68, 71–72. *See also* general theory of relativity; Newtonian gravitational theory
Great Chain of Being, 82
Great Depression, 279, 322
Grew, Nehemiah, 83
grue, 113–14, 128, 129–30, 130n, 139

hadrons, 271–72
Halley, Edmond, 215
hard core of research program, 173, 207–8, 209, 213n22
Harlow, Harry F., 276–77
Harman, Gilbert, 139
harmonic oscillator, 265–66, 269
heat: as elemental substance, 77, 159, 218; Maxwell's kinetic theory of, 77–79; Mill's example of, 65–66; phlogiston and, 159; speed of sound and, 192. *See also* thermodynamics
Hegel, Georg Wilhelm Friedrich, 229, 230, 292n4
Heisenberg, Werner, 217
Heisenberg's uncertainty principle, 302
heliocentrism, 155–57. *See also* Copernicanism
Helmholtz, Hermann: on anomalous dispersion, 241
Helmholtz resonator, 239–40
Hempel, C. G., 114–15, 131–37, 138, 139
Hermes Trismegistus, 33
Hermetic and Alchemical Writings (Paracelsus), 34, 332–34
Herodotus, Epicurus's letter to, 32, 330–32
Hipparchus, 75, 76
Hippocrates, 38–40, 86, 342–44
Hobbes, Thomas, 40–41, 87, 344–47

holistic view of theories, 171–74, 231; case studies on, 214–30. *See also* Duhem, Pierre; Kuhn, Thomas; Lakatos, Imre
homosexuality, 320–21
Hubbard, Ruth, 283, 325–26; case studies relating to, 315–23; excerpt from writings of, 294–306
Human Genome Project, 221, 274
Hume, David, 112–13, 115, 325; on cause and effect, 117–22, 123–26, 138; excerpt from *Enquiry Concerning Human Understanding*, 116–26; responses to, 138, 141, 148; Smith's friendship with, 169
humours, 39, 342–44
Hutton, James, 84–85, 163, 375–78
Huygens, Christiaan, 180
hydrogen: Bohr on (*see* Bohr model); Maxwell on molecules of, 359
hypotheses: accidental, 127, 128, 130; Aristotle on, 8; auxiliary, 147, 202, 203, 207, 208; Braithwaite on, 107, 108–11; corroboration of, 143; Duhem on testing of, 175–81; establishment of, 92, 110, 111; falsification of, 143; Giere on, 266–67, 268; lawlike, 127, 128, 129, 130; about models, 234; Newton on, 55; Spector on probability of, 246; Whewell on, 94, 100–4. *See also* confirmation of hypotheses; conjectures; predictions; theories
hypothetico-deductivism, 91–93, 171; case studies on, 155–69; evidence and, 112; Popper and, 142, 143; Spector on, 235, 236. *See also* Braithwaite, R. B.; Carnap, Rudolf; Whewell, William

Ibsen, Henrik, 292n9
ideal gas law, 233, 234, 247–48, 254n18
ideational approach, 227–28
index of refraction, 178
induction, 43, 325; Aristotle on, 5, 97; Bacon and, 44, 52, 112; Braithwaite on role of, 92, 111; Hume's acceptance of,

Mill's methods (*continued*)

Meyer, Oscar, 361

Mill, John Stuart, 44–45; excerpt from
  A System of Logic, 56–74; opposition to
  method of hypothesis, 91

Mill's methods, 45, 56–74; Boyle's work
  and, 80; Durkheim's use of, 88;
  Hutton's employment of, 85; Mendel's
  employment of, 82; Popper's use of
  deduction and, 142; Ptolemy and, 77;
  Quesnay's employment of, 90; Weber's
  approach and, 87

models: analogue, 241, 243, 258–59, 263,
  264; Black on, 233, 256–64; Giere on,
  234, 265–69; as heuristic fictions,
  262–63, 326; as icons, 257, 258; Lakatos
  on, 208–9; mathematical, 233, 234,
  259–61; modifications of a theory and,
  247–51; reality and, 238, 241–43, 245–47,
  254n15, 263, 266–67, 268–69, 326; in
  semantic view of theories, 231–34;
  Spector on, 232, 233, 235–55; Spector on
  Braithwaite's analysis of, 237–47, 248,
  250, 251, 253nn6–7, 254n17

"Models and Archetypes" (Black), 233,
  256–64

"Models and Theories" (Spector), 235–55

modernism, 281, 283

modus tollens, 92, 93, 143, 207–8

molecules, 77, 78, 157, 357–62

Molière, 212n10

momentum: Hume on, 120; Maxwell on
  diffusion of, 361; Whewell on concept
  of, 97

money: Aristotle on, 41–42, 349; Quesnay
  on, 89

moon: Eudoxus on orbit of, 75; Galileo's
  observations of, 155, 156; Newton's
  calculations of perigee, 192

moral reasoning, Hume on, 123

Morgan, Thomas Hunt, 160–62

motion: Aristotle on the heavens and,
  30–31, 75, 76, 327–29, 350; in Aristotle's

physics, 11, 12, 13, 14, 16, 185; Cartesian
  theory of, 201, 207; Giere on classical
  models of, 265–66, 267–68, 269;
  Whewell on forces and, 97. *See also*
  kinetic theory of gases; Newtonian
  mechanics; planetary motion

Nagel, Ernest, 238, 250

natural history: Kuhn on, 184; Linnaeus
  on, 372, 373

natural kind term, 139

natural law, Hobbes's theory of society and,
  40, 41

natural laws. *See* laws of nature

natural selection, 160, 222–24; vs.
  intelligent design, 319; sociobiology
  and, 299

nature: Aristotle on, 2–3, 11–14; Bacon on
  interpretation of, 49–52; Descartes
  on, 17; Newton on, 53; women's
  contributions to understanding of, 294,
  304

Nature of Man, The (Hippocrates), 39

necessary attributes, Aristotle on, 10

necessary truth, Aristotle on, 8

negative heuristic, 206–7

Neptune, discovery of, 270, 271, 315

Neptunism, 84. *See also* Woodward, John

Neumann, F. E., 176, 177

new riddle of induction, 113–14, 127–30

Newton, Isaac: inductivism of, 44–45,
  112; life of, 215. See also *Principia*
  (Newton)

Newtonian gravitational theory: Giere
  on, 267; Kepler's laws and, 74, 96, 208,
  214–15; Kuhn on, 192; Lakatos on,
  201–2, 207–9, 211n6, 212nn7–8, 212n10;
  in *Principia,* 54

Newtonian mechanics: Giere on standard
  textbooks of, 267–68; heat and, 77;
  Kuhn on, 182, 191; Lakatos on laws of,
  208; in *Principia,* 44, 53–54; Smith's
  economics and, 168; Whewell on

forces in, 96–97. *See also* Newtonian gravitational theory

Newtonian optics, 103, 177–78, 180, 182, 183, 184, 189

Nicod, Jean, 132

Nicod's criterion: Achinstein on, 139; Hempel on, 114, 132–34, 135, 136nn4–5, 137nn9–10

nitrogen, 158, 159, 160

No Free Lunch (Dembski), 319

non-Euclidean geometry, 199, 234

normal science, 172, 182–88, 189, 190, 193–94, 195, 196, 210

Novum Organum (Bacon), 44; Duhem's reference to, 180; excerpts from, 46–52

Novum Organum Renovatum (Whewell), 94–104

nucleus, Rutherford's discovery of, 157–58

objectivity, Hubbard on, 301, 302, 303, 304

observation: Boyle's use of, 79; Carnap on, 105; computer program needed for, 318; in hypothetico-deductive method, 92, 93; inductivism and, 43, 44; in Mill's deductive method, 73; Popper's view of, 143; predicted by model, 234; Whewell on, 102. *See also* confirmation of hypotheses; experience; experiments

observation terms, 235, 236–37, 244, 246, 248, 249, 251, 252–53, 253n7

Occam's razor, 44; grue paradox and, 139

"Of Dephlogisticated Air" (Priestley), 159–60

Ohm's law, 171–72

"On the Constitution of Atoms and Molecules" (Bohr), 249

On the Generation of Animals (Aristotle), 35, 37, 334–39

On the Heavens (Aristotle), 327–29

On the Revolutions of the Heavenly Spheres (Copernicus), 155, 190

operant conditioning, 166

operational definitions, 232

Oppenheim, Paul, 136n1

optics. *See* light

Osiander, Andrew, 155

oxygen, discovery of, 104, 158–60

papers. *See* case studies

Paracelsus, 32–34, 332–34; Boyle's departure from, 79, 80

paradigms, 172–74, 182–84, 185–86, 187, 188–96, 210, 231

paradoxes of confirmation, 135–36, 137n10

paradoxes of evidence, 112–15, 143, 325; readings on, 116–37; responses to, 138–40, 141. *See also* Goodman, Nelson; Hempel, C. G.; Hume, David

paranormal phenomena, 141

Parmenides, 31

Parsons, Talcott, 228, 278–79

partial interpretation thesis, 237, 239, 243, 247, 248, 249, 250–51, 253, 255n25

particle physics, 271–72, 316. *See also* atomic theory

patenting of genes, 318

Pauli, Wolfgang, 213n23, 217

Pavlov, Ivan Petrovich, 165–66

peer review, 295–96

Peirce, C. S., 254n21, 257

perception, measurement of, 86–87, 165, 378–79

perihelion of Mercury, 201, 270–71

permanent causes, Mill on, 64–66

phlogiston, 103, 158–60

photoelectric effect, 216, 271

physics: Bacon on, 51; Duhem on testing hypotheses in, 175–81; Popper on evolution of, 151. *See also* atomic theory; Einstein, Albert; light; Maxwell, James Clerk; motion; quantum theory

Physics (Aristotle): argument of, 2–3; excerpt from, 11–16; Kuhn's reference to, 182

physics track: paper 1 (ancient atomism), 31–32, 330–32; paper 2 (Maxwell's kinetic theory), 77–79, 357–62; paper 3 (Rutherford and the nucleus), 157–58; paper 4 (Bohr model), 216–18; paper 5 (standard model and top quark), 271–72; paper 6 (string theory), 316–17

physiocrats, 89, 168

Planck, Max, 183, 216, 217

planetary motion: Carnap on theories and, 105; case study on, 214–16; Lakatos on falsificationism and, 201–2, 211n6, 212nn7–8; Newton's successive analyses of, 208–9; perihelion of Mercury, 201, 270–71; Whewell on induction and, 96, 97, 98, 99, 102, 103, 104n2. *See also* Aristotle: on the heavens; Copernicanism; Galileo; gravitation; Kepler's laws; Ptolemy, Claudius

plants: Linnaeus's love of, 83; Mendelian genetics and, 81–82, 365–70

plate tectonics, 224–25

Plato, 2, 30, 82, 321

Platonic theory of light, 183

Pliny, 184

Pluto, 284, 315

Plutonism, 84–85. *See also* Hutton, James

Poincaré, H., 177

polarized light, 176, 177

Politics (Aristotle), 42, 347–49

Popper, Karl, 141–44, 145–54, 325; Lakatos and, 173, 198, 199, 203–4, 207, 210–11, 211nn4–5, 212n14

positive heuristic, 206, 208–10

positivism: dogma of meaning in, 147; society and, 87

Posterior Analytics (Aristotle), 2, 5–10

postmodernism, 283–84; string theory as, 317

power in science, 315, 317, 326

predictions: Braithwaite on, 106; Carnap on, 105; Duhem on hypothesis testing and, 175–76, 179–80; Goodman on, 128,

130; Hempel on, 134–35, 137n8; Lakatos on, 203, 209, 212n9; of mathematical models, 260; with models, 232, 234, 326; of paradigm, 185; Popper on, 143, 145–46, 152, 173; Whewell on, 94, 103–4. *See also* hypotheses

pressure. *See* ideal gas law; kinetic theory of gases; mercury: height of column of

prices, Smith on, 168–69

Priestley, Joseph, 158–60, 197, 360

primitive terms, 236, 244

Principia (Newton), 214; excerpt from, 53–55; Kuhn's reference to, 182; Lakatos's reference to, 209. See also *The Mathematical Principles of Natural Philosophy* (Newton)

Principles of Geology (Lyell), 164, 223

probabilism, 199–200

probability: inductive inference and, 44, 132, 138, 139, 146; Lakatos on, 199–200; Popper on, 146, 152, 153

profit, from research, 318

progress, scientific, 174, 326

progressive problemshift, 204, 206, 207, 211, 212n13

proof: Lakatos on, 198–99; in Whewell's analysis of induction, 98. *See also* logic

protective belt, 173, 174, 203, 207–10

Protestant Ethic and the Spirit of Capitalism (Weber), 166–68

Protestantism: capitalism and, 166–68; suicide and, 382–83

Proust, Joseph, 218

Prout, William, 213n23

pseudo-science, 141; Lakatos on, 204. *See also* demarcation, problem of

psychoanalysis, 225–27; homosexuality and, 321. *See also* Freud, Sigmund; mental illness

psychology: Braithwaite on scientific character of, 106; Popper's studies of, 141

psychology track: paper 1 (Hippocrates),

scientific method. *See* critical views of scientific theories; deductivism; falsificationism; holistic view of theories; hypothetico-deductivism; inductivism; paradoxes of evidence; semantic view of theories

syntactic view of theories, 1, 171, 231, 232.
  See also deductivism; falsificationism;
  hypothetico-deductivism; inductivism;
  paradoxes of evidence
synthetic *a priori* principles, 199
Systema Naturae (Linnaeus), 83, 371–75
System of Logic, A (Mill), 56–74
System of the Earth (Hutton), 84–85, 375–78

Tableau Économique (Quesnay), 89
tautology, 142, 145, 149
taxation, Quesnay on, 89–90, 383–87
taxonomy of life: Lamarck's contributions
  to, 163; Linnaeus and, 82–83, 371–75
technology/science distinction, 296
telescope: Galileo's discoveries with, 155–56;
  Lakatos's example of, 201, 211n6
Terman, Lewis, 277
theoretical hypothesis, Giere on, 266–67,
  268
theoretically progressive problemshift, 204
theoretical models: Black on, 261–64;
  Giere on, 265–69
"Theoretical Procedures in Science"
  (Carnap), 105
theoretical terms, 232–33, 234; Spector on,
  233, 235, 236–38, 244, 248, 249, 250–51,
  252–53, 253n7
theories: Carnap on, 105; Duhem on
  groups of, 175–81; Kuhn on novelties
  of, 188, 189–92; Lakatos on series of,
  204–6; Mill on verification of, 73–
  74; Popper on, 143–44, 145–54; social
  construction of, 265, 269; Spector on
  logical calculus of, 235–39, 243–53;
  three fundamental questions about,
  ix–x; Whewell on, 94, 99. *See also*
  confirmation of theories; critical views
  of scientific theories; falsification;
  holistic view of theories; hypotheses;
  models; semantic view of theories;
  syntactic view of theories
Theory of the Gene, The (Morgan), 161

thermodynamics: Boyle and, 80; classical
  dynamics and, 248, 250; Kuhn on,
  189; molecular motion and, 77–79.
  See also heat
Thomson, J. J., 157, 158
top quark, 271–72
Torricellian experiment, 72
tracks, x–xi, xv. *See also* astronomy track;
  chemistry track; economics track;
  evolutionary biology track; genetics
  track; geology track; physics track;
  psychology track; sociology track
trade: Aristotle on, 42, 348–49; Quesnay
  on, 89, 385–86; World Bank and IMF
  involvement in, 323
Trouble with Physics, The (Smolin), 316
truth: Aristotle on necessary truth, 8;
  confirmation of theories and, 326; in
  deductive logic, 2, 4, 43; Descartes
  on criterion of, 21–22, 23; Feyerabend
  on, 291; Giere on models and,
  265–66; in holistic view of theories,
  231; in hypothetico-deductivism, 236;
  Latour on, 307, 308, 312–13; objective,
  283; Popper's view of, 144, 146, 152, 153;
  positivism about, 147; in semantic view
  of theories, 234; socially constructed,
  283–84; Sorokin on, 228; Spector on
  models and, 249; in syntactic view of
  theories, 231; tautological, 142, 145, 149;
  Whewell on, 97, 102, 104

unemployment, 302
uniformitarianism, 163–64
universal generalizations: in Aristotle's
  deductivism, 2; Braithwaite on, 107–8,
  111, 111nn3–4; Hempel on confirmation
  of, 132–34, 135, 137nn9–10; in
  hypothetico-deductivism, 93; Lakatos
  on, 199; in Newton's reasoning, 44–45;
  raven paradox and, 114; in syllogisms,
  2; in syntactic view of theories, 171.
  See also laws of nature

unobservable objects: Maxwell on, 358;
Spector on, 252
Urban VIII (pope), 156

vacuum: Boyle's use of, 80, 363–65;
Descartes' denial of, 80
Vail, Tom, 320
van der Waals equation, 247–48
verification: Hempel on confirmation and,
131; Lakatos on, 210, 213n24; in Mill's
deductive method, 70, 73–74; Popper
on, 146, 147, 148; positivists on, 147; in
Whewell's analysis of induction, 98
void, of ancient atomists, 31, 331
Vulcan, 270, 315

Waismann, Friedrich, 147
Wallace, Alfred Russel, 223
Watson, James, 220–22, 234
wealth: Aristotle on, 41–42, 347–49;
scientific method and, 326; tax
structures and, 89–90, 383–87; Weber
on Protestant Ethic and, 166–68
Wealth of Nations, The (Smith), 169

Weber, Ernst Heinrich, 86–87, 165, 378–79
Weber, Max, 166–68, 227, 228, 279
Wegener, Alfred, 224–25
Werner, Abraham, 84, 85
Weyl, Hermann, 200
Whewell, William, 91–92, 325; excerpt from
Novum Organum Renovatum, 94–104
Whittaker, Edmund, 254n15
Wiener, O., 176, 177
Wilkins, Maurice, 220, 221
Wilson, E. O., 301
woman's nature, 297–300
women's health movement, 305
Woodward, John, 37–38, 84, 276, 339–41
Woolf, Virginia, 305
World Bank, 322–23
Wundt, Wilhelm, 86

Young, Thomas, 180, 183

Zenker, Wilhelm, 177
Zeno, 31
Zoological Philosophy (Lamarck), 163